T0331011

Radio Frequency Circuit Design

Radio Frequency Circuit Design

Second Edition

W. ALAN DAVIS
University of Texas Arlington

WILEY
A JOHN WILEY & SONS, INC., PUBLICATION

Published by John Wiley & Sons, Inc., Hoboken, New Jersey
Published simultaneously in Canada

For general information on our other products and services or for technical support, please contact our Customer Care Department within the United States at (800) 762-2974, outside the United States at (317) 572-3993 or fax (317) 572-4002.

Wiley also publishes its books in a variety of electronic formats. Some content that appears in print may not be available in electronic formats. For more information about Wiley products, visit our web site at www.wiley.com.

Library of Congress Cataloging-in-Publication Data is available:

ISBN 978-0-470-57507-9

Printed in the Singapore

10 9 8 7 6 5 4 3 2 1

In memory of Margaret
and to our children
Brent, Nathan, and Janelle

Contents

Preface to the Second Edition

Since the first edition of this book was published almost 10 years ago, radio frequency design techniques and applications have continued to rapidly expand. Readers of this second edition will find many changes from the first edition such as expansion of power amplifiers, oscillator phase noise, and impedance matching and deletion of other material. Some chapters and sections have been rearranged to provide a more logical flow. In particular, the chapter on noise now precedes the chapter on class A amplifiers. However, when this book is used in our course on radio frequency circuits, students are asked to do a design project using the software, Advanced Design System, from Agilent. It has been found helpful for students to start their project after understanding basic amplifier design and then treat the noise problem in their design subsequently. Throughout the book, design examples are given based on the text. Source code for the programs illustrated in the text are available at the website given in Chapter 1. These programs should be helpful to the working engineer in need of a quick solution and to the student wishing to understand some of the details in a computation.

I wish to acknowledge the many contributions made by Krishna K. Agarwal in the first edition of this book and the contributions to the class E power amplifier section by William Cantrell in this edition. I also wish to acknowledge the valuable suggestions given by the reviewers.

W. Alan Davis

Arlington, Texas
May 2010

Preface to the First Edition

The cellular telephone has become a symbol for the rapid change in the communications business. Within this plastic container reside the talents of engineers working in the areas of efficient power supplies, digital circuit design, analog circuit design, semiconductor device design, antennas, linear systems, digital signal processing, packaging, and materials science. All these talents are carefully coordinated at a cost that allows a wide cross section of the world's population to have available instant communication. The particular aspect of all these that is of primary focus in this text is in the area of analog circuit design with primary emphasis on radio frequency electronics. Topics normally considered in electronics courses or in microwave and antenna courses are not covered here. For example, there is no mention of distributed branch line couplers, since at 1 GHz their size would be prohibitive. On the other hand, topics such as transmission line transformers are covered because they fit so well into this frequency range.

This book is meant for those readers who have at least advanced standing in electrical engineering. The material in this text has been taught as a senior and graduate-level course in radio frequency circuit design at the University of Texas at Arlington. This class has continued to be popular for at least the last 20 years under the guidance of at least four different instructors, two of whom are the present authors. Because of the activity in the communications area, there has been ever greater interest in this subject. It is the intent of the authors, therefore, to update the current text offerings while at the same time avoiding simply reworking a microwave text.

The authors gratefully acknowledge the contribution of Michael Black, Raytheon Systems Company, to the phase lock loop discussion in Chapter 12.

W. Alan Davis
Krishna K. Agarwal

CHAPTER ONE

Information Transfer Technology

1.1 INTRODUCTION

The design of radio frequency (RF) circuits borrows from methods used in low frequency audio circuits as well as from methods used in design of microwave circuits. However, there are also important departures from audio and microwave frequency methods, so that design of radio frequency circuits requires some specialized techniques not found in these other frequency ranges. The radio frequency range for present purposes will be taken to be approximately somewhere between 300 MHz and 3 GHz. It is this frequency range where much of the present day activity in wireless communication occurs. In this range of frequencies, the engineer must be concerned with radiation, stray coupling, and frequency response of circuit elements that, from the point of view of lumped, low frequency analysis, might be expected to be independent of frequency. At the same time, the use of common microwave circuit elements such as quarter wave transformers is impractical because of the long line lengths required. The use of monolithic circuits have enabled many high frequency designs to be implemented with lumped elements, yet the frequency response of these "lumped" elements still must be carefully considered. The small size of lumped elements in integrated circuits has provided practical designs of filters, transformers, couplers, etc. in lumped element form. Therefore discussion of designs for low noise amplifiers, power amplifiers, oscillators, mixers, and phase lock loops will be addressed with both lumped and distributed elements. Several of the numerical examples given in the text use computer programs. Source code for these programs are available

Radio Frequency Circuit Design, Second Edition, by W. Alan Davis

on the web*. However, before getting into the details in the design of radio frequency circuits, it is important to understand that the purpose for these circuits is to transmit information.

1.2 INFORMATION AND CAPACITY

What exactly is information? *Random House Dictionary* 1966 states that "information" is "knowledge communicated or received concerning a particular fact or circumstance. ..." A narrower technical definition more closely aligns with the focus given here is that "information" is an "indication of the number of possible choices of messages, expressible as the value of some monotonic function of the number of choices, usually log to the base 2." *Information* then is a term for data that can be coded for digital processing.

Some examples of data that illustrate the meaning of information is helpful. If a signal were sent through a communication channel that never changed, then it would be conveying no information. There must be change to convey a message. If the signal consisted of 1 0 1 0 1 0 1 0 ... , there would be changes in the signal but still no information is conveyed because the next bit would be perfectly predictable. So while change is important, it is not the sole criterion for information. There is one last example. If a signal in an amplitude modulation system consists of purely random voltage fluctuations, then again no information is being transmitted. It is simply noise, and the receiver is no more knowledgeable after having heard it.

A communication system consists of a transmitter, a receiver, and a channel. The channel is capable of carrying only a certain limited amount of information. A water pipe can be seen as a rough analogy to a communication channel. The limitation in a communication channel is given the technical term *capacity*. It refers to the amount of information that is transmitted over a time interval of T seconds. The time interval can be broken up into short time intervals, each of duration τ. Clearly, the more distinct time intervals τ there are in the total time span T, the more information that can be transmitted. The minimum size of τ is determined by how well one pulse in one time frame can be distinguished from a pulse in a neighboring time frame. The limitation on how short a time frame can be is related to the channel bandwidth. In the water pipe analogy, the channel bandwidth corresponds to the pipe diameter.

In addition, the signal voltage will have a maximum amplitude that is limited by the available power in the system. This voltage range can be divided into many levels, each level representing a bit of information that is distinguished from another bit. The voltage range cannot be split indefinitely because of the noise that is always present in the system. Clearly, the more voltage intervals in a given time frame τ, the more information capacity there is in the system. Just as the flow of water through a pipe is limited by the amount of

*http://www-ee.uta.edu/online/adavis/rfsoftware

pressure on the water, by the friction on the walls of the pipe, and by the diameter of the pipe, so the capacity of a transmission system is limited by the maximum voltage level, by the noise in the system that tends to muddle the distinction between one voltage level and another, and by the bandwidth of the channel, which is related to the rise time of a pulse in the system.

In one of the time intervals, τ, there are n voltage levels. The smaller that τ is and the larger n is, the more information that can be transmitted through the channel. In each time interval, there are n possible voltage levels. In the next time interval there are also n possible voltage levels. It is assumed that the voltage level in each time frame is independent of what is going on in other time frames. The amount of information transmitted in a total of T seconds corresponds to the products of the possibilities in each interval:

$$n \cdot n \cdot n \cdot n \cdots n = n^{T/\tau} \tag{1.1}$$

The total information, H, transmitted intuitively is directly proportional to the total time span T, and is defined as the log of the above product. By convention, the base 2 logarithm is used.

$$H = T/\tau \log_2 n \tag{1.2}$$

The system capacity is simply the maximum *rate* of transmission (in bits/s) through a system:

$$C = H/T = 1/\tau \log_2 n \tag{1.3}$$

System capacity is inversely proportional to the minimum time interval over which a unit of information can be transmitted, τ. Furthermore, as the number of voltage levels increases, so does the capacity for more information.

Information can be transmitted through a channel in a variety of different forms, all giving the same amount of information. For example, suppose that a signal can take on any one of eight different voltage levels, 0,1, ... , 7, in a given time interval τ. But the eight-level signal could also equally be sent with just two levels, 0,1. However, for every interval that has eight possible levels, three intervals will be needed for the two-level signal. A convenient conversion between the two systems is shown in Table 1.1.

Clearly, a 16-level signal could be transmitted by a sequence of 4 binary signals, and a 32-level signal with a sequence of 5 binary signals, and so on. For n levels, $\log_2 n$ bits are needed. The information content of a signal is defined then to be the number of binary choices, or bits, that are needed for transmission. A system that is designed to transmit speech must be designed to have the capacity to transmit the information contained in the speech. While speech is not the total of what humans communicate, in a communication system, it is that with which engineers have to work. A decision must be made as to what level of fidelity the speech is to be transmitted. This translates to the bandwidth

TABLE 1.1 Eight-Level and Two-Level Systems

$n = 8$	$n = 2$
0	000
1	001
2	010
3	011
4	100
5	101
6	110
7	111

requirement of an analog system, or the number of voltage levels available in a given total voltage range. Ultimately the restriction is always present even if sophisticated coding techniques are used. The capacity of the system must be greater than or equal to the rate of information that is to be transmitted. Beyond this, system cost, power levels, and available transmission media must be considered.

1.3 DEPENDENT STATES

The definitions of the preceding section imply that the voltage level in each time interval, τ, is independent of the voltage level in other time intervals. However, one very simple example where this is not the case is the transmission of the English language. It is known in the English language that the letter *e* is much more likely to appear than the letter *z*. It is almost certain that the letter *q* will be followed by the letter *u*. So in transmitting a typical message in English, less information is being actually sent than there would be if each letter in the alphabet were equally likely to occur. A way to express this situation is in terms of probability. The total number of signal combinations that could occur in a message T seconds long if the value in each interval is independent of the others is $n^{T/\tau}$. On average, every possible message T seconds long would have a probability of occurrence of $1/n^{T/\tau}$.

The probability takes the form

$$P = \frac{\text{number of occurrences of a particular event}}{\text{total number of events}} \tag{1.4}$$

Information can be measured in terms of probability. The probability is $P = 1/n$ if there are n possible events specified as one of n voltage levels, and each of these events is equally likely. For any one event, the information transmitted is written $H_1 = -P\log_2 P$. For m intervals, each τ seconds long, there will be m

times more information. For m intervals, the information written in terms of probability is

$$H = \frac{T}{\tau}\log_2 n = -m\log_2 P \quad \text{bits} \tag{1.5}$$

Consider a binary system, where a number 0 occurs with a probability of p and the number 1 occurs with a probability of q. Knowing that $p + q = 1$, the information content of a message consisting of 0's and 1's is found. The total information is the sum of the information carried by the 0's and that of the 1's:

$$H = -\frac{T}{\tau}(p\log_2 p + q\log_2 q) \quad \text{bits} \tag{1.6}$$

If the probabilities of p and q were each 0.5, then the total information in T seconds is T/τ. If, for example, $p = 0.25$ and $q = 0.75$, then

$$H = -\frac{T}{\tau}(0.25\log_2 0.25 + 0.75\log_2 0.75) \quad \text{bits}$$

$$H = \frac{T}{\tau}(0.5 + 0.3113) = 0.8113\frac{T}{\tau} \quad \text{bits} \tag{1.7}$$

Hence, when there is a greater probability that an expected event will occur, there is less information. As p approaches 1 and q approaches 0, the near certainty of an event with probability p will give 0 information. Maximum information occurs when $p = q = 0.5$.

This scenario can be generalized for n signal levels in a given signal interval τ. Assume that each of these n signal levels, s_i, have a probability of occurrence of P_i where

$$P_1 + P_2 + \cdots P_n = \sum P_i = 1 \tag{1.8}$$

Assume further that the probability of finding a given signal level is independent of the value of the adjacent signal levels. The total information in T/τ intervals or in T seconds is

$$H = -\frac{T}{\tau}\sum_i^n P_i \log_2 P_i \quad \text{bits} \tag{1.9}$$

The capacity required to transmit this amount of information is then

$$C = -\frac{1}{\tau}\sum_i^n P_i \log_2 P_i \quad \text{bits/s} \tag{1.10}$$

In the case where each level is equally likely, $P_1 = P_2 = P_3 = \cdots P_n = 1/n$, then for the n level signal,

$$H = -\frac{T}{\tau}\sum_{i}^{n} P_i \log_2 P_i = \frac{T}{\tau}\log_2 n \quad \text{bits} \tag{1.11}$$

More details on information may be found in specialized texts; a short introduction is given by Schwartz [1]. In this study of radio frequency (RF) design the primary focus will be on the fundamental hardware design used in transmitters and receivers. Other topics that are of great interest to communication engineers such as programming digital signal processing chips, various modulation schemes, or electromagnetic propagation problems are more fully explored in specialized texts in those areas. In this book these areas will be referred to only as needed in illustrations of how systems may be implemented.

1.4 BASIC TRANSMITTER–RECEIVER CONFIGURATION

Analog RF and digital designs are both found in typical communication systems. There are many systems where digital signal processing is playing a large role along with advanced RF circuit design. A typical superheterodyne radio transmitter and receiver are shown in Fig. 1.1. An actual system would

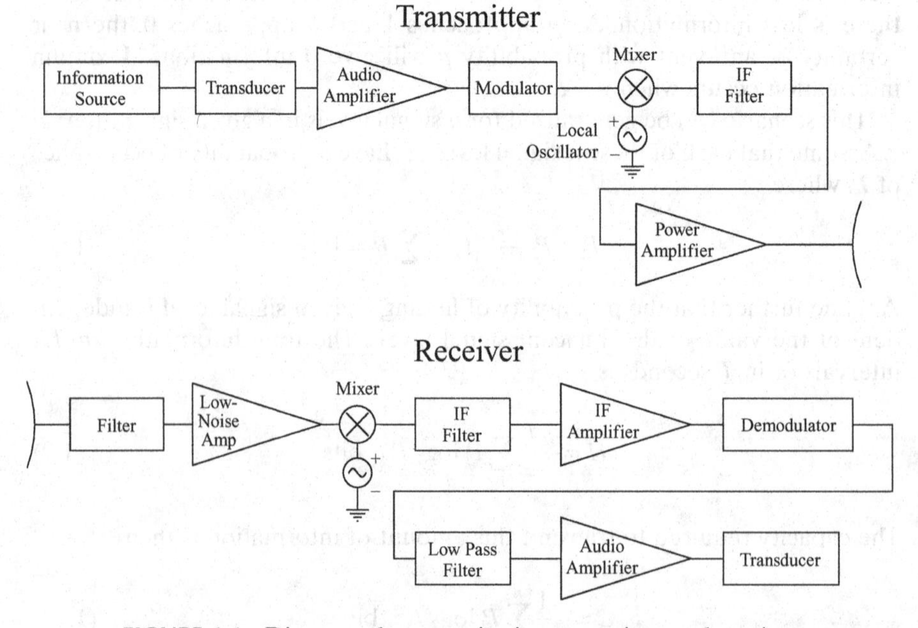

FIGURE 1.1 Diagram of communication transmitter and receiver.

be optimized for cost, noise immunity, fading, available bandwidth, bandwidth efficiency (the ratio of the throughput data rate per hertz in a given bandwidth), power efficiency (which measures the ability of a system to preserve the message under low-power conditions), intermodulation products, adjacent channel interference, and so on. The modulator and demodulator shown in the figure symbolize a large range of design options, often making use of digital techniques. Clearly, the circuits in Fig. 1.1 are only an outline of actual transmitters and receivers.

The transmitter in Fig 1.1 starts with some information source, which could be sound or a visual image. This is then converted to an electrical signal in the transducer, which may require amplification. The modulator codes the information and must be compatible with the demodulator. The modulator can be either analog or digital, and it comes in a wide variety of forms. It encodes the message in a certain way so as to meet the communication channel and receiver requirements. For example, if a video signal is being transmitted, the signal must carry information about the sweep time, intensity, and often color as well as the actual intelligence. The commonly used analog modulation techniques of amplitude modulation (AM), frequency modulation (FM), and phase modulation (PM) encode the carrier wave by changing its amplitude, frequency, or phase, respectively. Multiple signals can share the same channel if the signals are at different frequencies as in *frequency division multiple access* (FDMA), or at different time slots as in *time division multiple access* (TDMA), or with different digital codes as in *code division multiple access* (CDMA).

The mixer circuit is the first component in this discussion that breaks into the RF range, and it provides two necessary functions. First, it raises the carrier frequency that in AM and FM systems is distinct from neighboring transmitters. The second function of the modulator is that it translates the message information to a much higher frequency. This allows antennas to be made a manageable size since their mechanical size normally corresponds to the wavelength of the signal. A great deal of effort has gone into making electrically small antennas, but there are always design compromises. Chapter 11 is devoted to mixers.

The mixer is accompanied by a local oscillator that in some cases is carefully tuned to different frequencies or is fixed as in broadcast stations. The quality of an oscillator is judged on how low its phase noise is or how much its frequency will drift over time with temperature or age. Oscillators can be designed to be manually or electrically tuned to different frequencies. Techniques that are used to stabilize an oscillator include using high Q elements such as quartz crystals, dielectric resonators, or using a constant-temperature oven. Phase-lock loops can be used to stabilize a high frequency with a stable low-frequency oscillator. Design of oscillator circuits is considered in Chapter 10 and phase-lock loops in Chapter 12.

The filter that follows the mixer is required because the nonlinear multiplication process of the mixer produces unwanted frequencies. In addition, providing appropriate impedance levels to the mixer and the following amplifier often requires impedance matching. Radio-frequency filters and transformers are the primary subject in Chapters 3, 5, and 6 and are used in the design of amplifiers in Chapters 8 and 9.

The final stage of the transmitter before reaching the antenna is the power amplifier. Since this component uses the greatest amount of power, high efficiency becomes important. In FM systems, class C amplifiers are often used since in practice they can produce efficiencies as high as 70%. For AM systems, class A or B amplifiers are often used because of the required linearity of AM signal transmission. However, class A amplifiers typically have efficiencies of only 30 to 40%. In the transmission of digital modulated signals, linearity of the power amplifier becomes very important because of the need to minimize co-channel interference. In all these cases, it is clear that designing the amplifier for maximum power transfer so that the load impedance is conjugately matched to the amplifier output impedance would mean half the power would be dissipated in the transistor itself. The power amplifier must be designed for maximum efficiency where the internal output impedance is small relative to the external load.

The receiver is usually more complicated than the transmitter, and its purpose is to unravel the signal from the transmitter after the signal has acquired some noise and other distractions while going through the channel. If the received signal is strong enough, it can be put directly into the mixer. However, as will be seen in a later chapter, the overall noise response of the amplifier is greatly enhanced by using a low-noise amplifier for the front end. The design of the low-noise amplifier is described in detail in Chapter 8.

1.5 ACTIVE DEVICE TECHNOLOGY

The first RF vacuum devices made their appearance in the 1930s and today are still found to be the most reliable and efficient high-power amplifiers with power levels reaching up to 30 MW. Their demise is not likely to occur soon as is made evident by such things as the ubiquitous microwave oven. New device designs and new materials continue to improve the quality of vacuum tubes used in amplifiers and oscillators.

The solid-state entrance to the RF arena began with two-terminal diodes. These included the Gunn diode, the impact avalanche transit time (IMPATT) diode, the trapped plasma avalanche triggered transit (TRAPATT), the tunnel diode, and even the *pn* junction (varactor diode) used in parametric amplifiers. The three terminal GaAs *metal semiconductor field-effect transistor* (MESFET) soon displaced the diodes in most applications. Even though the MESFET did not have as low a noise figure as the parametric amplifier or the power (at the time) of an IMPATT, its stability and efficiency was superior. Furthermore, its noise level was low enough for many practical applications. Subsequent arrivals were the AlGaAs/GaAs *heterojunction bipolar transistor*

(HBT) and the *high electron mobility transistor* (HEMT) all based on GaAs or other III–V materials. These classes of devices in some cases still provide the best performance for a variety of high-power, high-frequency applications. Engineers are starting to make use of GaN and SiC for high-power RF applications. The wide band gap of GaN (3.4 eV), high break down voltage, high drift velocity, and high thermal conductivity of these materials make them attractive for high-power *heterojunction field-effect transistors* (HFET) devices. While self-heating and high flicker noise has been a problem with the GaN devices, some resolution with the flicker noise problem has been accomplished.

However, the world is made of silicon. Silicon has the advantage of being cheaper to manufacture than its GaAs cousins, has good thermal characteristics, and most important has an entrenched manufacturing infrastructure. Silicon soon surpassed its predecessor, germanium. Within a few years the *complementary metal–oxide semiconductor* (CMOS) technology found favor in digital circuits because of its ability to integrate a large number of transistors in a small space. The desire to integrate digital and analog applications on the same chip as well as to provide cost reduction relative to the GaAs devices has spawned much interest in RF CMOS designs. The progress in making small gate-length high-speed CMOS devices has provided the ability to make RF devices using CMOS technology. However, the mixed signal designs have required the sacrifice in the Early voltage, which is important in many analog circuits. *Laterally diffused metal–oxide semiconductor* (LDMOS) has also been used in power amplifiers. Their high gain, linearity, and reliability have made them the best choice in many cellular base station applications. More recently, the SiGe *heterojunction bipolar transistor* (HBT) has been found to have many advantages over straight CMOS. These include superior flicker noise, broadband noise, Early voltage, transconductance, and better tracking of V_{be} relative to the V_t of the MOSFET. The SiGe HBT does well with linearity, though not quite as well as the CMOS device.

In summary, there are a wide variety of devices available to the analog RF designer and with them a variety of specialized processing and circuit design techniques. It is the goal of the following chapters to provide basic circuit design techniques that can be applied to a wide variety of active devices.

PROBLEMS

1.1. A pulse train is being transmitted through a channel at the maximum channel capacity of 25×10^3 bits/s. The pulse train has 16 levels.

a. What is the pulse width?

b. The pulse width is doubled and sent back on the same channel. What number of levels is required?

1.2. A system can send out a signal at six different levels: 0, 1, 2, 3, 4, 5, each 1 ms long. The probability of each of these levels occurring is 1/8, 1/8, 1/16, 1/4, 3/8, 1/16, respectively. Each pulse value is independent of any previous pulse values. What is the total amount of information conveyed in 1 second?

REFERENCE

1. M. Schwartz, *Information Transmission, Modulation, and Noise*, 3rd ed., New York: McGraw-Hill, Chapter 1, 1980.

CHAPTER TWO

Resistors, Capacitors, and Inductors

2.1 INTRODUCTION

At radio frequencies, passive circuit elements typically have a more complicated model than those used in lower frequency designs. The simple resistor, capacitor, or inductor cannot be counted on to provide a pure resistance, capacitance, or inductance in high-frequency circuits. Usually the "lumped" element is best modeled as a combination of these pure elements. In addition, when the size of the element becomes larger than 0.1 wavelength in the circuit medium, the equivalent circuit may also include transmission lines.

2.2 RESISTORS

Integrated circuit resistors can be classified into three groups: (1) semiconductor films, (2) deposited metal films, and (3) cermets (a mixture of metal and dielectric materials). Of these, only the first two have found widespread use in high-frequency circuits. Semiconductor films can be fabricated by diffusion into a host semi-insulating substrate by depositing a polysilicon layer or by ion implantation of impurities into a prescribed region. Polysilicon, or polycrystalline silicon, consists of many small submicron crystals of silicon with random orientations.

Radio Frequency Circuit Design, Second Edition, by W. Alan Davis

2.2.1 Resistor Types

The resistance value of an integrated circuit resistor depends on the conductivity of the channel through which the current is flowing. In the diffused resistors in a semiconductor substrate, the conductivity is a function of the doping concentration and the carrier mobility. The conductivity is

$$\sigma = q(\mu_n n + \mu_p p) \tag{2.1}$$

It is usually expressed in the units of $(\Omega\text{-cm})^{-1}$. In this expression, q is the electronic charge (1.602×10^{-19}) C, μ_n and μ_p are the electron and hole mobilities (cm^2/V-s), and n and p are the number of free electrons and holes, respectively, available for conduction (cm^{-3}). At room temperature, it may be assumed that all the impurity atoms in the semiconductor are ionized. This means that for an n-type semiconductor, the number of available electrons is equal to the donor impurity concentration:

$$n_n \approx N_D \tag{2.2}$$

Similarly, for a p-type semiconductor, the number of holes equals the acceptor impurity concentration:

$$p_p \approx N_A \tag{2.3}$$

In either an n-type or p-type semiconductor, the relationship between the electron and hole concentrations is

$$np = n_i^2 \tag{2.4}$$

where $n_i = 1.45 \times 10^{10}$ cm^{-3} for silicon and 9.0×10^6 for gallium arsenide. This is called the mass action law. Thus, for an n-type semiconductor, the conductivity is

$$\sigma = q\left(\mu_n N_D + \mu_p \frac{n_i^2}{N_D}\right) \approx q\mu_n N_D \tag{2.5}$$

Typically, in integrated circuits, n-channel MOSFETs and *npn* bipolar transistors are preferred because of the much larger electron mobility over that of the hole mobility. The total number of processing steps required in a circuit design often dictates the choice of resistor channel type.

Ideally, the diffused resistor with conductivity, σ, can be represented by the rectangular block shown in Fig. 2.1. The resistance of the rectangular block is

$$R = \frac{L}{\sigma WT} \tag{2.6}$$

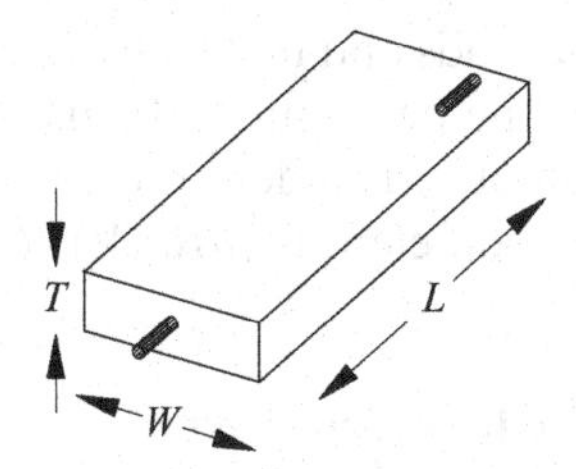

FIGURE 2.1 Diffused resistor of length L, width W, and height T.

TABLE 2.1 Resistor Materials

Resistor Type	Resistance	Temperature Coefficient	Voltage Coefficient
Diffused Si	10–100 Ω/□	1500 ppm/°C	200 ppm/V
Diffused GaAs	300–400 Ω/□	3000–3200 ppm/°C	—
Polysilicon	30–200 Ω/□	1500 ppm/°C	100 ppm/V
Ion implantation	0.5–2 kΩ/□	400 ppm/°C	800 ppm/V
AuGeNi (alloyed)	2 Ω/□	—	
Thin film Cr	13 μΩ-cm	3000 ppm/°C	
Thin film Ti	55–135 μΩ-cm	2500 ppm/°C	
Thin film Ta	180–220 μΩ-cm	−100 to +500 ppm/°C	
Thin film TaN	280 μΩ-cm	−180 to −300 ppm/°C	
Thin film Ni	7 μΩ-cm	—	
Thin film NiCr	60–600 μΩ-cm	200 ppm/°C	

Sources: From Pucel [1], Williams [2], and Allen and Holberg [3].

It is often convenient to separate the "processing" aspects from the "layout" aspects of the resistor. This is done by defining the sheet resistance in (Ω/□) as

$$R_{\square} = \frac{1}{\sigma T} \tag{2.7}$$

so that the total resistance is

$$R = R_{\square} \frac{L}{W} \tag{2.8}$$

The length-to-width ratio determines the resistance value once the conductivity and layer thickness is set.

Metal films are made by evaporation of the desired metal on a substrate and the desired pattern determined by photo lithography. Metal films are generally superior to the semiconductor films in that metal films are less sensitive to changes in temperature and voltage. Table 2.1 shows some of the main properties of a variety of methods and materials. The temperature and voltage coefficients are measures of the percentage change in resistance as a function of a change in a given parameter. The definition of temperature coefficient is $(dR/dT)/R$ and the voltage coefficient is $(dR/dV)/R$.

2.2.2 Resistance Determination from Layout

The layout shape of a resistor is typically simply a straight rectangular bar as shown in Fig. 2.1. However, it may at times be better to try to have different shapes in order to optimize the overall layout of a circuit. A convenient method for determining the resistance between two points on any shape is the method of curvilinear squares. Of course, computer-based numerical methods such as the finite-element technique can also be used. However, using paper and pencil, in just 20 minutes an answer can be obtained to within 10 to 20% accuracy.

A curvilinear rectangle may be defined "as any area which is bounded on opposite sides by two flux lines, and on the other sides by two equipotential lines" [4]. These rectangles can be divided and subdivided into squares of ever decreasing size. Then based on Eq. (2.8) the total resistance can be found by counting the squares.

Rather than estimating the "squareness" of a curvilinear square, circles can be drawn between two flow lines using a compass or a template. Each curvilinear square has four sides tangent to the inscribed circle.

The curvilinear square method is illustrated in Fig. 2.2 and is accomplished in the following way:

1. Draw flow lines between the two electrodes just as water would travel between the electrodes in a laminar flow. The spacing between two flow lines is less important than the shape of the flow lines. The flow lines should intersect the electrodes at right angles.
2. Between two adjacent flow lines, draw a series of circles tangent to the flow lines and to each other.
3. Draw equipotential lines between the circles orthogonal to the flow lines.
4. If there is more rectangle left over than an integral number of circles, then draw circles in the remaining rectangle in the orthogonal direction. This is continued until the last rectangle is sufficiently close to being a square.
5. Starting with the smallest square, count all the squares in series. Invert and add to the next largest row of squares going in the orthogonal direction. Continue inverting and adding to the next larger row of squares.

As Fig. 2.2 shows, the first step is to add the total number of the smallest squares. In this case the result is 2. Step 2 consists in inverting the result of step 1 and adding the remaining series squares, with the result $\frac{1}{2}+1=1.5$. In

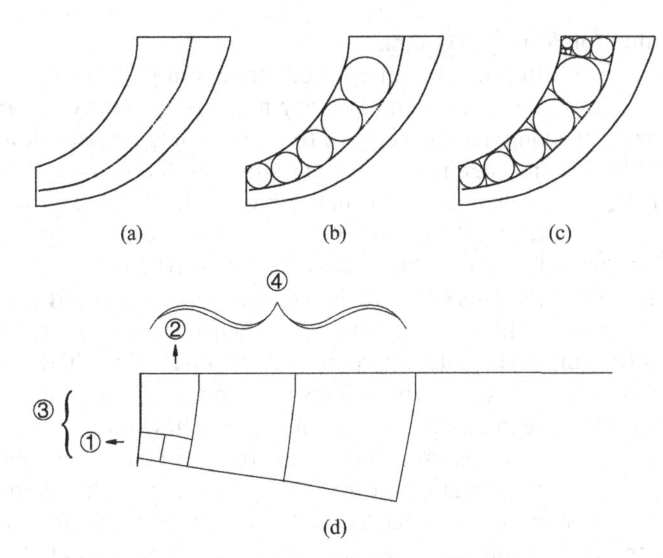

FIGURE 2.2 (*a*) Resistor shape with a flow line, (*b*) addition of tangential circles, (*c*) drawing best-fit curvilinear squares, and (*d*) expansion of the fractional curvilinear square from (*c*).

step 3, the result of step 2 is inverted and added to the remaining series squares. At the end of this step, the result is (1/1.5) + 2 = 2.67. Finally, step 4 gives 1/2.67 + 5 = 5.375. The resistance then in the indicated section of the resistor is $5.375R_{\square}$. These steps would be repeated for the other parallel flow lines to obtain the total resistance as a parallel combination. The most obvious application of this method to electrical engineers is in finding the resistance of an arbitrarily shaped resistor. However, it can also be applied in finding the magnetic reluctance in a magnetic circuit, capacitance, heat convection, and, of course, laminar fluid flow.

There are a couple of other details that should be considered in predicting resistance values. One is that the rectangular bars of resistance are not really rectangular bars. The bottom is rounded and a better estimate can be found by taking this into account. Another complication is that somewhere a semiconductor-diffused resistor is going to have to come in contact with a metal. The resulting Schottky barrier can cause an additional voltage drop. Normally, an ohmic contact is used for this interface. An ohmic contact is formed by heavily doping the semiconductor at the point of contact with the metal. This essentially promotes tunneling of electrons through the barrier. Nevertheless, there is still some residual resistance from the contact. Consequently, the previously given expression for resistance, Eq. (2.8), should be modified to incorporate the contact resistance, R_c:

$$R = R_{\square}\frac{L}{W} + \frac{2R_c}{W} \tag{2.9}$$

A typical value for R_c is 0.25 Ω-mm.

Active loads are often used in integrated circuits in place of passive loads where the required resistance value is fairly high. The primary advantage of the active load is its compact size relative to that of a large passive load. These are often used in common emitter *npn* transistor amplifiers or FET amplifiers as shown in Fig. 2.3. In the cases shown, the base–collector, the gate–drain of the enhancement mode MOSFET, or the gate–source of the depletion mode MOSFET are shorted together. An active load can also be made in GaAs with a "saturated resistor" [5]. This structure is essentially a GaAs MESFET without a gate, and it is simpler to construct than the usual depletion mode FET with gate shorted to source. The saturation current in GaAs is reached at a rather low saturation field of 3 kV/cm. This means that once saturation has occurred, there is a small increase in current with each increase in voltage. Consequently, a large effective resistance is obtained. The saturated resistor channel depth is effectively greater than that of the MESFET channel as shown in Fig. 2.4. Consequently, for a given resistance value, the width of the saturated resistor would have to be made narrower. Resistance values of 8 to 10 kΩ have been obtained [5]. However, the simpler processing of the saturated resistor has given improved reliability and repeatability of these devices.

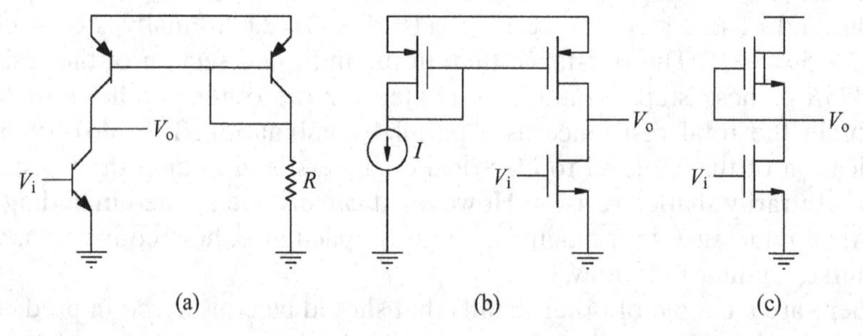

FIGURE 2.3 Active loads using (*a*) common emitter structure, (*b*) *p*-channel enhancement mode MOSFET load, and (*c*) *n*-channel depletion mode MOSFET load.

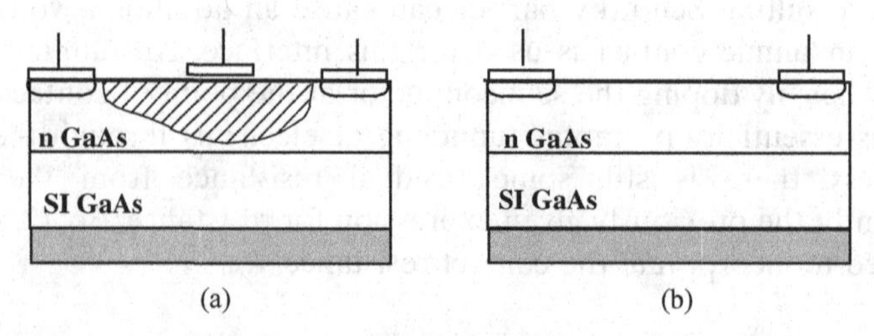

FIGURE 2.4 Charge distribution for (*a*) shorted gate-active load and (*b*) a saturated resistor.

2.3 CAPACITORS

Some of the most important parameters that need consideration in choosing a capacitance are (1) the capacitance value, (2) capacitance value tolerance, (3) loss or Q, (4) temperature stability, (5) mechanical packaging and size, and (6) parasitic inductance. These criteria are interdependent, so often the appropriate compromises depend on the constraints imposed by the particular application. This section will consider both hybrid and monolithic capacitor designs.

2.3.1 Hybrid Capacitors

Hybrid capacitors are available in both single-layer capacitors for high-frequency low-capacitance applications and multilayer capacitors for higher capacitance. Even for multilayer chip capacitors, the self-resonant frequency for a 0.1-pF capacitor is over 10 GHz and for a 1000-pF capacitor the self-resonant frequency of 250 MHz. These capacitors can be attached to printed circuit boards to provide high available capacitances with relatively low loss. Unlike low-frequency circuits, certain parasitic circuit elements must be accommodated in the overall design. The parasitic inductance is affected by the packaging since it is usually associated with the lead attachments to the capacitor and line length effects inside the capacitor. In low-frequency circuits, the effect of the inductance is so small that it can safely be neglected. However, at radio frequencies both the inductance and the metal losses often become significant. Consequently, the equivalent circuit for a chip capacitor as developed by chip capacitor manufactures is shown in Fig. 2.5 and can sometimes be simplified as simply a series *RLC* circuit. The additional parallel resistance, R_p, is added to this equivalent circuit to model resistive losses caused by dielectric loss. This parameter is the main loss at low frequencies in the hertz-to-kilohertz range, but at RF it becomes negligible when compared to R_s. The impedance of the circuit is

$$Z = R_s + \frac{j}{\omega C}(\omega^2 LC - 1) \tag{2.10}$$

Consequently, the effective capacitance is frequency dependent:

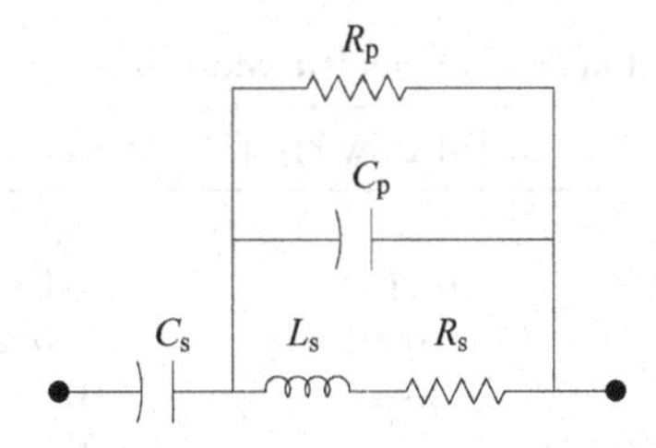

FIGURE 2.5 Typical equivalent circuit for a chip capacitor.

$$C_{\text{eff}} = \frac{C}{1-(\omega/\omega_0)^2} \tag{2.11}$$

where $\omega_0 = 1/\sqrt{LC}$ is the self-resonant frequency.

While loss in capacitors is usually less than that in inductors, capacitor loss can still be significant in circuit performance. Loss can be described in terms of dissipation factor (DF), loss tangent ($\tan\delta$), the equivalent series resistance (R_s), and Q_{cap}. Since the circuit Q is assumed to result from a series RLC configuration,

$$R_s = \frac{X_c}{Q_{\text{cap}}} \tag{2.12}$$

The loss terms then are related by

$$\tan\delta = \text{DF} = \frac{1}{Q_{\text{cap}}} \tag{2.13}$$

The angle $\delta = 90° - \theta$ where θ is the angle between the voltage and the current. In a lossless capacitor, $\theta = 90°$.

In a capacitor, the dielectric is the primary source of loss. An RF field can cause the dipole molecules in the dielectric to rotate at a rate proportional to the applied frequency and with a force proportional to electric field strength. The rotation of these molecular dipoles is converted to heat loss. When E is the electric field and f is the frequency, the energy dissipation is given by the following empirical expression [6]:

$$\mathcal{E} = E^2 f 55.5 \times 10^{-6} \varepsilon_r \tan\delta \quad \text{W/cm}^3 \tag{2.14}$$

Some of the most widely used dielectric materials for capacitors are shown in Table 2.2.

The $BaTiO_3$ $\varepsilon_r = 8000$ material provides the most compact capacitor. However, it has a relatively poor temperature coefficient, $\tan\delta$ shift with voltage, coefficient of expansion versus temperature, piezoelectric effects, and aging qualities because of its porosity.

TABLE 2.2 Loss Tangent ($\tan\delta$) of Dielectric Materials

Dielectric	$\tan\delta$ at Low Freq.	$\tan\delta$ at 100 MHz	$\tan\delta$ at 1 GHz
$BaTiO_3$ $\varepsilon_r = 8000$	—	0.1	—
$BaTiO_3$ $\varepsilon_r = 1200$	0.01	0.03	0.10
Ceramic (NPO) $\varepsilon_r = 30$	0.0001	0.002	0.10
Alumina $\varepsilon_r = 9.8$	—	0.0005	—
Porcelain (ATC 100) $\varepsilon_r = 15$	—	0.00007	—

The $BaTiO_3$ $\varepsilon_r = 1200$ capacitance varies by +15% from −55 to 125 °C. When the $BaTiO_3$ materials are heated to about the Curie point, the value for ε_r jumps up about 10 to 15%. After cooling and waiting 10 hours, the dielectric constant drops back down only 3% of its peak value, and after 10,000 hours, it drops down only 7% of its peak value. As the voltage changes over a range of 30 V, the loss tangent increases from 0.01 to 0.1 at low frequencies. There are four crystalline phases for $BaTiO_3$ as it is heated up. The crystal changes from orthorhombic to tetragonal to cubic (which is near the Curie point). At each of these changes, there is an abrupt change in the mechanical size of the crystal [7]. This has deleterious implications on solder joints of the capacitor.

The capacitance using NPO material varies with temperature ±30 ppm/°C. It moves in the negative direction, then in the positive direction exceeding the initial capacitance, and finally settling down near the original capacitance as the temperature rises. Hence, the name NPO.

The porcelain materials, such as ATC 100 from American Technical Ceramics, provide high Q, no piezoelectric effects, no aging effects (since it is not a porous material), and temperature coefficient of ±30 ppm/°C up to 125 °C. The coefficient of expansion of the porcelain capacitor is the same as alumina (Al_2O_3). For this reason when mounted on an alumina substrate the two will expand the same amount. The series resistance at 1 GHz varies with the value of capacitance as shown in Table 2.3.

For a 30-pF $BaTiO_3$, $\varepsilon_r = 1200$ capacitor operating at 300 MHz, the resistance can be as high as 1 Ω and result in 0.3- to 3-dB dissipation loss. In solid-state circuits that operate in high-current and low-voltage conditions, these losses can be quite significant. The generated heat further degrades the loss tangent, which increases the heat dissipation. Thermal runaway can occur causing self-destruction. Of the materials shown in Table 2.2, the porcelain material provides the best loss tangent, especially at frequencies in the 1- to 3 GHz range.

The frequency range of a chip capacitor can be extended by the simple expedient of turning it on its side (Fig. 2.6). Resonances appear to be the result

TABLE 2.3 Resistance of Porcelain Capacitors

C (pF)	R_s (Ω)
5.6	0.38
10	0.27
20	0.19
30	0.16
40	0.13
50	0.12
100	0.088

Source: Perna [7].

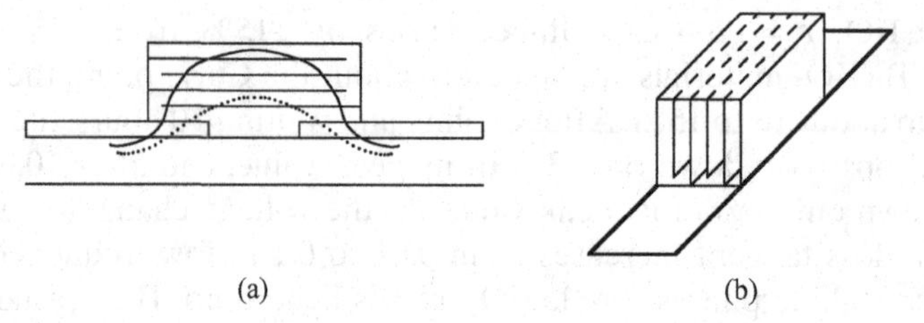

FIGURE 2.6 Metallic conductors in (*a*) horizontal and (*b*) vertical orientation.

of different path lengths of the path through the lower plates and upper plates of a multilayer capacitor. Turning the capacitor on its side tends to equalize the path lengths and eliminates all odd-order harmonic resonances [7].

2.3.2 Monolithic Capacitors

Capacitors in monolithic circuits are best avoided where possible because of the amount of real estate they occupy. Nevertheless, they are sometimes required. The capacitance tolerance is typically ±10%, and capacitance values range from 0.2 to 100 pF. There are four types of monolithic capacitors that might be used in integrated circuit designs: (1) open-circuit stub, (2) interdigital line, (3) metal–insulator–metal, and (4) varactor diode.

The open-circuit stub capacitance is simply an open-circuit transmission line whose length is less than $\lambda/4$. The capacitive susceptance is obtained from the transmission line equation:

$$B = Y_0 \tan\left(\frac{\omega l}{v_c}\right) \tag{2.15}$$

The value of the susceptance depends on the characteristic admittance, Y_0, of the transmission line, the length, l, of the transmission line, and the substrate material that governs the velocity of the wave, v_c. This open-circuit stub provides a shunt capacitance to ground. While the susceptance is not proportional to ω as in lumped capacitors, it is a good approximation when the argument of the tangent function is small. Line lengths can use a large amount of real estate at low frequencies, so typically the open-stub capacitor is most useful at frequencies greater than about 8 GHz.

The interdigital capacitor shown in Fig. 2.7, unlike the open stub, provides series capacitance. It is most useful for capacitances less than 1 pF, and at 12 to 14 GHz it typically has a Q of 35 to 50. The equivalent circuit shown in Fig. 2.7 includes series resistance and inductance, as well as some shunt capacitance to ground. The latter is caused by the metal–insulator–ground metal of the microstrip structure. The main series capacitance can be estimated from

$$C = (N_f - 1)C_g \ell \tag{2.16}$$

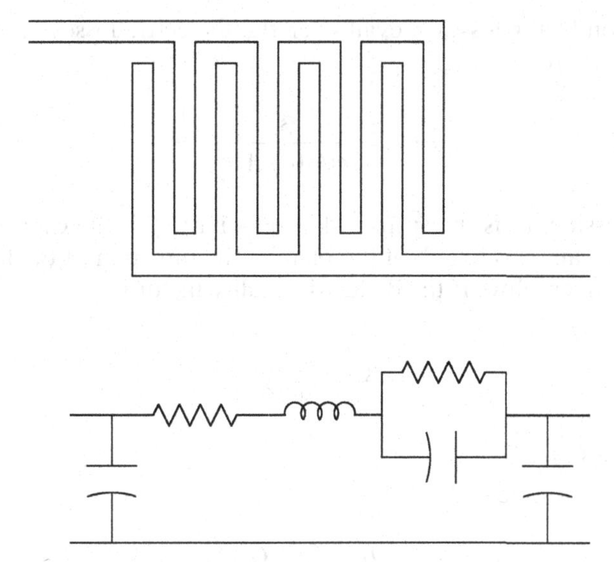

FIGURE 2.7 Interdigital capacitor layout and equivalent circuit.

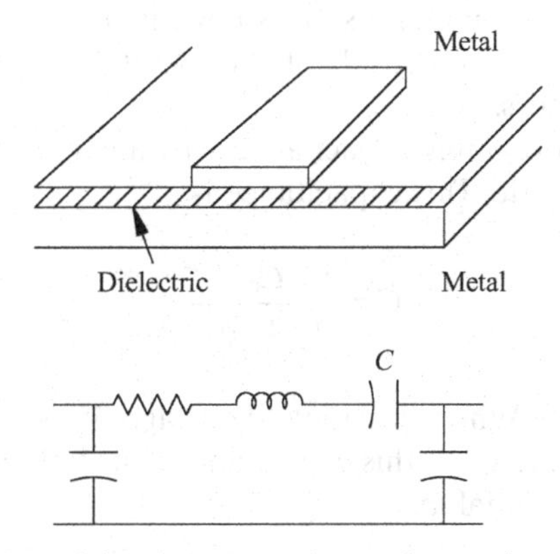

FIGURE 2.8 Metal–insulator–metal capacitor and equivalent circuit.

where N_f is the number of fingers, ℓ is the finger length, and C_g is the static gap capacitance per unit length between the fingers.

A third type of capacitor is the metal–insulator–metal capacitor (Fig. 2.8). Of the four monolithic capacitors, this is the most popular and is the most obvious. The dielectric thickness typically used is 0.1 to 0.4 μm. Losses can be reduced if the metal thickness is greater than 2 skin depths. The metal surface roughness should be as smooth as possible to reduce losses and avoid pin holes in the dielectric. Typically, the capacitance ranges from 50 to 300 pF/mm^2 [2].

When the conductor losses prevail over the dielectric losses, the conductor quality factor is [1]

$$Q_c = \frac{3}{2\omega R_s (C/A)\ell^2} \tag{2.17}$$

In this expression, R_s is the surface skin resistivity, C is the capacitance, A is the plate area, and ℓ is length of the plate in the direction of the longitudinal microwave current flow. If the dielectric quality factor is

$$Q_d = \frac{1}{\tan\delta} \tag{2.18}$$

then the total Q is

$$\frac{1}{Q_T} = \frac{1}{Q_d} + \frac{1}{Q_c} \tag{2.19}$$

The dielectric films used in monolithic capacitors tend to be much higher than that obtained in the hybrid capacitors described above. Some typical metal–insulator–metal dielectric materials are shown in Table 2.4. The variableness in the dielectric constant is a result of the variation in deposition methods, uniformity, and thickness.

The fourth way of obtaining capacitance is by means of the junction capacitance of a Schottky diode. This capacitance is

$$C = \frac{C_0}{(1 - V/\phi)^\gamma} \tag{2.20}$$

where $\gamma \approx \frac{1}{2}$ [8, p. 190]. When the applied voltage, V, is zero, the capacitance is C_0. A major disadvantage of this capacitance is its voltage dependence relative to the built-in potential, ϕ.

TABLE 2.4 Monolithic Capacitor Dielectric Materials

Dielectric	Nominal ε_r	Range of ε_r	Temperature Coefficient (*ppm*/°C)
SiO_2	5	4–5	50–100
Si_3N_4	7.5	5.5–7.5	25–35
Ta_2O_5	21	20–25	200–400
Al_2O_3	9	6–10	100–500
Polyimide	3.5	3–4.5	–500

Source: Pucel [1] and Williams [2].

2.4 INDUCTORS

Inductors operating at radio frequencies have a variety of practical limitations that require special attention. A tightly wound coil in addition to providing a self-inductance also has heat loss due to the nonzero wire resistance, skin effect losses, eddy current losses, and hysteresis losses when a magnetic material is used. Furthermore, two conductors close together at two different voltages will also exhibit an interelectrode capacitance. At radio frequencies these effects cannot be neglected as easily as they could at lower frequencies. The equivalent circuit is shown in Fig. 2.9. In this figure, the series resistance, R_s, represents the conductor loss as well as the skin effect losses. The parallel resistance, R_p, represents the effect of eddy current losses and the hysteresis loss in magnetic materials when present. The shunt capacitance, C_p, is the capacitance found between the coils. Straightforward circuit analysis gives the impedance for this equivalent circuit:

$$Z = \frac{R_p R_s + R_p L s}{s^2 L C_p R_p + s(R_s C_p R_p + L) + R_s + R_p} \tag{2.21}$$

If R_p is considered so large as to have negligible effect, and if the remaining series circuit $Q = 1/\omega R_s C$ is large, then the effective inductance is approximately

$$L_{\text{eff}} = \frac{L}{1 - \omega^2 L C_p} \tag{2.22}$$

and the effective resistance is

$$R_{\text{eff}} = \frac{R_s}{1 - \omega^2 L C_p} \tag{2.23}$$

Clearly, the presence of the capacitance dramatically increases the effective inductance and capacitance near the self-resonant frequency of the inductor.

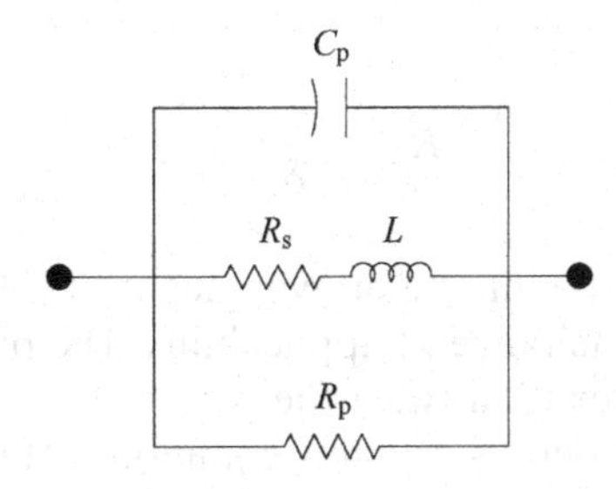

FIGURE 2.9 Simple equivalent circuit for an inductor.

The self-capacitance of the inductor is a function of the coil-length-to-coil-diameter ratio, ℓ/D, and has an optimum value [9, 10]. The following sections will describe in greater detail the origin of the parasitic circuit elements for a practical RF inductor and some design methods for RF inductors.

2.4.1 Resistance Losses

The direct current (dc) flowing through a wire with a cross-sectional area, A, will encounter half the resistance if the area is doubled. At radio frequencies, the alternating current (ac) tends to flow near the surface of the conductor because of the skin effect. This can be illustrated by an electric field impinging on a conductor whose resistance is not zero. The field will penetrate into the conductor and will exponentially decay as it penetrates deeper:

$$E(x) = E_0 e^{-x/\delta} \tag{2.24}$$

where

$$\delta = \sqrt{\frac{\rho}{\pi f \mu}} \tag{2.25}$$

In this equation f is the frequency, ρ is the resistivity, and μ is the permeability. Because of this skin depth, the resistance of a given wire with radius R will have a higher resistance at high frequencies than at direct current. The ac resistance is given by [9]

$$\begin{aligned} R_{ac} &= \frac{A_{tot}}{A_{skin}} R_{dc} \\ &= \frac{\pi R^2}{\pi R^2 - \pi (R-\delta)^2} R_{dc} \\ &= \left(\frac{\pi R^2}{2\pi \delta R - \pi \delta^2} \right) R_{dc} \end{aligned} \tag{2.26}$$

At high frequencies,

$$R_{ac} \approx \frac{R}{2\delta} R_{dc} \tag{2.27}$$

The possibility for R_{ac} to be infinite or even negative clearly indicates that Eq. (2.26) has gone beyond its range of applicability. The problem is that the skin depth has become greater than twice the wire radius. Listed in Table 2.5 are the resistivities and skin depths of a few common metals.

Another important loss mechanism is called the proximity effect. When one conductor supporting a changing magnetic field is brought close to another

TABLE 2.5 Common Conductors

Metal	Conductivity $(\Omega\text{-cm})^{-1}$	Skin Depth (cm)
Brass	1.57×10^5	$12.7f^{-1/2}$
Aluminum	3.54×10^5	$8.46f^{-1/2}$
Gold	4.27×10^5	$7.7f^{-1/2}$
Copper	5.8×10^5	$6.61f^{-1/2}$
Silver	6.14×10^5	$6.42f^{-1/2}$
Mu-Metal	1.58×10^8	$0.4f^{-1/2}$

conductor, currents will be induced on the second conductor in conformity with Faraday's law. These currents are called *eddy currents*, and they flow in closed paths so as to produce a magnetic field that is in opposition to the originally applied external field. These currents produce *joule heating*. This is exactly the condition that occurs in a tightly wound inductive coil. When many wires are close together, the loss problem is compounded and the eddy current losses can be quite significant. As an illustration of this, consider a coil with a length-to-diameter ratio of 0.7. If this coil is unwound and laid out as a straight wire, the losses would drop by a factor of 6 [9, p. 47].

2.4.2 Magnetic Materials

A recurring problem is the need for a large value of inductance. An obvious solution is to increase the flux density within an inductor coil with the addition of a magnetic material having high relative permeability μ_r. Most magnetic materials introduce losses that are unacceptable at radio frequencies. A variety of ferrite materials, however, have been found to have low loss at radio and microwave frequencies in comparison with most other magnetic materials. The relative permeability for ferrites is in the range $10 < \mu_r < 150$. Above the cutoff frequency, μ_r drops off quickly. The higher the permeability, the lower the cutoff frequency. Typically, for $\mu_r = 10$, $f_{\text{cutoff}} = 1\,\text{GHz}$. For $\mu_r = 150$, $f_{\text{cutoff}} = 20\,\text{MHz}$.

2.4.3 Solenoid Design up to 2 GHz [11]*

A design procedure for a single-layer solenoid is given below. The computer program, SOLENOID, follows the procedure outlined here and is described in Appendix A. The given parameters for the analysis of a solenoid are the form length, number of turns (n), and the form diameter. The pitch is defined as

$$\text{Pitch} = \frac{\text{form length}}{n} \quad \text{inches} \tag{2.28}$$

*Reprinted with permission, *Microwave Journal*, Vol. 39, pp. 70–76.

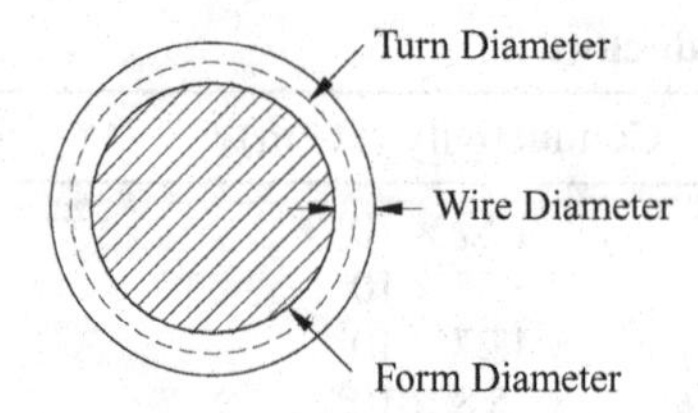

FIGURE 2.10 Inductor form cross section.

For maximum Q, the wire diameter should be 0.6 to 0.7 of the value for the pitch. The wire diameter is selected from the standard wire sizes. For a given American wire gauge (AWG) value, the wire diameter is

$$\text{Wire diameter} = \frac{0.005}{92^{(\text{AWG}-36)/39}} \quad \text{inches} \tag{2.29}$$

Another parameter is the turn diameter. It represents the diameter where the magnetic flux is generated. As shown in Fig. 2.10, turn diameter = form diameter + wire diameter. With these quantities now defined, the analysis recipe can be followed:

$$x = \frac{\text{turn diameter}}{\text{form length}} \tag{2.30}$$

$$K_\text{n} = \frac{1}{1+0.45x-0.005x^2} \tag{2.31}$$

When the wire diameter = 0, the current sheet correction factor, s, is set to 1. When the wire diameter > 0, the s correction factor is needed. The s factor and finally the inductance, is found by first finding a and b as shown below:

$$a = 2.3\log_{10}\left(1.73\frac{\text{wire diameter}}{\text{pitch}}\right) \tag{2.32}$$

$$b = 0.336\left(1-\frac{2.5}{n}+\frac{3.8}{n^2}\right) \tag{2.33}$$

$$s = 1-\frac{2(\text{form length})(a+b)}{\pi(\text{turn diameter})nK_\text{n}} \tag{2.34}$$

$$L_0 = \frac{[n\pi(\text{turn diameter})]^2\, 2.54\times10^{-9}}{\text{form length}} \tag{2.35}$$

$$L = L_0 K_\text{n} s \qquad \text{henries} \tag{2.36}$$

The value, L_0, is the inductance of a closely wound coil with a flat strip (wire diameter = 0). The value, K_n, is the Nagaoka correction factor and is used when the wire length is not much larger than the turn diameter. The value, s, is the current sheet correction factor and is needed when there is appreciable space between wire turns. Because L_0 is not dimensionless, the lengths must be given in terms of inches. An example given in [11] illustrates the use of these expressions:

Pitch = 0.0267 (n = 15)
Wire diameter = 0.0187 in.
Turn diameter = 0.2187 in.
$x = 0.5467$
$K_n = 0.8035$
$a = 0.1912$
$b = 0.2857$
$s = 0.954$
$L_0 = 674\,\text{nH}$
$L = 516.8\,\text{nH}$

A synthesis procedure is also available [11]. The goal is to design a given value of inductance. Only a finite number of form diameters are available, so the form diameter will also be considered as a given quantity. From this, the number of turns n and the form length, FLEN, is found.

The inductance is a function of n and FLEN, and L_d is the desired inductance. An iterative procedure is followed where

$$L_1 = L(n_1, \text{FLEN}) \tag{2.37}$$

$$L_2 = L(n_2 = n_1 \pm 1, \text{FLEN}) \tag{2.38}$$

$$n_3 = n_2 - (L_2 - L_d)\frac{n_2 - n_1}{L_2 - L_1} \tag{2.39}$$

This iteration loop is repeated until

$$L(n, \text{FLEN}) = L_d \tag{2.40}$$

The equality is obtained, although with a noninteger value for n. For printed circuit boards, n must be an integer. While the number of form diameters is limited, the form length can be cut to any desired length. Therefore the form length is adjusted to guarantee an integral n. The procedure is to increase n to the next higher integer value and adjust FLEN by an iterative scheme much like the previous one:

$$L_1 = L(n, \text{FLEN}_1) \tag{2.41}$$

$$L_2 = L(n, \text{FLEN}_2) \tag{2.42}$$

$$\text{FLEN}_3 = \text{FLEN}_2 - (L_2 - L_d)\frac{\text{FLEN}_2 - \text{FLEN}_1}{L_2 - L_1} \tag{2.43}$$

This iteration loop is repeated until

$$L(n, \text{FLEN}) = L_d \tag{2.44}$$

where n is an integer value.

Once L, n, and FLEN are known, the Q factor and the parasitic capacitance can be found using the formulas given in [11]. Using the value for x given in Eq. (2.30), a value for the capacitance is determined:

$$C = (\text{turn diameter})\frac{(0.301468x + 0.493025)x + 0.227858}{x} \quad \text{pF} \tag{2.45}$$

The coil resonant frequency is then simply

$$f_r = \frac{1}{2\pi\sqrt{LC}} \tag{2.46}$$

The value for Q is found from the empirical relationship for two cases where the turn diameter is in inches and f_r is in megahertz:

$$A = \begin{cases} [(58.6355x - 171.154)x + 200.674]x + 0.089708 & 0.2 < x < 1 \\ [(0.751186x - 9.49018)x + 42.506]x + 68.1191 & 1 < x < 5 \end{cases} \tag{2.47}$$

The value for Q is then obtained from the two-step formula below:

$$Q_0 = A(\text{turn diameter})\sqrt{f} \tag{2.48}$$

$$Q = Q_0\left[1 - \left(\frac{f}{f_r}\right)^2\right] \tag{2.49}$$

The procedure described above has been put in the form of the computer algorithm, SOLENOID. An example of the design of a 100-nH inductor is found in Appendix A using this program.

2.4.4 Older Solenoid Formulas

Empirical formulas for solenoid designs have been proposed by many others, one of these authors being Wheeler [12, 13]. In conformity with his notation, $2a \approx$ turn diameter, b = FLEN the form length, and n is the number of turns. In the 1928 study [12] where a and b are given in inches, the inductance is

TABLE 2.6 Solenoid Inductance Values

FLEN = b (in.)	Diameter = $2a$ (in.)	n	L_{1996} (μH) [11]	L_{1928} (μH) [12]	L_{1982} (μH) [13]
1	0.2	20	0.4643	0.3670	0.3693
1	0.2	25	0.6938	0.5734	0.5770
1	0.2	30	0.9693	0.8257	0.8309
0.4	0.2	10	0.2430	0.2041	0.2053
0.4	0.2	15	0.5168	0.4592	0.4619
0.4	0.2	20	1.520	1.3559	1.3648
1	0.4	20	1.520	1.3598	1.3648
1	0.4	25	2.324	2.1186	2.1325
1	0.4	30	3.297	3.0508	3.0609

$$L = \frac{a^2 n^2}{9a + 10b} \times 10^{-6} \quad \text{H} \tag{2.50}$$

A more accurate formula was published in 1982 [13]:

$$L = \mu_0 n^2 a \left[\ln\left(1 + \pi \frac{a}{b}\right) + \frac{1}{2.3 + 1.6(b/a) + 0.44(b/a)^2} \right] \quad \text{H} \tag{2.51}$$

This latter formula was taken to have an error <0.001 relative to a theoretical value. In this formula, $2.3 = 1/[\ln(8/\pi) - 0.5]$ and $0.44 = 6/(3\pi^2 - 16)$. These last two formulas do not account for the wire diameter nor the wire spacing.

A numerical comparison of these formulas with those given in Section 2.4.3, which were published in 1996, are shown in Table 2.6. For these particular designs, the values show that the 1996 inductance values range from 7.5 to 21% higher than those from the 1928 study.

2.4.5 Monolithic Spiral Inductors

Lumped monolithic inductors have been used in circuit designs as tuned loads for amplifiers, filters to reduce out-of-band signals and noise, and as a means of enhancing stage gain by tuning out device or parasitic capacitances at the center frequency. Planar inductors have been implemented in practical systems for many years using a variety of different substrates. They were examined early in the development of silicon integrated circuits, but they were abandoned because of process limitations and losses in the series resistance and substrate that effectively reduced their operating frequency. Now, however, technological improvements have made them available for mobile communications systems.

Small inductances in the nanohenry range can be fabricated using printed circuit techniques. These have typically been done in either a rectangular or circular spiral shape. Both are widely used, but the circular spiral design seems to provide greater inductance per unit area of real estate. In determining the inductance in either case, the self-inductance of the structure must be supplemented by the mutual inductance of neighboring turns as well as the mutual inductance of its mirror image associated with the ground plane for microstrip. There are, in addition, capacitances between turns of the spiral and capacitances to the ground. These capacitances are calculated from coupled microstrip line theory. A numerical implementation of the rectangular inductance based on [14] is provided. This study is heavily cited and does provide a comparison between the predicted and the measured inductance.

A comparison is made between square and circular inductors in [15] in which it is stated that square spirals provide less inductance than circular spirals for equivalent sized diameters, although the data seem ambiguous. This study uses a simple lumped-element equivalent circuit consisting of a series R–L circuit with shunt capacitances on either side to represent a single turn. However, the entire inductor is treated as a distributed circuit.

A design of a square inductor is described in [16], which is modeled like the one in [15] except that an additional resistance is added in series to the shunt capacitors to ground. A comparison is made with measured data and the design is incorporated into a low-pass filter design.

In an effort to increase the desired Q for an inductor, the ground plane under the square spiral is removed in [17]. Excellent agreement is obtained up to 5 GHz.

An extensive study of over 100 inductors was made in [18]. Comparisons were made between square, octagonal, and circular spirals. Empirically determined equivalent circuits were obtained based on measured data. The basic conclusions were that the resistance of circular or octagonal shaped spiral is 10% lower than that for a square spiral inductor for the same inductance. Furthermore, it is better to maximize line spacing rather than maximizing line width to achieve high Q.

The capacitance itself becomes a major part of the inductor model. An effort is made in [19] to predict the distributed capacitances of circular spiral inductors by means of a Green's function analysis. Good agreement between predicted and measured values are obtained.

An actual determination of an equivalent circuit model for a spiral inductor was obtained in [20]. The computer program is posted on the Web.[†] The "circular" spiral is a p-sided polygon of n turns with a total of np sections. Each section is modeled as shown in Fig. 2.11. The analysis includes the effects of the internal impedance of each section as well as the magnetic and electric coupling to neighboring segments and the substrate. The primary advantages of using this analysis tool is the speed of computation (unlike a three-dimensional field simulator), optimization, and the ability to analyze spiral transformers as well as inductors with various metalizations and shapes. The

[†]ASITIC is found at http://rfic.eecs.berkeley.edu/~niknejad/asitic.html.

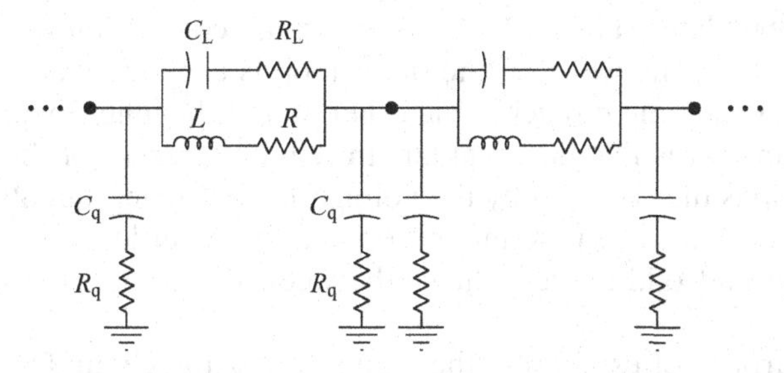

FIGURE 2.11 Spiral inductor modeled by a cascade of equivalent circuit sections [21].

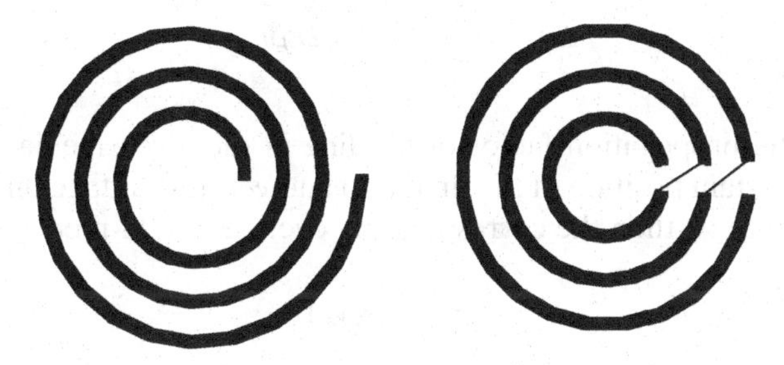

FIGURE 2.12 Spiral inductor can be approximated by concentric circular coupled lines.

geometrical shape of the inductor depends on the area of the spiral, metal width, metal spacing, the number of turns, and frequency of operation. The appropriate choice for these parameters are aided by the ASITIC program.

An alternate approach approximates an n turn circular spiral as a set of n concentric circular microstrips (Fig. 2.12). Each of these circular microstrips are modeled by an equivalent circuit shown in Fig. 2.11 where $R_L = R_q = 0$ [15]. The total equivalent circuit of the circular spiral is simply the cascade of each of the circular sections. The series resistance, R, represents the resistive loss in the conductor. The resistance is be proportional to $\sqrt{f}$ because of the skin effect. The capacitances C_{q1} and C_{q2} are the capacitances to the ground plane, and C_L is the total coupling capacitance between neighboring turns. What is lacking here, but considered in [20], is that mutual coupling occurs for line segments that are not parallel. What is helpful though are the closed-form equations given in [21] for various types of inductor elements.

The capacitances are determined from coupled line theory in which each line can be excited with the same voltage (even mode) or equal but opposite signed voltages (odd mode). The actual capacitance is a linear combination of the even- and odd-mode capacitances. The percentage of the even-mode and odd-mode capacitances between two adjacent turns of the spiral may be found based on the following approximations. For a given pair of adjacent concentric circular lines, assume there is a small break between the excitation of one loop and its end, 360° later (Fig. 2.12). There is a small connection from this point to the next loop. If the wave on the inner loop is excited by a 1-V source of $1 \times e^{j0°}$, then by the time it reaches the other end of the inner loop, the voltage is $1 \times e^{j\Delta\phi}$, where $\Delta\phi$ is the electrical length (circumference) of the inner loop. The outer loop is then excited by the voltage $1 \times e^{j\Delta\phi}$. Consequently, there is a voltage difference between the inner loop and the outer loop. The percentage of even-mode and odd-mode voltages between the two loops is a function of $\Delta\phi$.

For the purpose of estimating the value for $\Delta\phi$, the circumferences of the two circles will be assumed to be the average of the two circles:

$$\Delta\phi = 2\pi\frac{r_1 + r_2}{2}\beta = 2\pi\beta r_{\text{avg}} \tag{2.52}$$

where β is the propagation factor of the line in the given media. If v_1 is the voltage at a certain position of the first loop and v_2 is the voltage on the second loop adjacent to v_1, then the corresponding even- and odd-mode voltages are

$$v_{\text{e}} = \tfrac{1}{2}(v_1 + v_2) \tag{2.53}$$

$$v_{\text{o}} = (v_1 - v_2) \tag{2.54}$$

No information is lost in doing this since the original voltages v_1 and v_2 are easily recovered if v_e and v_o are known. The percentage of even- and odd-mode capacitances are proportional to the even- and odd-mode voltages:

$$\%C_{\text{e}} = \frac{v_{\text{e}}}{v_{\text{e}} + v_{\text{o}}} \tag{2.55}$$

$$\%C_{\text{o}} = \frac{v_{\text{o}}}{v_{\text{e}} + v_{\text{o}}} \tag{2.56}$$

In the equivalent circuit for a single turn of the spiral shown in Fig. 2.11, the percentage of even-mode excitation determines the relative amount of even- and odd-mode capacitance components. For the even mode,

$$C_{\text{L}} = 0 \tag{2.57}$$

$$C_{qi} = C_{\text{m}} + C_{\text{f}} + C_{\text{f}}' \tag{2.58}$$

and for the odd mode

$$C_{\text{L}} = C_{\text{ga}} + C_{\text{ge}} \tag{2.59}$$

$$C_{qi} = C_{\text{m}} + C_{\text{f}}' \quad i = 1, 2 \tag{2.60}$$

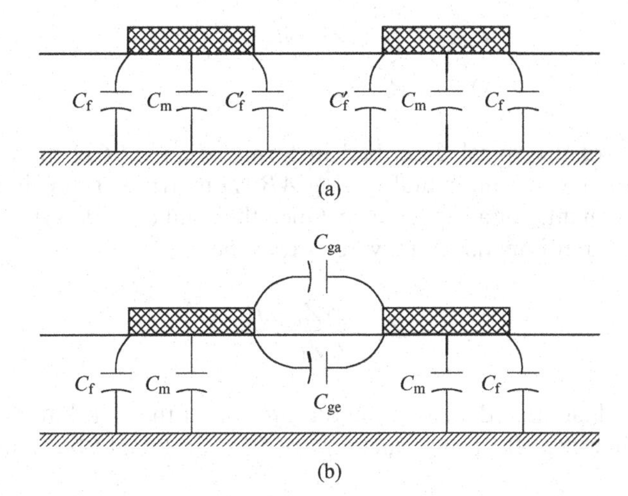

FIGURE 2.13 (*a*) Even- and (*b*) odd-mode excitation of microstrip lines.

In these expressions, C_{ga} and C_{ge} represent the gap capacitances between the lines through the air and through the dielectric, respectively. The capacitance, C_m, represents the parallel-plate capacitance between the spiral conductor and the ground plane. This is modified by the fringing capacitance, C_f', between each of the two lines to ground (which is nonzero only for the even-mode excitation) and the fringing capacitance, C_f, on the other side of the conductors. The even- and odd-mode capacitances are added together in proportion to their even- and odd-mode voltages:

$$C_L = 0 + \%C_o(C_{ga} + C_{ge}) \tag{2.61}$$

$$C_{qi} = \%C_e(C_f + C_m + C_f') + \%C_o(C_m + C_f') \tag{2.62}$$

Detailed formulas for the circuit elements in Fig. 2.13 are found in [21], and some of these are summarized in Appendix B.

Once the equivalent π circuit for the spiral section is known, the entire spiral inductor is modeled by cascading each of these sections. The C_L, R, and L are combined into the single impedance Z_p:

$$Z_p = \frac{R + sL}{s^2 LC_L + sC_L R + 1} \tag{2.63}$$

The *ABCD* parameters described in Section 4.2 are used to cascade the individual π circuits. Thus,

$$A = 1 + sC_{q2}Z_p \tag{2.64}$$

$$B = Z_p \tag{2.65}$$

$$C = s^2 C_{q1} C_{q2} Z_p + s(C_{q1} + C_{q2}) \tag{2.66}$$

$$D = 1 + sC_{q1} Z_p \tag{2.67}$$

Each section of the spiral described in terms of an *ABCD* matrix may be cascaded together by simply multiplying *ABCD* matrices. The *C* in Eq. (2.66) is a matrix element, not a capacitance. Once the total cascaded *ABCD* matrix is found, the input impedance may be determined:

$$Z_{in} = \frac{AZ_L + B}{CZ_L + D} \tag{2.68}$$

The Z_L is the load impedance on the output side of the spiral. If Z_L is a short to ground, then the effective inductance of the spiral might be estimated by

$$L_{eff} = \frac{\Im\{Z_{in}\}}{\omega} \tag{2.69}$$

In the cascade analysis, the capacitance, C_{q2}, from one section is the same as the C_{q1} of the subsequent section, and hence ought not to be counted twice in evaluating the cascaded equivalent circuit. One approach is to simply choose $C_{q2} \rightarrow C_{q2}/2$ and $C_{q1} \rightarrow C_{q1}/2$ except, of course, for the innermost and outermost coupled line section.

2.5 CONCLUSIONS

The basic components used in an RF transceiver are the same old resistors, inductors, and capacitors. However, at these high frequencies, these components may look and act a lot differently than their low-frequency counterparts. These differences arise from such things as the increasing importance of stray parasitic reactances, skin effect losses, and frequency dependence of materials.

PROBLEMS

2.1. Calculate the resistance of a 1-m long copper wire over a frequency range of 100 MHz to 1 GHz when (a) the diameter of the wire is 31.2 mils (AWG #20) and (b) when it is 10.0 mils (AWG #30). Plot your results of ac resistance versus frequency.

2.2. The diagram in Fig. 2.14 shows a piece of material with two terminals. The material is 1 cm thick.

a. Determine the resistance between the two terminals if $R_\square = 2\,\Omega/\square$.

b. Determine the capacitance between the two terminals if $\varepsilon = 10 \times 10^{-14}$ F/cm.

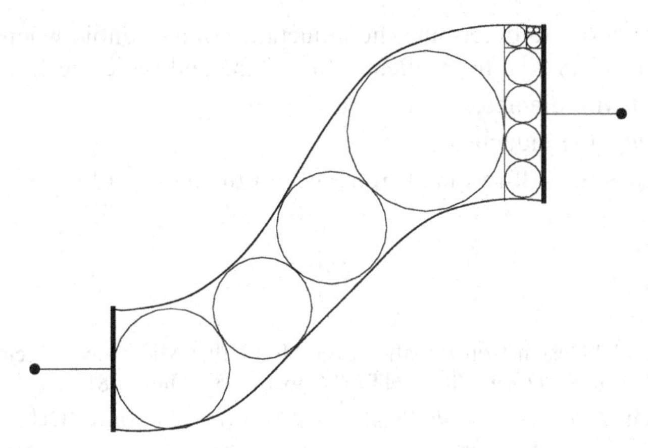

FIGURE 2.14 Cross-section geometry for Problem 2.2.

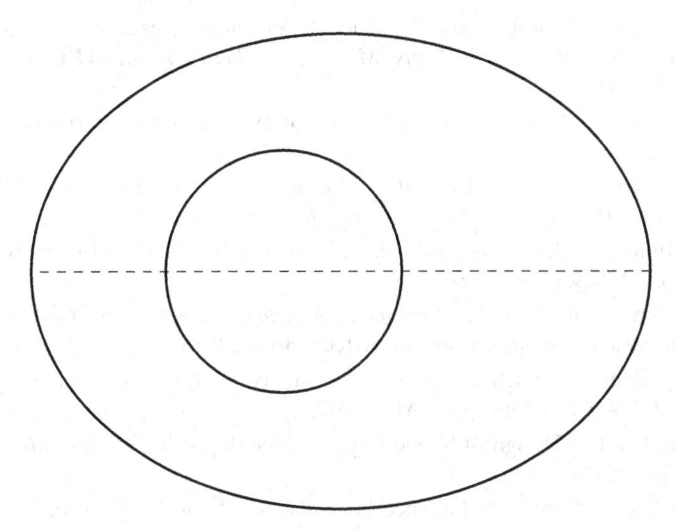

FIGURE 2.15 Cross-section geometry for Problem 2.3.

2.3. A cross section of a conductor is shaped in an irregular shape shown in Fig. 2.15 where the inner circle is one electrode and the outer ellipse is the other electrode. If the conducting material between these two electrodes is made of a material in which $R_\square$ is $4\,\Omega/\square$, what is the resistance between the left and right side of the conductors? If the space between the two ends is filled with a dielectric where $\varepsilon_r = 2$, what is the capacitance between the two terminals? To do this derive an expression for $C_\square$ similar to $R_\square$. You may wish to enlarge the drawing to achieve better accuracy.

2.4. A capacitance is modeled as a series RLC circuit. If $C = 20\,\text{pF}$ and $R = 5\,\Omega$, what is the Q for this capacitor at 20 Mrad/s?

2.5. You are asked to determine the inductance of a solenoid when the form length is 1.5 in., the form diameter is 0.3 in., and there are 12 turns.

a. What wire diameter would you choose?

b. What is the inductance?

c. What is the self-resonant frequency of the inductor?

REFERENCES

1. R. A. Pucel, "Design Considerations for Monolithic Microwave Circuits," *IEEE Trans. Microwave Theory Tech.*, **MTT-29**, pp. 513–534, Jun. 1981.
2. R. Williams, *Gallium Arsenide Processing Techniques*, Boston: Artech, 1990.
3. P. E. Allen and D. R. Holberg, *CMOS Analog Circuit Design*, New York: Oxford University Press, p. 66, 1987.
4. A. D. Moore, *Fluid Mapper Patterns*, Ann Arbor: Overbeck, 1956.
5. C. P. Lee, B. M. Welch, and R. Zucca, "Saturated Resistor Load for GaAs Integrated Circuits," *IEEE Trans. Microwave Theory Tech.*, **MTT-30**, pp. 1007–1013, Jul. 1982.
6. V. F. Perna, Jr. "A Guide to Judging Microwave Capacitors," *Microwaves Mag.*, pp. 40–42, Aug. 1970.
7. V. F. Perna, Jr., "Chip Capacitor Dielectric Effects on Hybrid Microwave Amplifiers," *Proc. Int. Soc. Hybrid Microelectron.*, Oct. 13, 1971.
8. R. S. Muller and T. I. Kamins, *Device Electronics for Integrated Circuits*, 2nd ed., New York: Wiley, p. 190, 1986.
9. P. L. D. Abrie, *The Design of Impedance-Matching Networks for Radio-Frequency and Microwave Amplifiers*, Boston: Artech House, 1985.
10. R. G. Medhurst, "High Frequency Resistance and Capacity of Single-Layer Solenoids," *Wireless Eng.*, p. 35, Mar. 1947.
11. P. R. Geffe, "The Design of Single-Layer Solenoids for RF Filters," *Microwave J.*, **39**, pp. 70–76, Dec. 1996.
12. H. A. Wheeler, "Simple Inductance Formulas for Radio Coils," *Proc. IRE*, **16**, pp. 1398–1400, 1928.
13. H. A. Wheeler, "Inductance Formulas for Circular and Square Coils," *Proc. IEEE*, **70**, pp. 1449–1450, Dec. 1982.
14. H. M. Greenhouse, "Design of Planar Rectangular Microelectronic Inductors," *IEEE Trans. Parts, Hybrids, Packaging*, pp. 101–109, Jun. 1974.
15. M. Parisot, Y. Archambault, D. Pavlidis, and J. Magarshack, "Highly Accurate Design of Spiral Inductors for MMIC's with Small Size and High Cut-off Frequency Characteristics," *1984 IEEE MTT-S Digest*, pp. 106–110, 1984.
16. N. M. Nguyen and R. G. Meyer, "Si IC-Compatible Inductors and LC Passive Filters," *IEEE J. Solid State Circuits*," **25**, pp. 1028–1031, Aug. 1999.
17. R. G. Arnold and D. J. Pedder, "Microwave Lines and Spiral Inductors in MCM-D Technology," *IEEE Trans. Components, Hybrids, and Manufact. Tech.*, **15**, pp. 1038–1043, Dec. 1992.

18. S. Chaki, S. Andoh, Y. Sasaki, N. Tanino, and O. Ishihara, "Experimental Study on Spiral Inductors," *1995 IEEE MTT-S Digest*, pp. 753–756, Apr. 1995.
19. Z. Jiang, P. S. Excell, and Z. M. Hejazi, "Calculation of Distributed Capacitances of Spiral Resonators," *IEEE Trans. Microwave Theory Tech.*, **45**, pp. 139–142, Jan. 1997.
20. A. M. Niknejad, R. G. Meyer, "Analysis, Design, and Optimization of Spiral Inductors and Transformers for Si RF IC's," *IEEE J. Solid-State Circuits*, **33**, pp. 1470–1481, Oct. 1998.
21. E. Pettenpaul, H. Kapusta, A. Weisgerber, H. Mampe, J. Luginsland, and I. Wolff, "CAD Models of Lumped Elements on GaAs up to 18 GHz," *IEEE Trans. Microwave Theory Tech.*, **MTT-36**, pp. 294–304, Feb. 1988.

CHAPTER THREE

Impedance Matching

3.1 INTRODUCTION

A major part of RF design is matching one part of a circuit to another to provide maximum power transfer between the two parts. Even antenna design can be thought of as matching impedance of free space to a transmitter or receiver. This chapter describes a few techniques that can be used to match between two real impedance levels. While some comments will be made relative to matching to a complex load, the emphasis will be on real impedance matching. The first part of this chapter will discuss the circuit quality factor, Q. The Q factor is useful in certain matching circuit designs.

3.2 THE Q FACTOR

The circuit Q factor is defined as the ratio of stored to dissipated power in the following form:

$$Q = \frac{2\pi(\text{max instantaneous energy stored})}{\text{energy dissipative per cycle}} \tag{3.1}$$

For a typical parallel RLC circuit, the Q becomes

$$Q = \frac{\omega C}{G} \tag{3.2}$$

where G is $1/R$. For a series RLC circuit,

Radio Frequency Circuit Design, Second Edition, by W. Alan Davis

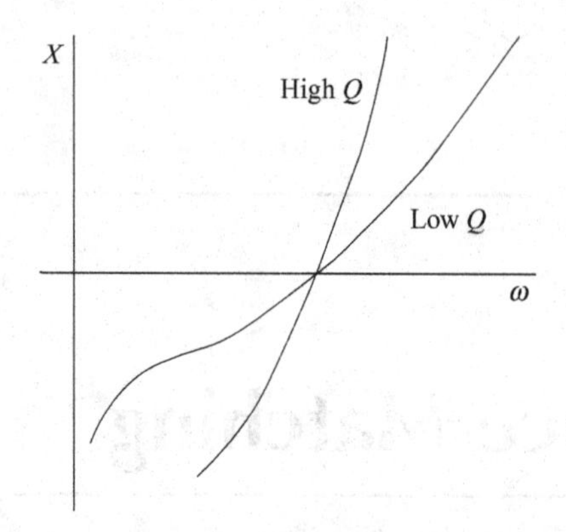

FIGURE 3.1 Reactance slope related to Q.

$$Q = \frac{\omega L}{R} \tag{3.3}$$

It should be emphasized that Q is defined at circuit resonance. If the circuit reactance is plotted as a function of frequency, the slope of the reactance at resonance is a measure of Q (Fig. 3.1). This is explicitly given as

$$Q = \frac{\omega_0}{2G} \frac{dB}{d\omega}\bigg|_{\omega_0} \tag{3.4}$$

where B is the susceptance and G the conductance. Alternately,

$$Q = \frac{\omega_0}{2R} \frac{dX}{d\omega}\bigg|_{\omega_0} \tag{3.5}$$

where R and X are the resistance and reactance of the circuit. For a series RLC circuit this latter formula will result in the solution given by Eq. (3.3). On the other hand, the Q of a complicated circuit can be readily obtained from Eq. (3.4) or (3.5), numerically if necessary.

3.3 RESONANCE AND BANDWIDTH

The minimum insertion loss or maximum transmission of a parallel RLC circuit occurs at the resonant frequency of the circuit. When this circuit is excited by a current source, and the output is terminated with an open circuit, the transfer function is

$$\frac{V_{\text{out}}}{I_{\text{in}}} = \frac{1}{(1/R) + j\omega C - (j/\omega L)} \tag{3.6}$$

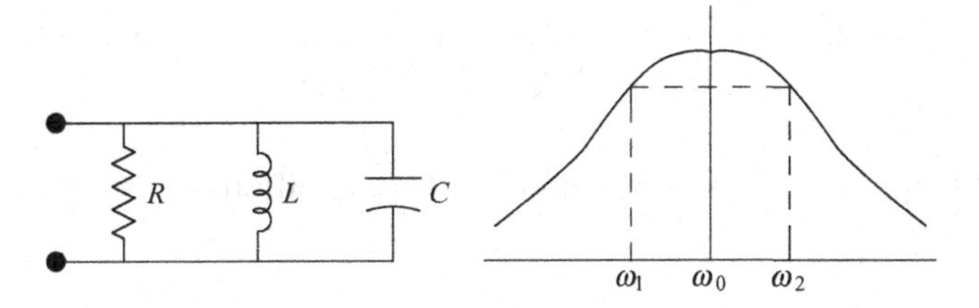

FIGURE 3.2 Simple parallel resonant circuit.

This is shown in Fig. 3.2. The bandwidth is often defined when the output voltage, V_{out}, drops from the resonant value by $\sqrt{2}$ (or −3 dB). This occurs when the denominator of the transfer function increases from $1/R$ at resonance to

$$\left|\frac{1}{R}+j\omega C-\frac{j}{\omega L}\right|=\frac{\sqrt{2}}{R} \tag{3.7}$$

Equation (3.7) is a quadratic equation in ω^2:

$$\omega^4 C^2 L^2 R^2 - \omega^2\left(2CLR^2 + L^2\right) + R^2 = 0 \tag{3.8}$$

Since the resonant frequency is $\omega_0 = 1/\sqrt{LC}$, and the parallel Q from Eq. (3.2) is $\omega_0 C/G = R/\omega_0 L$, then Eq. (3.8) can be written in terms of Q and ω_0:

$$\omega^4 - \omega^2\omega_0^2\left(2+\frac{\omega_0^2 L^2}{R^2}\right)+\omega_0^4 = 0 \tag{3.9}$$

$$\omega^4 - \omega^2\omega_0^2\left(2+\frac{1}{Q^2}\right)+\omega_0^4 = 0 \tag{3.10}$$

The two solutions for ω^2 are

$$\omega^2 = \omega_0^2\left\{1+\left[\frac{1}{2Q^2}\pm\frac{1}{Q}\sqrt{1+\frac{1}{4Q^2}}\right]\right\}$$

$$\omega^2 = \omega_0^2\left\{1+\frac{1}{4Q^2}\pm\frac{1}{Q}\sqrt{1+\frac{1}{4Q^2}}+\frac{1}{4Q^2}\right\}$$

$$\omega^2 = \omega_0^2\left\{\sqrt{1+\frac{1}{4Q^2}}\pm\frac{1}{2Q}\right\}\left\{\sqrt{1+\frac{1}{4Q^2}}\pm\frac{1}{2Q}\right\} \tag{3.11}$$

This has been written as a product of two equal terms, so that the original quartic equation has two pairs of equal roots. Taking the square root of Eq. (3.11) provides the two 3-dB frequencies of the resonant circuit:

$$\omega_{1,2} = \omega_0 \left\{ \sqrt{1 + \frac{1}{4Q^2}} \pm \frac{1}{2Q} \right\} \quad (3.12)$$

The 3-dB bandwidth of the resonant circuit is the difference between the two 3-dB frequencies:

$$\Delta\omega = \omega_2 - \omega_1 = \frac{1}{RC} \text{ rad/s} \quad (3.13)$$

The response is clearly not symmetrical about the resonant frequency ω_0. The resonant frequency can be found by taking the geometric mean of the two solutions of Eq. (3.12).

$$\omega_1\omega_2 = \omega_0^2 \left[\sqrt{1 + \frac{1}{4Q^2}} - \frac{1}{2Q} \right] \left[\sqrt{1 + \frac{1}{4Q^2}} + \frac{1}{2Q} \right]$$

$$\omega_0 = \sqrt{\omega_1\omega_2} \quad (3.14)$$

However, for narrow bandwidths, the arithmetic mean of the two 3-dB frequencies can be used with small error.

3.4 UNLOADED Q

In real physical reactive elements there are always some resistive losses. The loss in a capacitor or an inductor can be described in terms of its Q. For example, if a lossy inductor is placed in parallel with a lossless capacitor, the Q of the resulting parallel circuit is said to be the circuit Q of the inductor. The inductor Q_{ind} then is

$$Q_{ind} = \omega_0 CR = \frac{R}{\omega_0 L} \quad (3.15)$$

or

$$R = X_L Q_{ind} \quad (3.16)$$

Similarly, for a lossy capacitor, its resistive component could be expressed in terms of the capacitor Q_{cap}. If the inductor, capacitor, and a load resistance R_L are placed in parallel, then the total resistance is R_T:

$$\frac{1}{R_T} = \frac{1}{R_L} + \frac{1}{Q_{ind}X_L} + \frac{1}{Q_{cap}X_C} \quad (3.17)$$

At resonance, $X_L = X_C$, so

$$\frac{X_L}{R_T} = \frac{X_L}{R_L} + \left[\frac{1}{Q_{ind}} + \frac{1}{Q_{cap}}\right] \tag{3.18}$$

The unloaded Q, Q_u, is the Q associated with the reactive elements only (i.e., without the load). The bracketed term is the unloaded Q:

$$\frac{1}{Q_u} = \frac{1}{Q_{ind}} + \frac{1}{Q_{cap}} \tag{3.19}$$

3.5 *L* CIRCUIT IMPEDANCE MATCHING

There are four possible configurations that will provide impedance matching with only two reactive elements. In each case, the design of the matching circuits is based on the Q factor, a concept that will become even more important in designing broadband matching circuits [1]. Two of the circuits will be described as the *series connection* since the reactive element closest to the load resistance is a series reactance (Figs. 3.3*a* and 3.3*b*). The circuits with a shunt reactance closest to the load resistance are called the *shunt connection* (Figs. 3.3*c* and 3.3*d*). For the series connection in which the series reactance is an inductance, the total input admittance is given as follows:

$$\begin{aligned} Y_{in} &= j\omega C + \frac{1}{R + j\omega L} \\ &= \frac{R}{R^2 + (\omega L)^2} + j\left[\omega C - \frac{\omega L}{R^2 + (\omega L)^2}\right] \end{aligned} \tag{3.20}$$

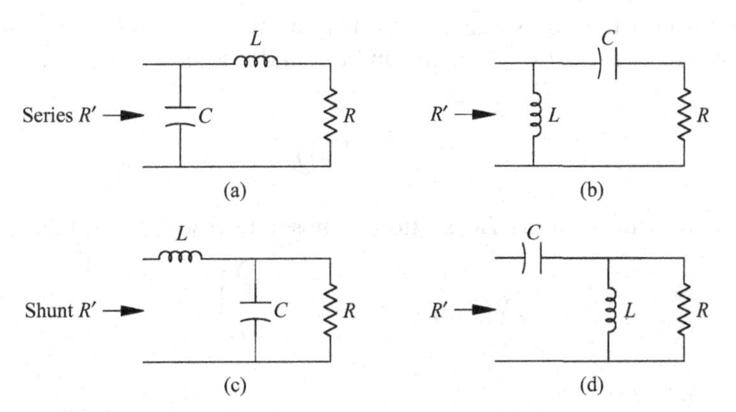

FIGURE 3.3 Four possible *L* matching circuits.

Resonance occurs when the total shunt susceptance $jB = 0$. Thus,

$$C = \frac{L}{R^2 + (\omega_0 L)^2} \tag{3.21}$$

Solution of this for the resonant frequency gives the following expression for the resonant frequency:

$$\omega_0 = \sqrt{\frac{1}{LC} - \frac{R^2}{L^2}} \tag{3.22}$$

The effect of the load resistor is to modify the resonant frequency somewhat. The conductive part of Y_{in} at this frequency (where $B = 0$) can be found. Its reciprocal is the input resistance, R', of the circuit:

$$\begin{aligned} R' &= \frac{R^2 + \omega_0^2 L^2}{R} \\ &= R(1 + Q_1^2) \end{aligned} \tag{3.23}$$

The subscript 1 for Q is present to emphasize this is the first resonator closest to the load. More complicated circuits might have several pertinent Q factors to consider.

At center frequency, the reactance of the series part (i.e., not the capacitance part) will change with changing frequency. Its value can be found from the input admittance expression and is the amount the reactance changes because of the series inductance. This reactance change is

$$jX_1' = j\frac{R^2 + (\omega_0 L)^2}{\omega_0 L} \tag{3.24}$$

If X_1 represents the series reactance, which in this case is $\omega_0 L$, then the reactance change of the series element can be found also in terms of Q:

$$jX_1' = jX_1\left(1 + \frac{1}{Q_1^2}\right) \tag{3.25}$$

The second element in the LC section is chosen to resonate out this X_1':

$$jX_2 = -jX_1' = -jX_1\left(1 + \frac{1}{Q_1^2}\right) \tag{3.26}$$

or with Eq. (3.23)

TABLE 3.1 *L* Matching Circuit Design Where X_1, B_1 are Reactance or Susceptance Closest to the Load *R*

Circuit	R'	jX_2	Q_1
Series	$R(1+Q_1^2)$	$-jX_1(1+1/Q_1^2)$ or $-jR'/Q_1$	X_1/R
Shunt	$R/(1+Q_1^2)$	$-jX_1/(1+1/Q_1^2)$ or $-jR'Q_1$	B_1/G

$$jX_2 = \frac{-jR'}{Q_1} \tag{3.27}$$

In the typical synthesis problem, R' and R are known. Equation (3.23) gives the necessary value of Q_1, Eq. (3.3) gives the required L, and Eq. (3.26) or (3.27) gives the required C from X_2. This procedure is summarized in Table 3.1.

A similar procedure can be applied for the shunt connection in which the shunt capacitance is closest to the load resistance. The input impedance is expressed as follows:

$$\begin{aligned} Z_{in} &= j\omega L + \frac{1}{G + j\omega C} \\ &= \frac{G}{G^2 + (\omega C)^2} + j\left[\omega L - \frac{\omega C}{G^2 + (\omega C)^2}\right] \end{aligned} \tag{3.28}$$

For resonance, the series reactance $X = 0$. Solution for the resonant frequency for the shunt connection is

$$\omega_0 = \sqrt{\frac{1}{LC} - \frac{G^2}{C^2}} \tag{3.29}$$

Substituting this back into the input impedance expression gives the input resistance:

$$\begin{aligned} R' &= \frac{1/G}{1 + (\omega_0 C/G)^2} \\ &= \frac{R}{1 + Q_1^2} \end{aligned} \tag{3.30}$$

The reactance associated with the capacitance is

$$jX_1' = \frac{-j\omega_0 C}{G^2 + (\omega_0 C)^2}$$

Since $jX_1 = 1/j\omega C$,

$$jX_1' = \frac{jX_1}{1+1/Q_1^2} \tag{3.31}$$

$$= R'Q_1 \tag{3.32}$$

Since $jX_2 = -jX_1'$, the values in Table 3.1 are obtained. The sign of jX_2 must always be the opposite of jX_1.

The major feature that should be recognized, whether dealing with elaborate lumped circuits or microwave circuits, if the impedance level needs to be raised, a series connection is needed. If the impedance needs to be lowered, a shunt connection is needed. Furthermore, since the design is based on a resonance condition, the two reactances in the circuit must be of the opposite type. This means two inductors or two capacitors will not work.

3.6 π TRANSFORMATION CIRCUIT

In the previous L matching circuit, the value for Q is completely determined by the transformation ratio. Consequently, there is no independent control over the value of Q that is related to the circuit bandwidth. Addition of a third circuit element gives flexibility to design for bandwidth. If a design begins with a shunt L matching circuit, then addition of another shunt susceptance on the other side of the series element provides the necessary circuit flexibility to be able to choose the circuit Q as a design parameter. The resulting π matching circuit is shown in Fig. 3.4. In this circuit B_1 and X_2 both act as the impedance transforming elements while the third, B_3, is the compensation element that tunes out the excess reactance from the first two elements. As in the L matching circuit, the first shunt element, B_1, reduces the resistance level by a factor of $1/(1+Q_1^2)$ and X_2 increases the resistance level by $1+Q_2^2$ where Q_2 is a Q factor related to the second element. The final transformation ratio can be $R'' < R$ or $R'' > R$, depending on which Q is larger as shown in the diagram of Fig. 3.5. To make $R'' < R$, make $Q_1 > Q_2$. The maximum Q, $Q_{max} = Q_1$, will be the major factor that determines the bandwidth.

Now consider design of a circuit where $R'' < R$. Then the first shunt transformation gives

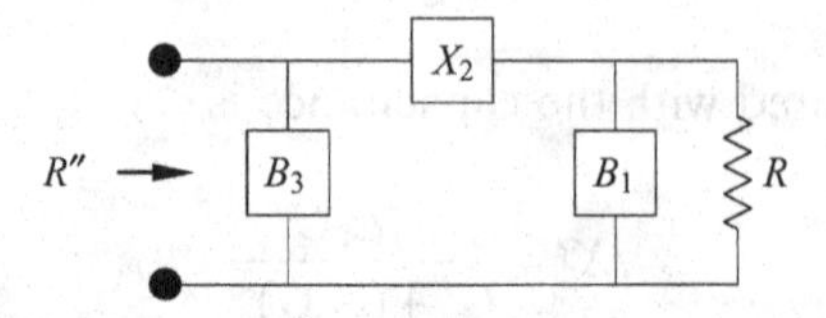

FIGURE 3.4 π Impedance transformation circuit.

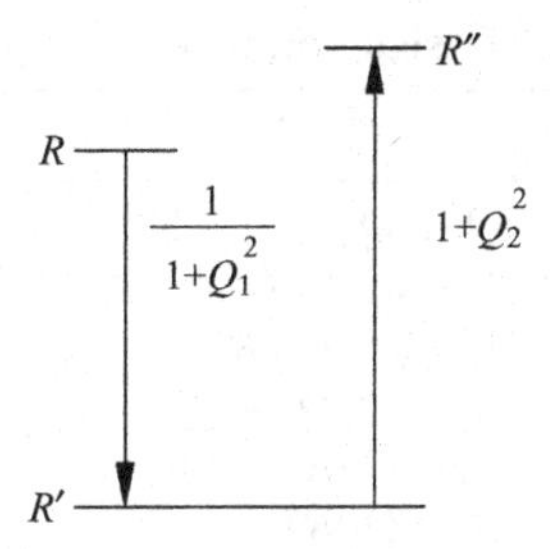

FIGURE 3.5 Diagram showing two-step transformation.

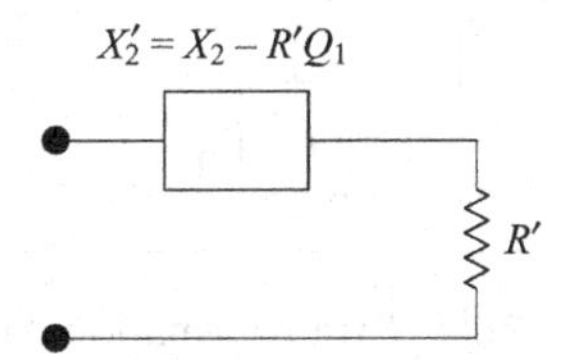

FIGURE 3.6 Equivalent series reactance after first transformation.

$$R' = \frac{R}{1+Q_1^2} \tag{3.33}$$

$$X' = -R'Q_1 \tag{3.34}$$

The incremental reactance, X', is to be added to the series arm. This results in the circuit in Fig. 3.6, which shows that R has been transformed to R' with a modified series reactance. This series reactance will act to increase the resistance level from R' to R''. The second transformation Q is

$$Q_2 = \frac{X_2 - R'Q_1}{R'} = \frac{R'(X_2/R - Q_1)}{R'}$$

or

$$\frac{X_2}{R'} = Q_1 + Q_2 \tag{3.35}$$

The X_2, B_3 combination is a series L section with "load" of R'. Consequently,

$$R'' = R'(1+Q_2^2) \tag{3.36}$$

$$X'' = -\frac{R''}{Q_2} \tag{3.37}$$

TABLE 3.2 π Matching Circuit Design Formulas

Step Number	$R'' < R$	$R'' > R$
1	$Q_1 = Q_{max}$	$Q_2 = Q_{max}$
2	$R' = R/(1+Q_1^2)$	$R' = R''/(1+Q_2^2)$
3	$1+Q_2^2 = R''/R'$	$1+Q_1^2 = R/R'$
4	$X_2 = R'(Q_1 + Q_2)$	$X_2 = R'(Q_1 + Q_2)$
5	$B_1 = Q_1/R$	$B_1 = Q_1/R$
6	$B_3 = Q_2/R''$	$B_3 = Q_2/R''$

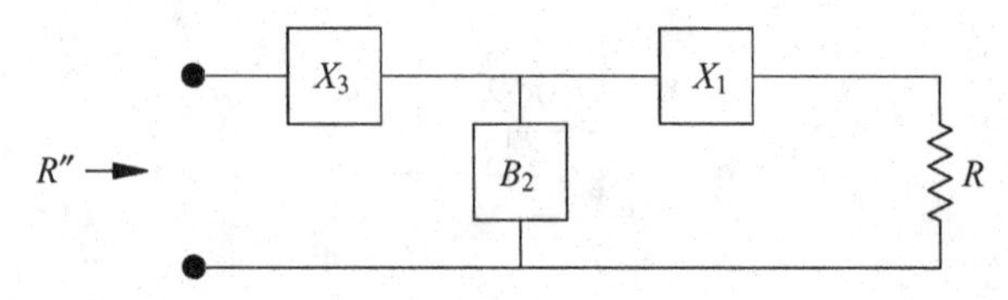

FIGURE 3.7 *T* transformation circuit.

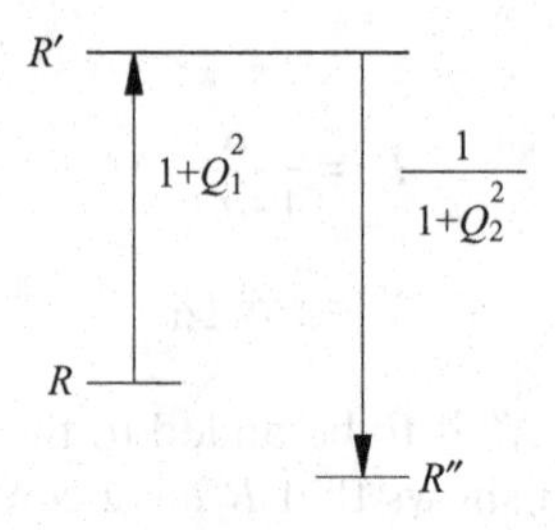

FIGURE 3.8 Diagram showing impedance transformation for *T* circuit.

A summary for the design process is shown in Table 3.2. To make $R'' < R$, make $Q_1 > Q_2$ where $Q_1 = Q_{max}$ and follow the design steps in the first column of Table 3.2. For $R'' > R$, use the second column.

3.7 *T* TRANSFORMATION CIRCUIT

The *T* transformation circuit is the dual to the π transformation circuit and is shown in Fig 3.7. In this circuit, however, the series reactance X_1 first raises the resistance level to R', and the remaining shunt susceptance lowers the resistance level as indicated in Fig. 3.8. The design formulas are derived in the same way as the π circuit formulas and are summarized in Table 3.3.

TABLE 3.3 ***T* Matching Circuit Design Formulas**

Step Number	$R'' > R$	$R'' < R$
1	$Q_1 = Q_{max}$	$Q_2 = Q_{max}$
2	$R' = R(1+Q_1^2)$	$R' = R''(1+Q_2^2)$
3	$1+Q_2^2 = R'/R''$	$1+Q_1^2 = R'/R$
4	$X_1 = Q_1R$	$X_1 = Q_1R$
5	$B_2 = (Q_1 + Q_2)/R'$	$B_2 = (Q_1 + Q_2)/R'$
6	$X_3 = Q_2R''$	$X_3 = Q_2R''$

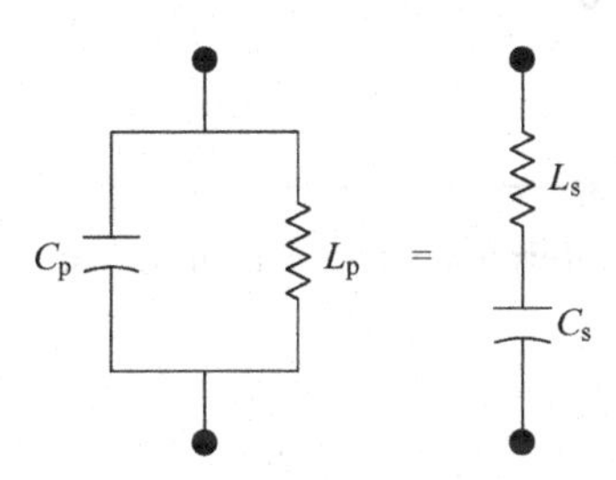

FIGURE 3.9 Parallel *RC* to series *RC* conversion.

3.8 TAPPED CAPACITOR TRANSFORMER

The tapped capacitor circuit is another approximate method for obtaining impedance-level transformation. The description of this design process will begin with a parallel *RC* to series *RC* conversion. Then the tapped *C* circuit will be converted to an *L*-shaped matching circuit. The Q_1 for an equivalent load resistance, R_{eqv}, will be found. Finally, a summary of the circuit synthesis procedure will be given.

3.8.1 Parallel-to-Series Conversion

Shown in Fig. 3.9 is a parallel *RC* circuit that will be forced to have the same impedance as the series *RC* circuit, at least at one frequency. The conversion is, of course, valid for only a narrow frequency range, so that this method is fundamentally limited by this approximation.

The impedance of the parallel circuit is

$$Z_p = \frac{R_p}{1+sC_pR_p} \tag{3.38}$$

The Q for a parallel circuit is $Q_p = \omega C_pR_p$. The equivalent series resistance and reactance in terms of Q_p are

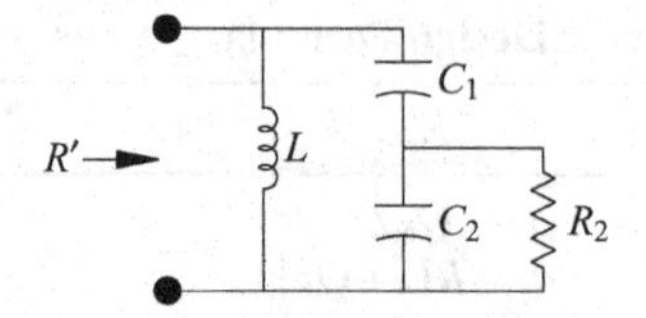

FIGURE 3.10 Tapped C transformation circuit.

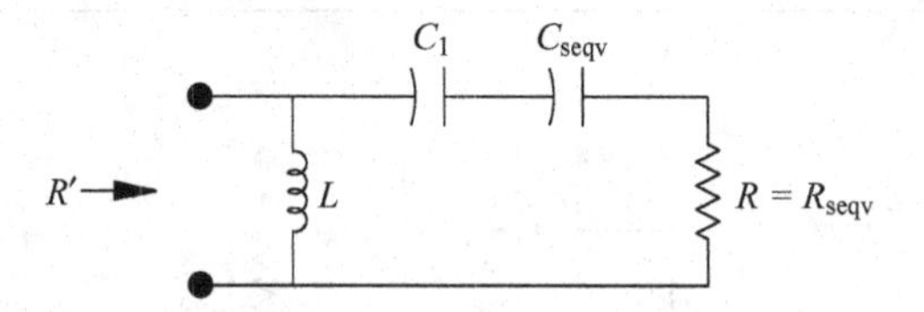

FIGURE 3.11 Intermediate equivalent transformation circuit.

$$R_{\mathrm{seqv}} = \frac{R_{\mathrm{p}}}{1+Q_{\mathrm{p}}^2} \tag{3.39}$$

$$X_{\mathrm{seqv}} = -\frac{X_{\mathrm{p}}Q_{\mathrm{p}}^2}{1+Q_{\mathrm{p}}^2} \tag{3.40}$$

3.8.2 Conversion of Tapped C Circuit to an L-Shaped Circuit

The schematic of the tapped C circuit is shown in Fig. 3.10 where R' is to be matched to R_2. The parallel R_2C_2 section is converted to a series $R_{\mathrm{eqv}}C_{\mathrm{eqv}}$, as indicated in Fig. 3.11. Making use of Eqs. (3.39) and (3.40),

$$C_{\mathrm{seqv}} = C_2\left(\frac{1+Q_{\mathrm{p}}^2}{Q_{\mathrm{p}}^2}\right) \approx C_2 \quad \text{for high } Q_{\mathrm{p}} \tag{3.41}$$

$$R_{\mathrm{seqv}} = R = \frac{R_2}{1+Q_{\mathrm{p}}^2} \tag{3.42}$$

where $Q_{\mathrm{p}} = \omega_0 C_2 R_2$. Considering R' as the load, and using the L circuit transformation for a shunt circuit in Table 3.1,

$$R_{\mathrm{seqv}} = \frac{R'}{1+Q_1^2} \tag{3.43}$$

This is the transformed resistance looking through C_1 toward the left. Looking toward the right through C_{seqv} and again using the parallel-to-series conversion Eq. (3.39),

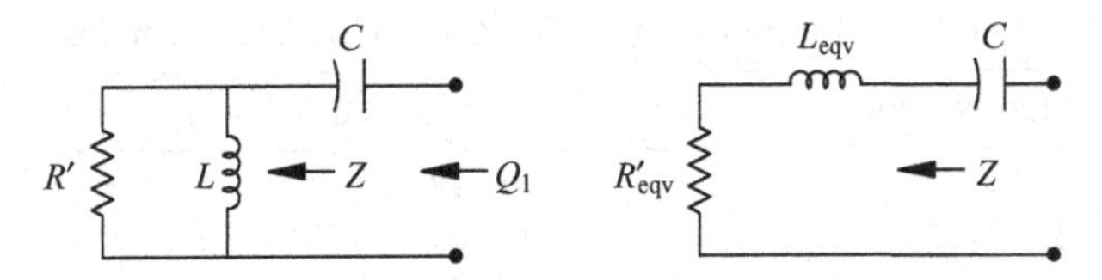

FIGURE 3.12 Equate the left- and right-hand circuits.

$$R_{\text{seqv}} = \frac{R_2}{1+Q_{\text{p}}^2} \tag{3.44}$$

These two expressions for R_{seqv} can be equated and solved for Q_{p}:

$$Q_{\text{p}} = \left[\frac{R_2}{R'}(1+Q_1^2)-1\right]^{1/2} \tag{3.45}$$

3.8.3 Calculation of Circuit Q

An approximate value for Q can be found by equating the impedances of the two circuits in Fig. 3.12:

$$Z = \frac{R'\omega^2 L^2 + jR'^2\omega L}{R'^2 + (\omega L)^2} = R'_{\text{eqv}} + j\omega L_{\text{eqv}} \tag{3.46}$$

If the Q of the right-hand circuit is approximately that of the left-hand circuit in Fig. 3.12, then

$$Q_1 = \frac{\omega_0 L_{\text{eqv}}}{R'_{\text{eqv}}} = \frac{\omega_0 R'^2 L}{R'\omega_0^2 L^2} = \frac{R'}{\omega_0 L} \tag{3.47}$$

The variable C represents the total capacitance of C_1 and C_{seqv} in series as implied in Fig. 3.11 and represented in Fig. 3.12. For a high-Q circuit, circuit analysis gives the resonant frequency:

$$\omega_0^2 = \frac{1}{LC - L^2/R'^2} \approx \frac{1}{LC} \tag{3.48}$$

As a result, the approximate value for Q_1 can be found:

$$Q_1 = \omega_0 R'C = \frac{f_0}{\Delta f} \tag{3.49}$$

Here Δf is the bandwidth in hertz and f_0 is the resonant frequency.

TABLE 3.4 Tapped *C* Matching Circuit Design Formulas

Step Number	Tapped *C* Formula
1	$Q_1 = f_0/\Delta f$
2	$C = Q_1/(\omega_0 R') = 1/(2\pi\,\Delta R')$
3	$L = 1/\omega_0^2 C$
4	$Q_p = [(R_2/R')(1+Q_1^2)-1]^{1/2}$
5	$C_2 = Q_p/\omega_0 R_2$
6	$C_{seqv} = C_2(1+Q_p^2)/Q_p^2$
7	$C_1 = C_{seqv}C/(C_{seqv} - C)$

3.8.4 Tapped *C* Design Procedure

The above ideas are summarized in Table 3.4, which provides a design procedure for the tapped *C* matching circuit. Similar expressions could be found for a tapped inductor transforming circuit, but such a circuit is typically less useful because high *Q* inductors are more difficult to obtain than capacitors.

3.9 PARALLEL DOUBLE-TUNED TRANSFORMER

Each of the above described *T*, π, or tapped *C* matching circuits provide some control over the bandwidth. Where precise control over the bandwidth is required, a double-tuned circuit allows controlling bandwidth by specifying two different frequencies where maximum transmission occurs. For a small pass band, the midband dip in the transmission coefficient can be made small. Furthermore, the double-tuned circuit is especially useful when a large difference in impedance levels is desired, although its high end frequency range is limited. The filter transmission gain is shown in Fig. 3.13.

The double-tuned circuit consists of a coupled coil transformer with resonating capacitances on the primary and secondary side. This circuit is shown in Fig. 3.14. The transformer is described by its input and output inductance as well as the coupling coefficient *k*. The turns ratio for the transformer is

$$n:1 = \sqrt{\frac{L_{11}}{k^2 L_{22}}}:1 \tag{3.50}$$

The circuit in Fig. 3.14 can be replaced by an equivalent circuit using an ideal transformer (Fig. 3.15*a*). Since an ideal transformer has no self-inductance, the inductances and coupling factor, *k*, must be added to the ideal transformer. The final circuit topology is shown in Fig. 3.15*b*). Looking toward the right through the ideal transformer, the circuit values in Fig. 3.15*b*) are

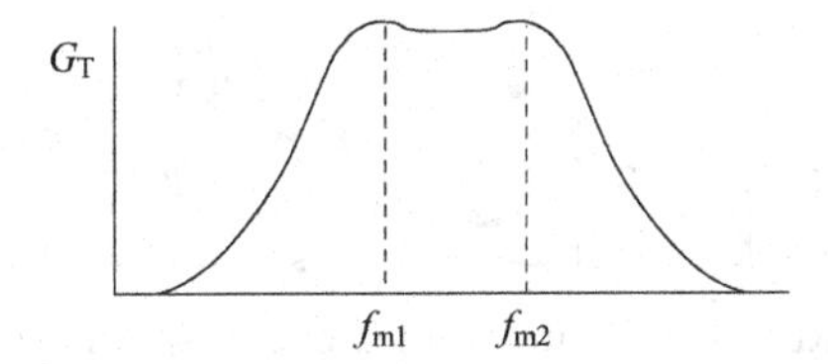

FIGURE 3.13 Double-tuned transformer response.

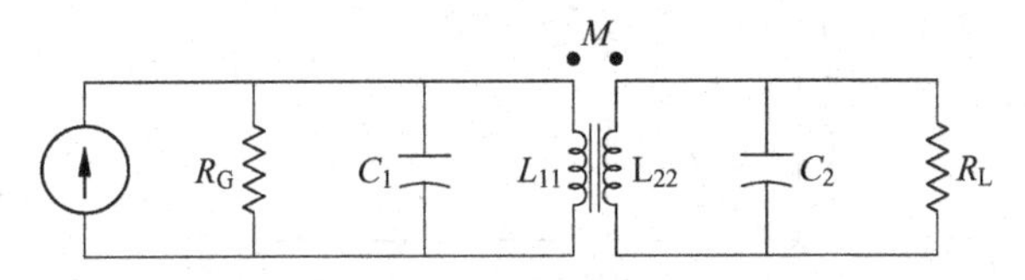

FIGURE 3.14 Real transformer with resonating capacitances.

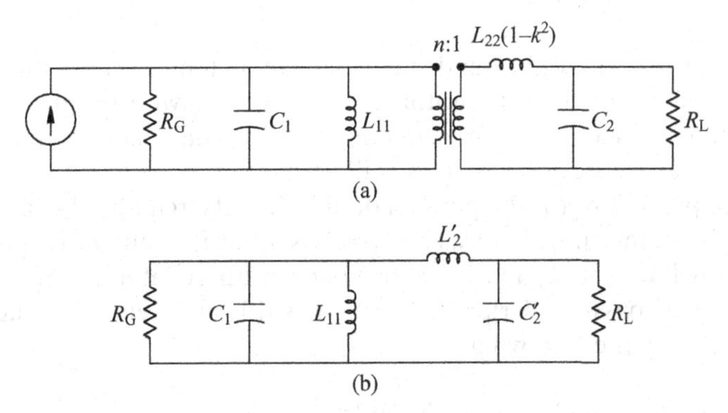

FIGURE 3.15 (*a*) Alternate equivalent circuit with ideal transformer and (*b*) final equivalent circuit.

$$L'_2 = L_{11}\left(\frac{1}{k^2} - 1\right) \tag{3.51}$$

$$C'_2 = \left(\frac{k^2 L_{22}}{L_{11}}\right) C_2 \tag{3.52}$$

$$R'_L = \frac{L_{11}}{k^2 L_{22}} R_L \tag{3.53}$$

The circuit elements will be chosen to give an exact match at the two frequencies, f_{m1} and f_{m2}. The circuit in Fig. 3.15*b* can be conceptually split into two (Fig. 3.16). The resistance R_1 with the parallel resonant circuit will never

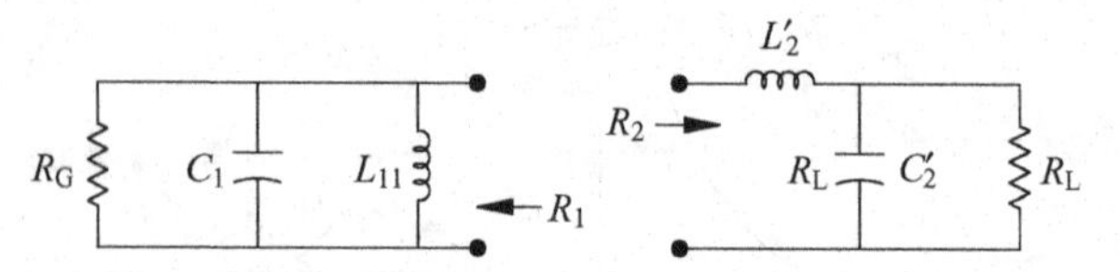

FIGURE 3.16 Double-tuned circuit split into two.

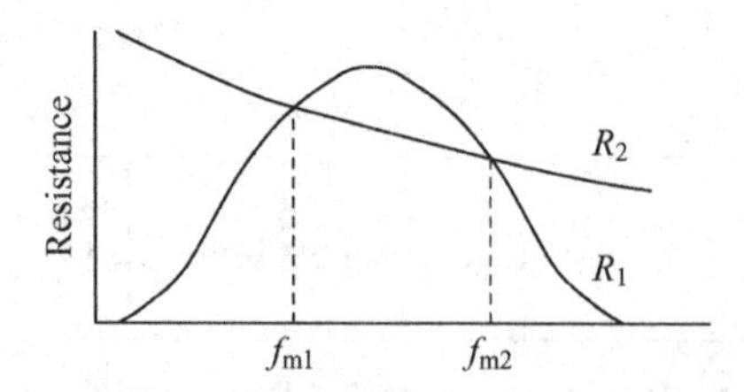

FIGURE 3.17 Plot of left- and right-hand resistance values vs. frequency.

be larger than R_G. The right-hand side is an L matching circuit with the reactance of the shunt element monotonically decreasing with frequency. Hence, R_2 monotonically decreases. Consequently, if R_L is small enough, there will be two frequencies where $R_1 = R_2$. This is illustrated in Fig. 3.17.

A design procedure for the parallel double-tuned circuit has been reviewed in [1] and is summarized below.The typical synthesis problem is to design a circuit that will match R_G and R_L over a bandwidth, Δf, at a center frequency, f_0, with a given pass-band ripple. The bandwidth and center frequency are approximated by the following:

1. Determine f_{m1} and f_{m2} from Δf and f_0:

$$\Delta f \approx \sqrt{2}(f_{m2} - f_{m1}) \tag{3.54}$$

$$f_0 \approx \sqrt{f_{m1} f_{m2}} \tag{3.55}$$

The minimum pass-band gain for the filter is dependent on the difference between the match frequencies: The larger the distance between f_{m1} and f_{m2}, the larger the dip in the center of the pass-band characteristic:

$$G_{T,\min} = \frac{4 f_{m2}/f_{m1}}{(f_{m2}/f_{m1})^2 + 2 f_{m2}/f_{m1} + 1} \tag{3.56}$$

Equation (3.56) provides an approximation to the minimum gain at the center of the pass band, so that it predicts whether a chosen ripple factor can be met.

2. Determine the actual transducer gain for the given ripple factor:

$$G_T = 10^{-\text{ripple factor (dB)}/10} \tag{3.57}$$

3. Determine the resistance ratio r. If $G_T > G_{T,\min}$, then the pass-band ripple specification can be met.

$$r = \frac{1+|1-G_T|^{1/2}}{1-|1-G_T|^{1/2}} \tag{3.58}$$

4. Calculate the Q_2 at the two matching frequencies:

$$Q_{2\text{-m1}}^2 = r\frac{f_{m1}}{f_{m2}} - 1 \tag{3.59}$$

$$Q_{2\text{-m2}}^2 = r\frac{f_{m2}}{f_{m1}} - 1 \tag{3.60}$$

5. Solve the following simultaneous equations for L_2' and C_2':

$$-\omega_{m1}L_2' + \frac{1}{\omega_{m1}C_2'} = |Q_{2\text{-m1}}|\frac{R_G}{1+Q_{2\text{-m1}}^2} \tag{3.61}$$

$$+\omega_{m2}L_2' - \frac{1}{\omega_{m2}C_2'} = |Q_{2\text{-m2}}|\frac{R_G}{1+Q_{2\text{-m2}}^2} \tag{3.62}$$

6. Find the value for R_L':

$$R_L' = \frac{1+Q_{2\text{-m1}}^2}{\omega_{m1}^2 C_2'^2 R_G} \tag{3.63}$$

7. Calculate the input susceptance of the right-hand side where $G_L' = 1/R_L'$:

$$B_{m1} = \text{Im}\left\{\frac{1}{j\omega_{m1}L_2' + 1/(G_L' + j\omega_{m1}C_2')}\right\} \tag{3.64}$$

$$B_{m2} = \text{Im}\left\{\frac{1}{j\omega_{m2}L_2' + 1/(G_L' + j\omega_{m2}C_2')}\right\} \tag{3.65}$$

8. Solve the following simultaneous equations for L_{11} and C_1:

$$\frac{1}{\omega_{m1}L_{11}} - \omega_{m1}C_1 = |B_{m1}| \tag{3.66}$$

$$\frac{1}{\omega_{m2}L_{11}} - \omega_{m2}C_1 = -|B_{m2}| \tag{3.67}$$

9. Find the transformer coupling coefficient, and hence L_{22} and C_2:

$$k = \frac{1}{\sqrt{1 + L_2'/L_{11}}} \tag{3.68}$$

$$L_{22} = \frac{L_{11}R_L}{k^2 R_L'} \tag{3.69}$$

$$C_2 = \frac{L_{11}}{k^2 L_{22}} C_2' \tag{3.70}$$

This procedure has been coded into the program DBLTUNE, and an example of its use is given in Appendix B.

The parallel double-tuned transformer makes use of a coupled coil. Sometimes coupled-coil transformers can be implemented in an integrated circuit with coupled spiral coils. This was referenced in Section 2.4.5 relative to using the ASITIC program.

3.10 CONCLUSIONS

This chapter has provided a variety of circuits that can be used to transform one impedance level to another. Impedance matching is required in many places within a transceiver, especially in the amplifiers. However, the circuit designs described in this chapter do not provide accurate bandwidth specifications, nor do they make use of transmission lines. These topics will be considered later in Chapters 5 and 6.

PROBLEMS

3.1. The graph in Fig. 3.18 shows the susceptance of a circuit as well as the frequency response.

a. From these graphs determine the Q of the circuit.

b. Determine the equivalent circuit that would approximate this frequency response. Give numerical values for the resistive and reactive components.

3.2. Design an impedance transforming network that matches a generator resistance, $R_G = 400\,\Omega$ to a load resistance $R_L = 20\,\Omega$. The center frequency for the circuit is $f_0 = 6\,\text{MHz}$. The desired ripple (where appropriate) is to be less than 0.25 dB. In some cases, the ripple factor will not be able to be controlled in the design. The problem is to design four different transformation circuits with the above specifications, and for each design do an analysis using SPICE. See Appendix G, Sections G.1 and G.2.

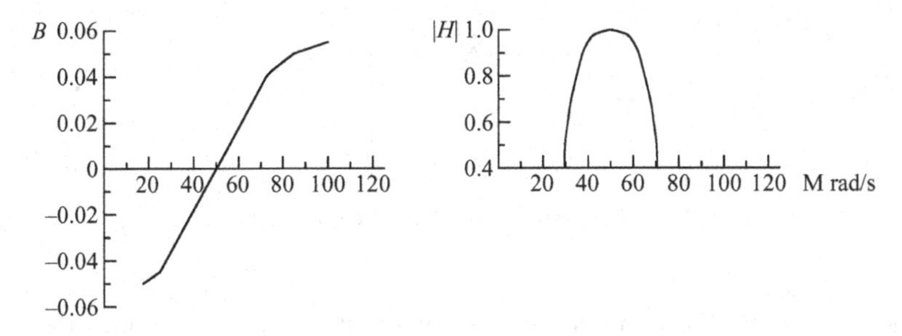

FIGURE 3.18 Susceptance and frequency response for Problem 3.1.

a. Design a two-element L matching circuit and check the results with SPICE.

b. Design a three-element tapped capacitor matching circuit with a bandwidth $\Delta f = 50\,\text{kHz}$ and check the results with SPICE to determine the actual bandwidth.

c. Design a three-element π matching circuit with a bandwidth of $\Delta f = 50\,\text{kHz}$ and check the results with SPICE to determine the actual bandwidth.

d. Design a double-tuned transformer matching circuit with a bandwidth of $\Delta f = 50\,\text{kHz}$ and check the results with SPICE to determine the actual bandwidth.

e. Repeat part (d) for a 3-dB bandwidth of 2 MHz. Again check the results using SPICE.

3.3. The π matching circuit shown in Fig. 3.4 is used to match the load, $R = 1000\,\Omega$ to $R'' = 80\,\Omega$. If the intermediate resistance level is $R' = 20\,\Omega$, determine the following:

a. What is Q_1?

b. What is Q_2?

c. What is B_1, the first susceptance nearest R?

d. What is the estimated 3-dB bandwidth for this circuit in terms of the center frequency, f_0?

3.4. The tapped capacitor transformer is to be used in a narrow band of frequencies around $\omega = 4 \times 10^9\,\text{rad/s}$. In designing the matching circuit, the tapped C circuit is converted to an L matching circuit. If R_2 in Fig. 3.10 is $50\,\Omega$, $C_2 = 8\,\text{pF}$, and $C_1 = 5.0\,\text{pF}$, then what is the total capacitance for the L matching circuit?

3.5. A lossless π matching circuit has a load resistance $R = 340\,\Omega$. The center frequency is $\omega_0 = 20 \times 10^6\,\text{rads/s}$ and the bandwidth is $\Delta\omega = 5 \times 10^6\,\text{rad/s}$. It is also known that the series element in the π circuit is $L_2 = 6\,\mu\text{H}$.

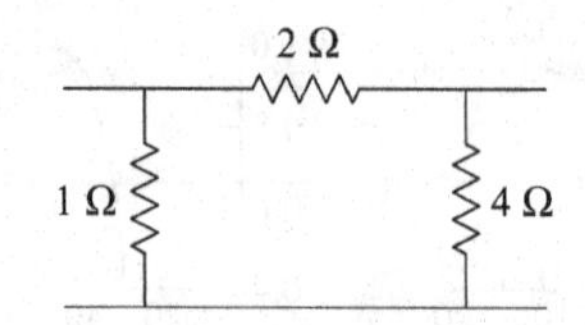

FIGURE 3.19 Circuit for Problem 3.6.

a. Determine the matching generator resistance that is smaller than the load.

b. Determine the susceptance at the load side.

c. Determine the susceptance at the generator side.

3.6. Determine the impedances that would match both sides of the two-port circuit in Fig. 3.19.

3.7. Show that the part of the circuit in Fig. 3.14 consisting of L_{11}, L_{22}, and M is equivalent to the part of the circuit in Fig. 3.15*a* consisting of L_{11}, L_{22}, n, and k. This can be done by equating corresponding z parameters of both circuits. Recall that $k^2 = k_1 k_2$, $M = k\sqrt{L_{11}L_{22}}$, and $n = L_{11}/M$.

REFERENCE

1. P. L. D. Abrie, *Design of RF and Microwave Amplifiers and Oscillators*, Norwood, MA: Artech, 1999.

CHAPTER FOUR

Multiport Circuit Parameters and Transmission Lines

4.1 VOLTAGE–CURRENT TWO-PORT PARAMETERS

A linear n-port network is completely characterized by n independent excitation variables and n dependent response variables. These variables are the terminal voltages and currents. There are four ways of arranging these independent and dependent variables for a two-port circuit that are particularly useful, especially when considering feedback circuits. They are the impedance parameters (z matrix), admittance parameters (y matrix), hybrid parameters (h matrix), and the inverse hybrid parameters (g matrix). These four sets of parameters are defined as:

$$\begin{bmatrix} v_1 \\ v_2 \end{bmatrix} = \begin{bmatrix} z_{11} & z_{12} \\ z_{21} & z_{22} \end{bmatrix} \begin{bmatrix} i_1 \\ i_2 \end{bmatrix} \tag{4.1}$$

$$\begin{bmatrix} i_1 \\ i_2 \end{bmatrix} = \begin{bmatrix} y_{11} & y_{12} \\ y_{21} & y_{22} \end{bmatrix} \begin{bmatrix} v_1 \\ v_2 \end{bmatrix} \tag{4.2}$$

$$\begin{bmatrix} v_1 \\ i_2 \end{bmatrix} = \begin{bmatrix} h_{11} & h_{12} \\ h_{21} & h_{22} \end{bmatrix} \begin{bmatrix} i_1 \\ v_2 \end{bmatrix} \tag{4.3}$$

$$\begin{bmatrix} i_1 \\ v_2 \end{bmatrix} = \begin{bmatrix} g_{11} & g_{12} \\ g_{21} & g_{22} \end{bmatrix} \begin{bmatrix} v_1 \\ i_2 \end{bmatrix} \tag{4.4}$$

Two networks connected in series (Fig. 4.1) can be combined by simply adding the z parameters of each network together. This configuration is called the

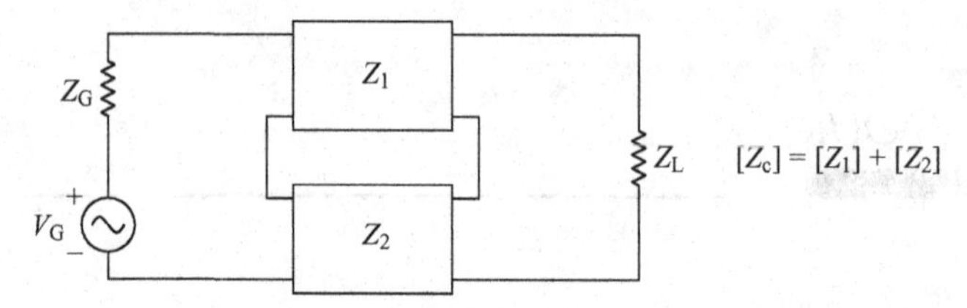

FIGURE 4.1 Series–series connection.

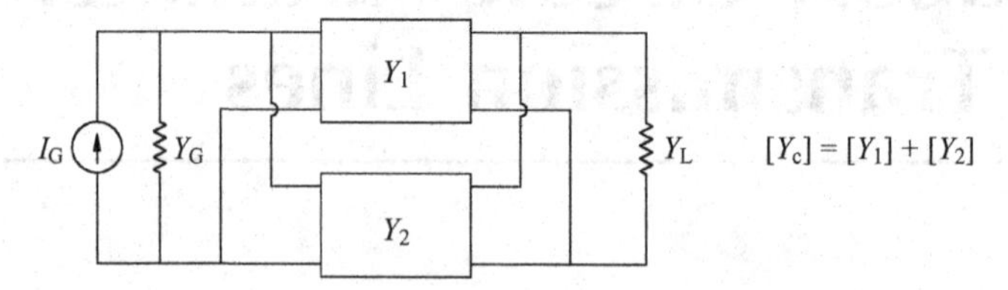

FIGURE 4.2 Shunt–shunt connection.

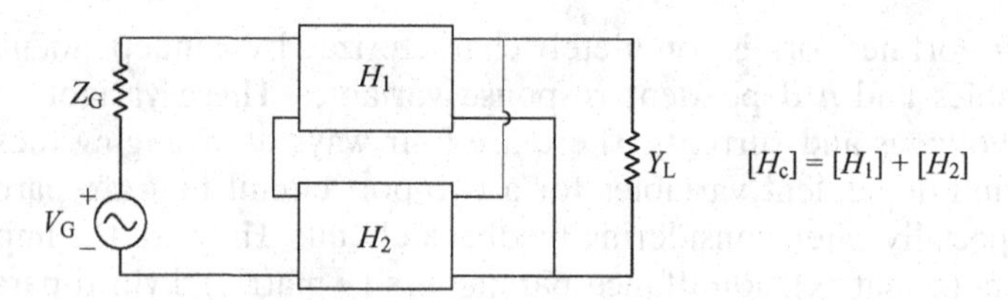

FIGURE 4.3 Series–shunt connection.

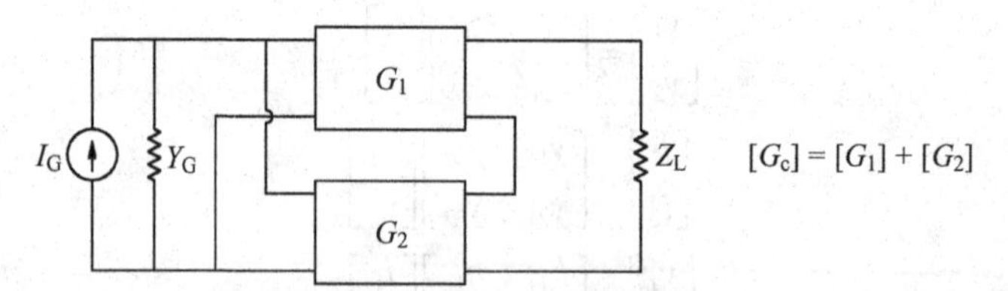

FIGURE 4.4 Shunt–series connection.

series–series connection. In the *shunt–shunt* configuration shown in Fig. 4.2, the two circuits can be combined by adding their y matrices together. In the *series–shunt* configuration (Fig. 4.3), the composite matrix for the combination is found by adding the h parameters of each circuit together. Finally, the circuits connected in the *shunt–series* configuration (Fig. 4.4) can be combined by adding the g parameters of the respective circuits. In each case the independent variables for the particular configuration are the same for each of the

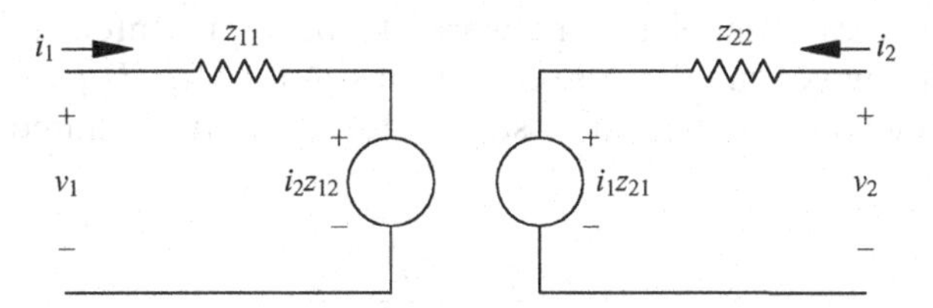

FIGURE 4.5 Equivalent circuit for z parameters.

individual circuits; thus matrix addition is valid most of the time. The case where the matrix addition is not valid occurs when, for example, in Fig. 4.1 a current going in and out of port 1 of circuit 1 is not equal to the current going in and out of port 1 of circuit 2. These pathological cases will not be of concern here, but further information is found in [1, pp. 188–191] where a description of the Brune test is given.

Any of the four types of circuit parameters described above can be represented by an equivalent circuit with controlled sources. As an example, the impedance (or z) parameters can be represented as shown in Fig. 4.5. The input port 1 side is represented by a series resistance of value z_{11} together with a current-controlled voltage source with gain z_{12} in series. The controlling current is the port 2 current. If the current at port 1 is i_1 and the current at port 2 is i_2, then the voltage at port 1 is

$$v_1 = i_1 z_{11} + i_2 z_{12}$$

A similar representation is used for the port 2 side.

The individual impedance parameters are found for a given circuit by setting i_1 or i_2 to 0 and solving for the appropriate z parameter. The z parameters are sometimes termed the *open-circuit parameters* for this reason. The y parameters are sometimes called the *short-circuit parameters* because they are found by shorting the appropriate port. The conversion of these parameters is summarized in Appendix D.

4.2 *ABCD* PARAMETERS

Two-port networks are often cascaded together, and it would be useful to be able to describe each network in such a way that the product of the matrices of each individual network would describe the total composite cascaded network. The *ABCD* parameters have the property of having the port 1 variables being the independent variables and the port 2 variables being the dependent ones:

$$\begin{bmatrix} v_1 \\ i_1 \end{bmatrix} = \begin{bmatrix} A & B \\ C & D \end{bmatrix} \begin{bmatrix} v_2 \\ -i_2 \end{bmatrix} \tag{4.5}$$

This allows the cascade of two networks to be represented as the matrix product of the two circuits expressed in terms of the *ABCD* parameters. The *ABCD* parameters can be expressed in terms of the commonly used z parameters:

$$A = \left.\frac{v_1}{v_2}\right|_{i_2=0} = \frac{z_{11}}{z_{21}} \tag{4.6}$$

$$B = -\left.\frac{v_1}{i_2}\right|_{v_2=0} = \frac{\Delta_z}{z_{21}} \tag{4.7}$$

$$C = \left.\frac{i_1}{v_2}\right|_{i_2=0} = \frac{1}{z_{21}} \tag{4.8}$$

$$D = -\left.\frac{i_1}{i_2}\right|_{v_2=0} = \frac{z_{22}}{z_{21}} \tag{4.9}$$

where

$$\Delta_z \triangleq z_{11}z_{22} - z_{21}z_{12}$$

In addition, if the circuit is reciprocal so that $z_{12} = z_{21}$, then the determinate of the *ABCD* matrix is unity, namely

$$AD - BC = 1 \tag{4.10}$$

4.3 IMAGE IMPEDANCE

A generator impedance is said to be matched to a load when the generator can deliver the maximum power to the load. This occurs when the generator impedance is the complex conjugate of the load impedance. For a two-port circuit, the generator delivers power to the circuit, which in turn has a certain load impedance attached to the other side (Fig. 4.6). Consequently, maximum power transfer from the generator to the input of the two-port circuit occurs when it has the appropriate load impedance, Z_L. The optimum generator impedance depends on both the two-port circuit itself and its load impedance. In addition, the matched load impedance at the output side will depend on the two-port itself as well as the generator impedance on the input side. Both sides are matched simultaneously when the input side is terminated with an impedance equal to its image impedance, Z_{I1}, and the output side is terminated with a load impedance equal to Z_{I2}. The actual values for Z_{I1} and Z_{I2} are determined completely by the two-port circuit itself and are independent of the loading on either side of the circuit. Terminating the two-port circuit in this way will guarantee maximum power transfer from the generator into the

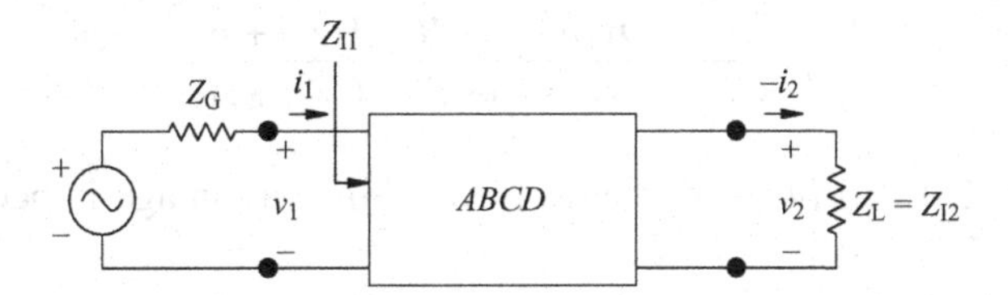

FIGURE 4.6 Excitation of a two-port circuit at port 1.

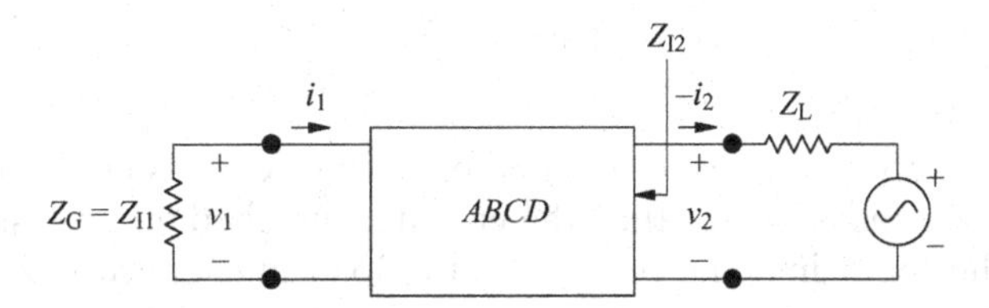

FIGURE 4.7 Excitation of a two-port circuit at port 2.

input side and maximum power transfer from a generator at the output side (if it exists).

The volt–ampere equations for a two-port circuit are given in terms of their *ABCD* parameters as

$$v_1 = Av_2 - Bi_2 \tag{4.11}$$

$$i_1 = Cv_2 - Di_2 \tag{4.12}$$

If the input port is terminated by $Z_{I1} = v_1/i_1$, and the output port by $Z_{I2} = v_2/(-i_2)$, then both sides will be matched. Taking the ratio of Eqs. (4.11) and (4.12) gives

$$\begin{aligned} Z_{I1} &= \frac{v_1}{i_1} = \frac{Av_2/(-i_2)+B}{Cv_2/(-i_2)+D} \\ &= \frac{AZ_{I2}+B}{CZ_{I2}+D} \end{aligned} \tag{4.13}$$

The voltage and current for the output side in terms of these parameters of the input side are found by inverting Eqs. (4.11) and (4.12):

$$v_2 = Dv_1 - Bi_1 \tag{4.14}$$

$$i_2 = Cv_1 - Ai_1 \tag{4.15}$$

If the output port is excited by v_2 as shown in Fig. 4.7, then the matched load impedance is the same as the image impedance:

$$Z_{I2} = \frac{v_2}{i_2} = \frac{Dv_1/(-i_1) + B}{Cv_1/(-i_1) + A} = \frac{DZ_{I1} + B}{CZ_{I1} + A} \tag{4.16}$$

Equations (4.13) and (4.16) can be solved to find the image impedances for both sides of the circuit:

$$Z_{I1} = \sqrt{\frac{AB}{CD}} \tag{4.17}$$

$$Z_{I2} = \sqrt{\frac{DB}{AC}} \tag{4.18}$$

When a two-port circuit is terminated on each side by its image impedance, so that $Z_G = Z_{I1}$ and $Z_L = Z_{I2}$, then the circuit is matched on both sides simultaneously. The input impedance is Z_{I1} if the load impedance is Z_{I2} and vice versa.

The image impedance can be written in terms of the open-circuit z parameters and the short-circuit y parameters by making the appropriate substitutions for the *ABCD* parameters (see Appendix D):

$$Z_{I1} = \sqrt{\frac{z_{11}}{y_{11}}} \tag{4.19}$$

$$Z_{I2} = \sqrt{\frac{z_{22}}{y_{22}}} \tag{4.20}$$

Therefore, an easy way to remember the values for the image impedances is

$$Z_{I1} = \sqrt{z_{oc1} z_{sc1}} \tag{4.21}$$

$$Z_{I2} = \sqrt{z_{oc2} z_{sc2}} \tag{4.22}$$

where z_{oc1} and z_{sc1} are the input impedances of the two-port circuit when the output port is an open circuit or a short circuit, respectively.

As an example consider the simple T circuit in Fig. 4.8. The input impedance when the output is an open circuit is

$$z_{oc1} = Z_a + Z_b \tag{4.23}$$

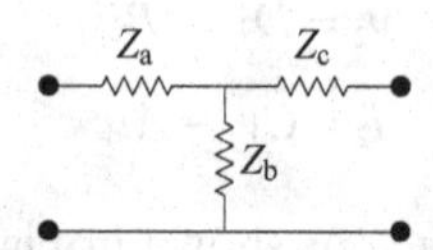

FIGURE 4.8 Example T circuit.

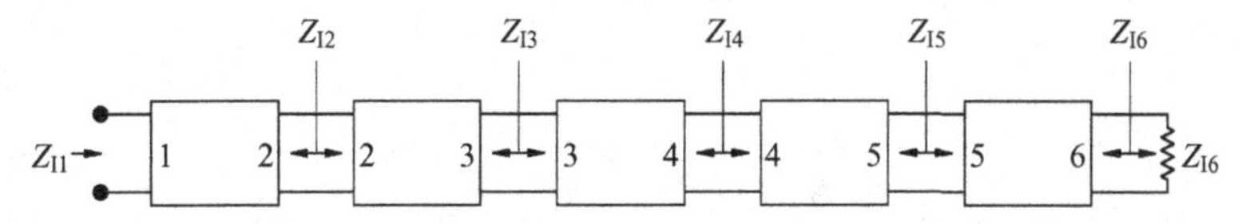

FIGURE 4.9 Chain of matched two-port circuits.

and the input impedance when the output is a short circuit is

$$z_{sc1} = Z_a + Z_b \| Z_c \tag{4.24}$$

The image impedance for the input port for this circuit is

$$Z_{I1} = \sqrt{(Z_a + Z_b)(Z_a + Z_b \| Z_c)} \tag{4.25}$$

and similarly for the output port

$$Z_{I2} = \sqrt{(Z_c + Z_b)(Z_c + Z_b \| Z_a)} \tag{4.26}$$

The output side of the two-port circuit can be replaced by another two-port circuit whose input impedance is Z_{I2}. This is possible if Z_{I2} is the image impedance of the second circuit and the load of the second circuit is equal to its output image impedance, say Z_{I3}. A cascade of two-port circuits where each port is terminated by its image impedance would be matched everywhere (Fig. 4.9). A wave entering from the left side could propagate through the entire chain of two-port circuits without any internal reflections. There, of course, could be some attenuation if the two-port circuits contain lossy elements.

The image propagation constant, γ, for a two-port circuit is defined as

$$e^{\gamma} = \sqrt{\frac{v_1 i_1}{v_2(-i_2)}} = \frac{v_1}{v_2}\sqrt{\frac{Z_{I2}}{Z_{I1}}} \tag{4.27}$$

If the network is symmetrical so that $Z_{I1} = Z_{I2}$, then $e^{\gamma} = v_1/v_2$. For the general unsymmetrical network, the ratio v_1/v_2 is found from Eq. (4.11) as

$$\begin{aligned}
\frac{v_1}{v_2} &= \frac{Av_2 - Bi_2}{v_2} \\
&= A + \frac{B}{Z_{I2}} \\
&= A + B\sqrt{\frac{AC}{BD}} \\
&= \sqrt{\frac{A}{D}}\left(\sqrt{AD} + \sqrt{BC}\right)
\end{aligned}$$

Similarly,

$$\frac{i_1}{(-i_2)} = CZ_{I2} + D$$

$$= \sqrt{\frac{D}{A}}\left(\sqrt{AD} + \sqrt{BC}\right)$$

The image propagation constant is obtained from Eq. (4.27):

$$e^{\gamma} = \sqrt{\frac{v_1 i_1}{v_2(-i_2)}} = \sqrt{AD} + \sqrt{BC} \tag{4.28}$$

Also,

$$e^{-\gamma} = \sqrt{AD} - \sqrt{BC} \tag{4.29}$$

When the circuit is reciprocal, $AD - BC = 1$. Now if Eqs. (4.28) and (4.29) are added together and then subtracted from one another, the image propagation constant can be expressed in terms of hyperbolic functions.

$$\cosh\gamma = \sqrt{AD} \tag{4.30}$$

$$\sinh\gamma = \sqrt{BC} \tag{4.31}$$

If n represents the square root of the image impedance ratio, the $ABCD$ parameters can then be written in terms of n and γ.

$$n \triangleq \sqrt{\frac{Z_{I1}}{Z_{I2}}}$$

$$= \sqrt{\frac{A}{D}} \tag{4.32}$$

$$A = n\cosh\gamma \tag{4.33}$$

$$B = nZ_{I2}\sinh\gamma \tag{4.34}$$

$$C = \frac{\sinh\gamma}{nZ_{I2}} \tag{4.35}$$

$$D = \frac{\cosh\gamma}{n} \tag{4.36}$$

Hence, from the definition of the $ABCD$ matrix, Eq. (4.5), the terminal voltages and currents can be written in terms of n and γ:

$$v_1 = v_2 n\cosh\gamma - i_2 nZ_{I2}\sinh\gamma \tag{4.37}$$

$$i_1 = \frac{v_2}{nZ_{I2}} \sinh\gamma - \frac{i_2}{n} \cosh\gamma \tag{4.38}$$

Division of these two equations gives the input impedance of the two-port circuit when it is terminated by Z_L:

$$Z_{in} = \frac{v_1}{i_1} = n^2 Z_{I2} \frac{Z_L + Z_{I2} \tanh\gamma}{Z_L \tanh\gamma + Z_{I2}} \tag{4.39}$$

This is simply the transmission line equation for a lumped-parameter network when the output is terminated by $Z_L = v_2/(-i_2)$. A clear distinction should be drawn between the input impedance of the network, Z_{in}, which depends on the value of Z_L, and the image impedance Z_{I2}, which depends only on the two-port circuit itself. For a standard transmission line, $Z_{I1} = Z_{I2} = Z_0$ where Z_0 is the characteristic impedance of the transmission line. Just as for the image impedance, the characteristic impedance does not depend on the terminating impedances, but is a function of the geometrical features of the transmission line itself. When the lumped-parameter circuit is lossless, $\gamma = j\beta$ is pure imaginary and the hyperbolic functions become trigonometric functions:

$$Z_{in} = n^2 Z_{I2} \frac{Z_L + jZ_{I2} \tan\beta}{Z_{I2} + jZ_L \tan\beta} \tag{4.40}$$

where β is real. For a lossless transmission line of electrical length $\theta = \omega\ell/v$,

$$Z_{in} = Z_0 \frac{Z_L + jZ_0 \tan\theta}{Z_0 + jZ_L \tan\theta} \tag{4.41}$$

where ω is the radian frequency, ℓ is the length of the transmission line, and v is the velocity of propagation in the transmission line medium.

4.4 TELEGRAPHER'S EQUATIONS

A transmission line consists of two conductors that are spaced considerably less than a quarter wavelength apart. The transmission line is assumed to support only a transverse electromagnetic (TEM) wave. The transmission line might support higher order modes at higher frequencies, but it is assumed here that only the TEM wave is present. This assumption applies to the vast number of two conductor transmission lines used in practice. A transmission line may take a wide variety of forms: Here it will be represented as a two-wire transmssion line (Fig. 4.10). This line is represented as having a certain series inductance per unit length, L, and a certain shunt capacitance per unit length, C (Fig. 4.11). The inductance for the differential length is thus $L\,dz$, and the

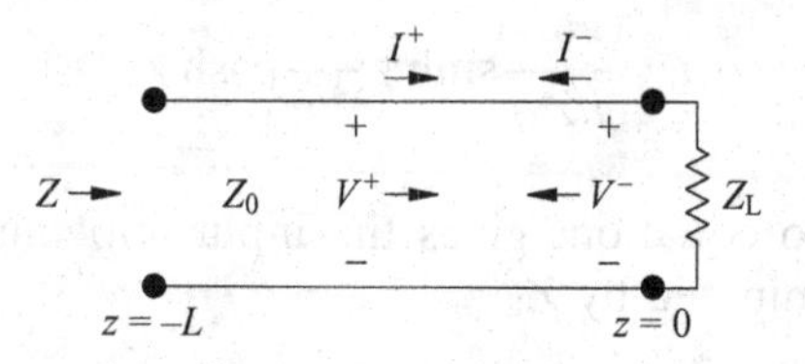

FIGURE 4.10 Two-wire representation of transmission line.

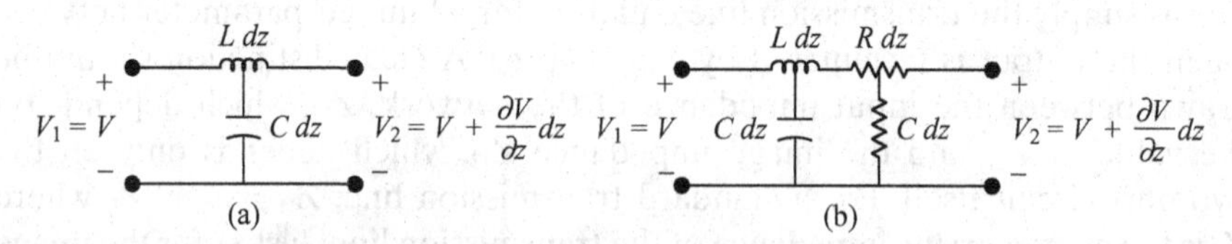

FIGURE 4.11 Circuit model of a differential length of transmission line where (*a*) is the lossless line and (*b*) is the lossy line.

capacitance is $C\,dz$. If the incoming voltage and current wave entering port 1 is $V = v_1$ and $I = i_1$, respectively, then the voltage at port 2 is

$$v_2 = V + \frac{\partial V}{\partial z} dz$$

so that the voltage difference between ports 1 and 2 is

$$v_2 - v_1 = \frac{\partial V}{\partial z} dz = -L\, dz \frac{\partial I}{\partial t} \tag{4.42}$$

The negative sign for the derivative indicates the voltage is decreasing in going from port 1 to port 2. Similarly, the difference in current from port 1 to port 2 is the current going through the shunt capacitance:

$$i_2 - i_1 = \frac{\partial I}{\partial z} dz = -C\, dz \frac{\partial V}{\partial t} \tag{4.43}$$

The telegrapher's equations are obtained from (4.42) and (4.43):

$$\frac{\partial V}{\partial z} = -L \frac{\partial I}{\partial t} \tag{4.44}$$

$$\frac{\partial I}{\partial z} = -C \frac{\partial V}{\partial t} \tag{4.45}$$

Differentiation of Eq. (4.44) with respect to z and Eq. (4.45) with respect to t and then combining produces the voltage wave equation:

$$\frac{\partial^2 V}{\partial z^2} = \frac{1}{v^2}\frac{\partial^2 V}{\partial t^2} \tag{4.46}$$

In similar fashion the current wave equation can be found:

$$\frac{\partial^2 I}{\partial z^2} = \frac{1}{v^2}\frac{\partial^2 I}{\partial t^2} \tag{4.47}$$

The velocity of the wave is

$$v = \frac{1}{\sqrt{LC}} \tag{4.48}$$

The solution for these two wave equations given below in terms of the arbitray functions F_1 and F_2 can be verified by substitution back into Eqs. (4.46) and (4.47):

$$V(z,t) = F_1\left(t - \frac{z}{v}\right) + F_2\left(t + \frac{z}{v}\right) \tag{4.49}$$

$$I(z,t) = \frac{1}{Z_0}\left[F_1\left(t - \frac{z}{v}\right) - F_2\left(t + \frac{z}{v}\right)\right] \tag{4.50}$$

The most useful function for F_1 and F_2 is the exponential function $\exp[j(\omega t \pm \beta z)]$ where $\beta = \omega/v$. The term, Z_0, is the same characteristic impedance used in Eq. (4.41) for the transmission line. For the telegrapher's equations it is

$$Z_0 = \sqrt{\frac{L}{C}} = Lv = \frac{1}{Cv} \tag{4.51}$$

The units for L and C are given in terms of henries per unit length and farads per unit length. These are to be distinguished from L and C used in lumped-element circuit theory.

4.5 TRANSMISSION LINE EQUATION

The transmission line equation was determined in Section 4.3 for a cascade of lumped-element matched circuits. It is the input impedance of a transmission line terminated with a load, Z_L, and it can also be found directly from analysis of a transmission line itself. The transmission line is characterized by its mechanical length, ℓ, and its characteristic impedance, Z_0. The characteristic impedance of a transmission line is a function only of the geometry and dielectric constant of the material between the conductors and is independent of its terminating impedances. This is similar to the image impedance for lumped-element circuits. The input impedance of the transmission line depends on ℓ, Z_0, and Z_L. When terminated with a nonmatching impedance, a standing wave is set up in the transmission line where the forward- and backward-going voltages and currents are as indicated in Fig. 4.10. At the load,

$$V_L = V^+ + V^- \tag{4.52}$$

$$I_L = I^+ - I^- \tag{4.53}$$

Since the forward current wave is $I^+ = V^+/Z_0$ and the reverse current wave is $I^- = V^-/Z_0$, the current at the load is

$$I_L = \frac{V^+ - V^-}{Z_0} = \frac{V_L}{Z_L} \tag{4.54}$$

Replacing V_L above with Eq. (4.52), the voltage reflection coefficient can be determined:

$$\Gamma = \frac{V^-}{V^+} = \frac{Z_L - Z_0}{Z_L + Z_0} \tag{4.55}$$

If the transmssion line is lossy, the reflection coefficent is actually

$$\Gamma = \frac{V^-}{V^+} = \frac{Z_L - Z_0^*}{Z_L + Z_0} \tag{4.56}$$

The phase velocity of the wave is a measure of how fast a given phase moves down a transmission line. This is illustrated in Fig. 4.12 where $e^{j\omega t}$ time dependence is assumed. If time progresses from t_1 to t_2, then in order for $e^{j(\omega t - \beta z)}$ to have the same phase at each of these two times, the wave must progress in the forward direction from z_1 to z_2. Consequently,

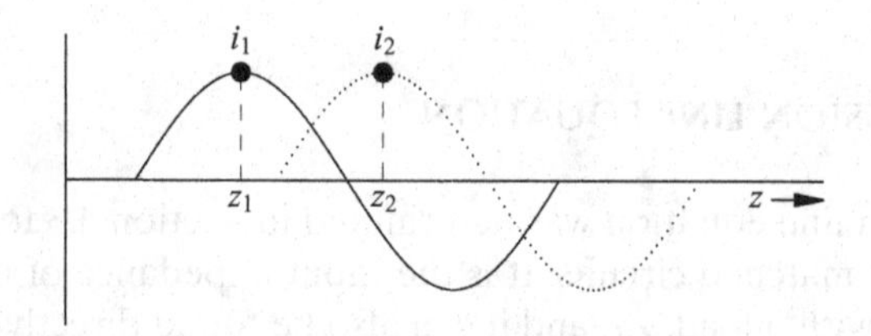

FIGURE 4.12 Forward directed propagating wave.

$$0 = \beta(z_2 - z_1) - \omega(t_2 - t_1)$$

giving the phase velocity

$$v = \frac{\Delta z}{\Delta t} = \frac{\omega}{\beta} \tag{4.57}$$

This is to be distinguished from the group velocity,

$$v_g = \frac{d\omega}{d\beta}$$

which is a measure of velocity of energy flow. For low loss media, $v_g v = c^2/\varepsilon$ where c is the velocity of light in a vacuum. The negative-going wave, of course, has a phase velocity of $-\omega/\beta$.

This traveling wave corresponds to the solution of the lossless telegrapher's equations. The total voltage at any position, z, along the transmission line would be the sum of the forward- and backward-going waves:

$$V(z) = V^+ e^{-j\beta z} + V^- e^{+j\beta z} \tag{4.58}$$

The total current at any point z is by Kirchhoff's law the difference of the two currents:

$$I(z) = \frac{1}{Z_0}\left(V^+ e^{-j\beta z} - V^- e^{+j\beta z}\right) \tag{4.59}$$

At the input to the line (or left side) where $z = -\ell$, the ratio of Eqs. (4.58) and (4.59) gives the input impedance:

$$Z_{\text{in}} = Z_0 \frac{V^+ e^{-j\beta z} + V^- e^{+j\beta z}}{V^+ e^{-j\beta z} - V^- e^{+j\beta z}} \tag{4.60}$$

$$= Z_0 \frac{e^{-j\beta z} + \Gamma e^{+j\beta z}}{e^{-j\beta z} - \Gamma e^{+j\beta z}} \tag{4.61}$$

At the position $z = -\ell$

$$Z_{\text{in}} = Z_0 \frac{Z_L + jZ_0 \tan\beta\ell}{Z_0 + jZ_L \tan\beta\ell} \tag{4.62}$$

If the propagation constant is the complex quantity $\gamma = \alpha + j\beta$, then

$$Z_{\text{in}} = Z_0 \frac{Z_L + Z_0 \tanh\gamma\ell}{Z_0 + Z_L \tanh\gamma\ell} \tag{4.63}$$

A few special cases illustrates some basic features of the transmission line equation. If $z = 0$, $Z_{in}(0) = Z_L$ no matter what Z_0 is. If $Z_L = Z_0$, then $Z_{in}(z) = Z_0$ no matter what z is. For a quarter wavelength line, $Z_{in}(z = \lambda/4) = Z_0^2/Z_L$. The input impedance for any length of line can be readily calculated from Eq. (4.62) or by using the Smith chart.

4.6 SMITH CHART

The Smith chart, as shown in Fig. 4.13, is merely a plot of the transmission line equation on a set of polar coordinates. The reflection coefficient is really an

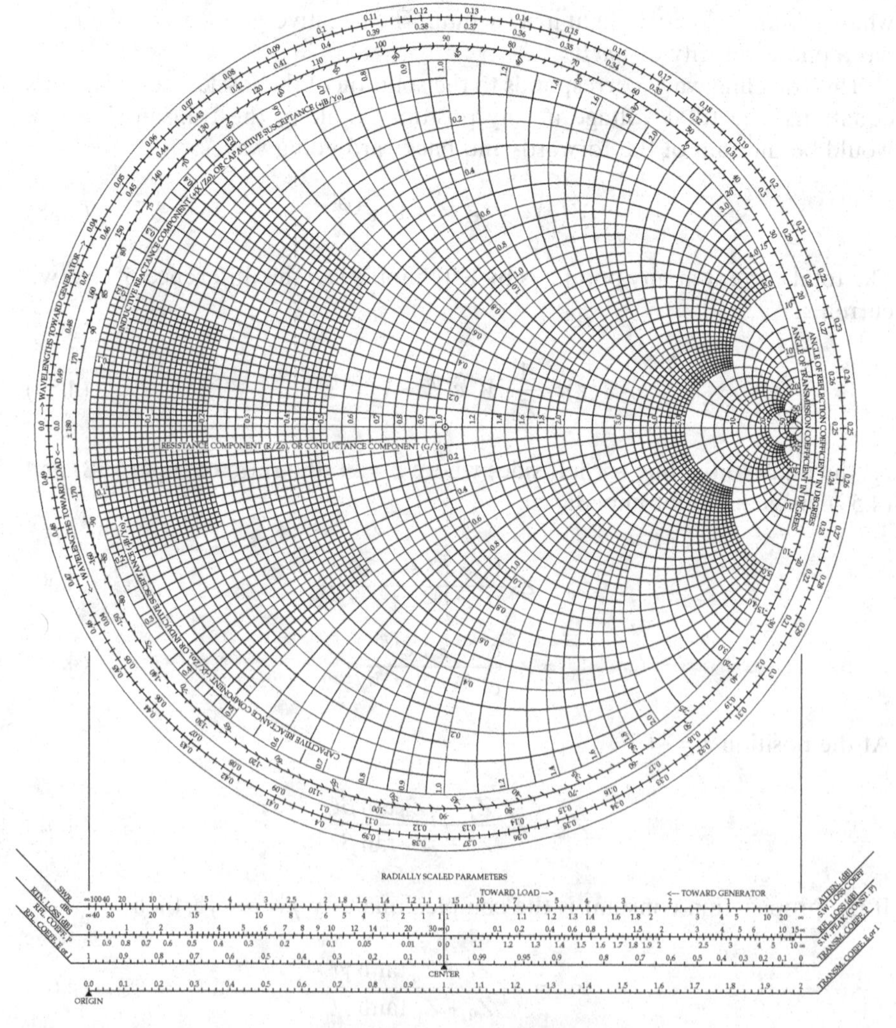

FIGURE 4.13 Smith chart.

alternate way of expressing the input impedance relative to some standard value (Z_0), which is typically 50 Ω. The reflection coefficient, Γ, has a magnitude between 0 and 1 and a phase angle between 0° and 360°. The equations describing the radii and centers of the circles of the coordinates of the Smith chart are found by solving the normalized version of Eq. (4.60):

$$\zeta = r + jx = \frac{Z_{in}}{Z_0} = \frac{1 + \Gamma e^{-2j\beta\ell}}{1 - \Gamma e^{-2j\beta\ell}} \tag{4.64}$$

Solution of the real part of Eq. (4.64) gives the center of the resistance circles as $(r/(1 + r), 0)$ with a radius of $1/(1 + r)$. Solution of the imaginary part gives the center of the reactance circles as $(1, 1/x)$ with a radius of $1/x$ [2, pp. 121–129].

The Smith chart can be used as a computational tool, and it often gives insight where straight equation solving will not. It is also a convenient plotting tool of measured or calculated data, since any passive impedance will fall within its boundaries.

4.7 TRANSMISSION LINE STUB TRANSFORMER

In Chapter 3 a variety of lumped-element impedance transformers were described. Impedance transformation can also be achieved using a simple transmission line circuit with a shunt or series stub. The shunt stub design is the most practical for the most common types of transmission lines. When the transmission line and stub lengths are a small fraction of a wavelength, the circuit can be practical for the upper RF range. Impedance matching can be done using either short-circuit or open-circuit stubs. Impedance matching can be done using shunt or series stubs. Finally, it can be done with one, two, or three stubs as described by Collin [3].

4.7.1 Matching a Real Load Impedance

Consider the shunt stub circuit shown in Fig. 4.14. The load admittance, Y_L, is to be matched to the generator side whose admittance is $Y_G = Y_0$. A distance, d, is determined so that the admittance looking toward the load just to the right side of the stub is

$$Y'_{in} = Y_0 + jB \tag{4.65}$$

When Y_L is real as is assumed here, then $d_2 = 0$ and $\Gamma_L = \pm|\Gamma_L|$ in Fig. 4.14. At the intersection of the main line and the shunt stub, the shunt stub is designed to produce a shunt admittance of $-jB$, which cancels the susceptance of Y'_{in} in leaving

$$Y_{in} = Y'_{in} - jB = Y_0 + jB - jB = Y_0 \tag{4.66}$$

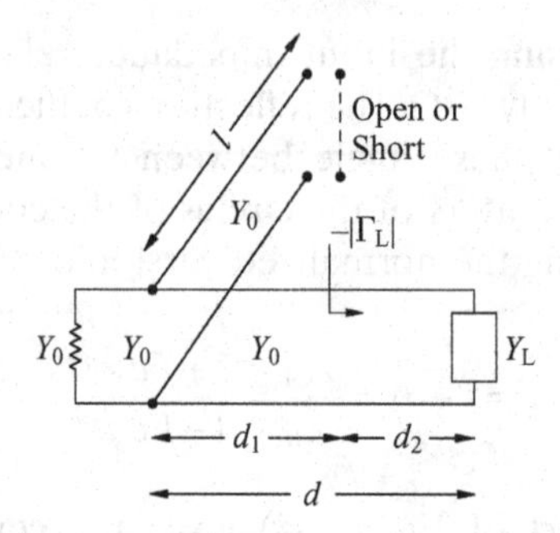

FIGURE 4.14 Open- or short-circuit stub-matching circuit with Y_L complex. When Y_L is real, $d_2 = 0$ and $\Gamma_L = \pm|\Gamma_L|$.

Because of the periodicity of the tangent function, there are many solutions for d and shunt stub length ℓ. The admittance repeats for each added half wavelength, $n\lambda/2$. Usually it is wise to choose the smallest values for d and ℓ in order to minimize frequency sensitivity. The end of the stub may be either a short or an open circuit to ground. When the stub length is shorter than a quarter wavelength, then

$$B = \begin{cases} \tan(\beta\ell) & \text{open circuit} \\ -\cot(\beta\ell) & \text{short circuit} \end{cases} \tag{4.67}$$

Knowing the sign of the required susceptance determines whether to use an open- or a short-circuited stub to achieve the minimum distance, ℓ.

The characteristic admittance used in each of the transmission lines can be any realizable value, but for simplicity the value is chosen to be the system characteristic admittance, Y_0. Then the only two design parameters that need to be found are d and ℓ.

The load admittance is assumed to be real, so that $Y_L = G_L$. This restriction will be removed later. At a distance d from the load,

$$Y'_{in} = Y_0 + jB = Y_0 \frac{G_L + jY_0 \tan\theta}{Y_0 + jG_L \tan\theta} \tag{4.68}$$

$$Y_0(G_L + jY_0 \tan\theta) = (Y_0 + jB)(Y_0 + jG_L \tan\theta) \tag{4.69}$$

where $\theta = \beta d$. The real and imaginary parts of this equation are

$$Y_0 G_L = Y_0^2 - BG_L \tan\theta \tag{4.70}$$

$$jY_0^2 \tan\theta = jY_0(B + G_L \tan\theta) \tag{4.71}$$

which are, respectively,

$$\tan\theta = \frac{Y_0(Y_0 - G_L)}{BG_L} \tag{4.72}$$

$$B = (Y_0 - G_L)\tan\theta \tag{4.73}$$

Substitution of Eq. (4.73) into Eq. (4.72) gives

$$\tan\theta = \pm\sqrt{\frac{Y_0}{G_L}} \tag{4.74}$$

A useful alternative form makes use of the trigonometric double-angle formula

$$\tan^2\theta = \frac{1-\cos 2\theta}{1+\cos 2\theta} \tag{4.75}$$

so that

$$\cos 2\theta = \frac{G_L - Y_0}{G_L + Y_0} \tag{4.76}$$

$$d = \frac{\lambda}{4\pi}\arccos\left(\frac{G_L - Y_0}{G_L + Y_0}\right) \tag{4.77}$$

The ambiguity in the sign for the $\sqrt{G_L}$ in Eq. (4.74) is used to determine the stub length ℓ. The + sign is chosen when $0 < \beta d_1 < \pi/2$ or when $0 < d_1 < \lambda/4$. The − sign is chosen when $\lambda/4 < d_1 < \lambda/2$. From Eqs. (4.73) and (4.74)

$$B = \sqrt{Y_0}\,\frac{Y_0 - G_L}{\sqrt{G_L}} \tag{4.78}$$

The transmission line equation gives the input admittance for a short-circuit stub as

$$Y_{\text{stub}} = -jB = -jY_0\cot\beta\ell \triangleq -jY_0\cot\phi \tag{4.79}$$

so that from Eq. (4.78)

$$\cot\phi = \frac{Y_0 - G_L}{\sqrt{G_L Y_0}} \tag{4.80}$$

or

$$\ell = \frac{\lambda}{2\pi}\arctan\left(\frac{\sqrt{G_L Y_0}}{Y_0 - G_L}\right) \tag{4.81}$$

The design is complete since the characteristic admittance and length of each line section have been found. The design of the open-circuit stub is similar and will be given after considering matching to a complex load.

4.7.2 Matching a Complex Load Impedance

The voltage across the transmission line varies sinusoidally as a function of the position along the line. At a certain point, the voltage will reach a minimum as inferred by Eq. (4.58):

$$V(z) = V^{+}\left(e^{-j\beta z} + \Gamma e^{j\beta z}\right) \tag{4.82}$$

Since $\Gamma = |\Gamma| \exp(j\psi)$, then at a certain point, d_2, Γ will be negative and real (Fig. 4.14):

$$\Gamma(d_2) = -|\Gamma| \tag{4.83}$$

Here the admittance is maximum and real:

$$Y(d_2) = Y_0 \frac{1-\Gamma}{1+\Gamma} = Y_0 \frac{1+|\Gamma|}{1-|\Gamma|} \tag{4.84}$$

$$= Y_0 S \tag{4.85}$$

The variable, S, is the voltage standing-wave ratio, which is a scalar ratio of the maximum to minimum voltage on the transmission line. Moving from the load admittance, Y_L, a distance d_2 will give an admittance looking toward the load a real value of $Y_0 S$ for the "load" at position d_2. The equations for the real load admittance can now be used where $G_L = Y_0 S$. From Eqs. (4.76), (4.77), and (4.81)

$$2\theta_1 = \arccos\left(\frac{SY_0 - Y_0}{SY_0 + Y_0}\right) \tag{4.86}$$

$$d_1 = \frac{\lambda}{4\pi} \arccos\left(\frac{S-1}{S+1}\right) \tag{4.87}$$

$$\ell = \frac{\lambda}{2\pi} \arctan\left(\frac{\pm\sqrt{S}}{S-1}\right) \tag{4.88}$$

Choose the $+\sqrt{S}$ when $0 < d_1 < \lambda/4$ and $-\sqrt{S}$ when $\lambda/4 < d_1 < \lambda/2$. The total distance from the complex load, Y_L, to the stub is

$$d = d_1 + d_2 \tag{4.89}$$

where d_1 is the same solution for d in Section 4.7.1 (Eq. 4.76).

The above designs have been based on the use of a short-circuit stub. However, the open circuit is often easier to implement. If the susceptance from the load at d turns out to be negative, the open-circuit stub will provide the shortest line length to get the positive susceptance needed for cancellation. The input admittance for an open-circuit stub is $jY_0 \tan \phi$. Then by analogy with Eqs. (4.80) and (4.81),

$$\tan\phi = -\frac{Y_0 - G_L}{\sqrt{G_L Y_0}} \tag{4.90}$$

$$\ell = \frac{\lambda}{2\pi}\arctan\left(\frac{G_L - Y_0}{\sqrt{G_L Y_0}}\right) \tag{4.91}$$

or for a complex load

$$\ell = \frac{\lambda}{2\pi}\arctan\left(\frac{S-1}{\sqrt{S}}\right) \tag{4.92}$$

As an example, design a matching circuit for a load impedance of $Z_L = 30 + j40\,\Omega$ with $Z_0 = 50\,\Omega$ using a short-circuit stub. The reflection coefficient at the load is

$$\Gamma_L = |\Gamma_L|e^{j\psi} = \frac{Z_L - Z_0}{Z_L + Z_0} = j0.5$$

The magnitude of Γ_L is 0.5 and the angle ψ is $\pi/2$. Since,

$$\begin{aligned}\Gamma(d_2) &= \frac{V^-}{V^+}\left(\frac{e^{j\beta z}}{e^{-j\beta z}}\right)\\ &= \Gamma_L e^{-2j\beta d_2} && (4.93)\\ &= |\Gamma_L|e^{j\psi}e^{-2j\beta d_2} && (4.94)\\ &= -|\Gamma_L|\\ &= |\Gamma_L|e^{\pm j\pi} && (4.95)\end{aligned}$$

The sign change in Eq. (4.93) is a result of placing the origin at the load and moving in the $-z$ direction toward the generator a distance of d_2. The minus sign is used to ensure positive line lengths for d_2. Solving Eqs. (4.94) and (4.95) for d_2 gives

$$d_2 = \frac{\lambda}{4\pi}(\psi + \pi) \tag{4.96}$$

In the present example, $\psi = \pi/2$ so that $d_2 = 0.375\lambda$. From Eq. (4.85) the standing-wave ratio, $S = 3$, from Eq. (4.87) $d_1 = 0.0833\,\lambda$ so that $d = d_1 + d_2 = 0.4583\,\lambda$, and from Eq. (4.88) $\ell = 0.205\,\lambda$.

4.8 COMMONLY USED TRANSMISSION LINES

Because TEM transmission lines have neither an elecrtric nor magnetic field component in the direction of propagation, the characteristic impedance can be found from Eq. (4.51) and electrostatics. Since the velocity of propagation in the given media is presumably known, all that is necessary is to calculate the electrostatic capacitance between the conductors. When the geometry is particularly nasty and the solution is needed quickly, the field mapping approach described in Section 2.2.2 can be used.

4.8.1 Two-Wire Transmission Line

The two-wire transmission line commonly used, for example, between a TV antenna and the receiver, consists of two round conductors each with a radius of a and separated by a distance b (Fig. 4.15). The dielectric surrounding the wires has a dielectric constant of ε. The field theory analysis, such as that given in [2], shows that the characteristic impedance of the two-wire line is

$$Z_0 = \frac{\eta}{\pi}\operatorname{arccosh}\left(\frac{b}{2a}\right) \tag{4.97}$$

where

$$\eta = \sqrt{\frac{\mu}{\varepsilon}} \tag{4.98}$$

While the field analysis for a given structure may bring some challenges, the good news is that once Z_0 is known, the rest of the problem can be solved by circuit theory. Losses in the two-wire transmission line stem from the lossy dielectric between the conductors and the resistive losses experienced by the current as it flows along the conductor. Since a voltage wave is attenuated as it goes down a line by $\exp(-\alpha z)$, the power loss is proportional to $\exp(-2\alpha z)$ where

$$\alpha = \alpha_{\mathrm{d}} + \alpha_{\mathrm{c}} \tag{4.99}$$

The dielectric and conductor losses are

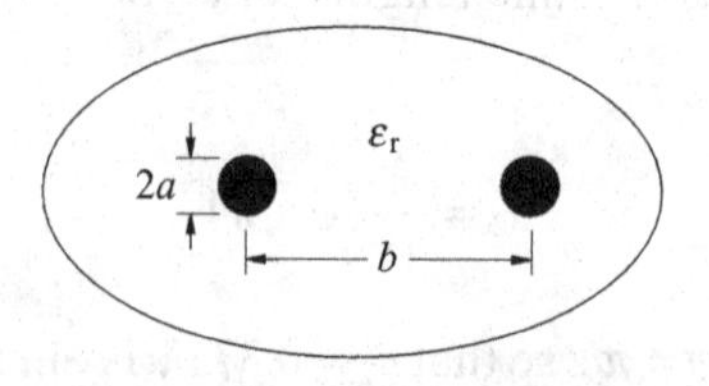

FIGURE 4.15 Two-wire transmission line.

$$\alpha_{\rm d} = \frac{\sigma_{\rm d}\eta}{2} \tag{4.100}$$

$$\alpha_{\rm c} = \frac{1}{2a}\sqrt{\frac{\omega\mu}{2\sigma_c}}\frac{1}{\eta\,\text{arccosh}(b/2a)} \tag{4.101}$$

where $\sigma_{\rm d}$ and $\sigma_{\rm c}$ are the conductivities of the dielectric and conductor, respectively. The two-wire line is inexpensive and widely used in ultrahigh frequency (UHF) applications.

4.8.2 Parallel Strip Transmission Line

The parallel strip transmission line consists of two separate conductors of width b and separated by a distance a (Fig. 4.16). This is a rectangular waveguide without the side walls. It is fundamentally distinct from the rectangular waveguide, which is not a TEM transmission line. The Maxwell equations for a plane wave for this system are the exact analog to the telegrapher's equations:

$$\frac{\partial E_x}{\partial z} = -\mu\frac{\partial H_y}{\partial t} \tag{4.102}$$

$$\frac{\partial H_y}{\partial z} = -\varepsilon\frac{\partial E_x}{\partial t} \tag{4.103}$$

The voltage between the strips is the integral of the electric field:

$$V = -\int_0^a E_x\,dx = -aE_x \tag{4.104}$$

The magnetic field in the y direction will produce a current in the conductor that will travel in the z direction according to Ampère's law:

$$I = -\int_0^b H_y\,dy = -H_y b \tag{4.105}$$

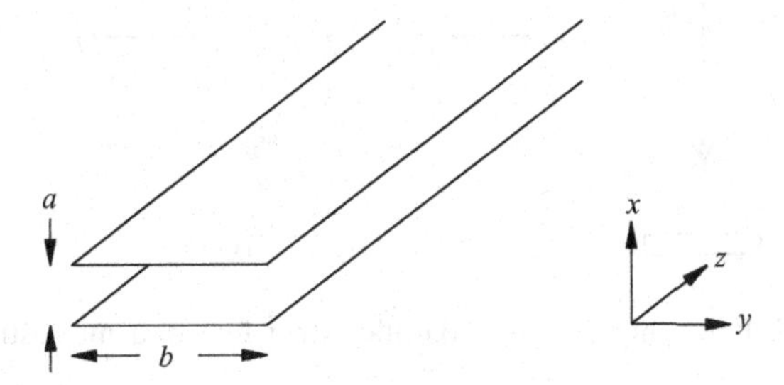

FIGURE 4.16 Parallel strip transmission line.

Substitution of Eqs. (4.104) and (4.105) into Eq. (4.44) gives

$$\frac{\partial E_x}{\partial z} = -\frac{Lb}{a}\frac{\partial H_y}{\partial t} \tag{4.106}$$

Comparison of this with Eq. (4.102) indicates that

$$L = \frac{\mu a}{b}. \tag{4.107}$$

A similar substitution of Eqs. (4.104) and (4.105) into Eq. (4.45) and comparison with Eq. (4.103) indicates that

$$C = \frac{\varepsilon b}{a} \tag{4.108}$$

so that the characteristic impedance for the parallel strip guide is

$$Z_0 = \sqrt{\frac{L}{C}} = \frac{a}{b}\sqrt{\frac{\mu}{\varepsilon}} \tag{4.109}$$

While this solution illustrates the major feactures of the parallel strip transmission line, the neglect of the fringing fields makes this expression in practice useless. A more uselful expression was developed for the parallel strip line [4], which was later used to determine the characteristic impedance for microstrip by the method of images (Fig. 4.17). Conformal transformations of the structure were made that included an approximation of the fringing fields. As a

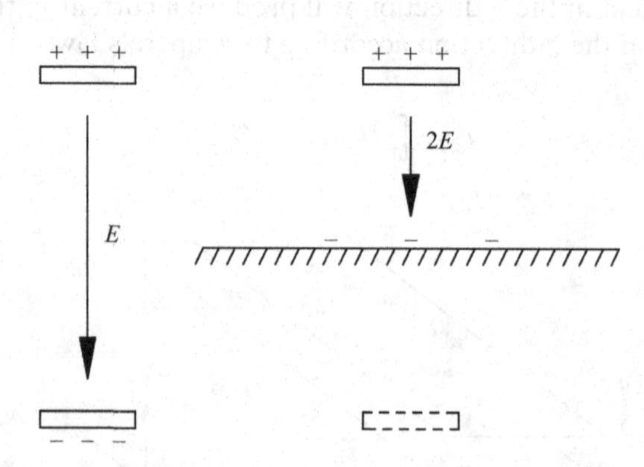

FIGURE 4.17 Relationship between parallel strip line and microstrip by method of images.

result two formulas were developed, one for wide and the other for narrow strips. The analysis formulas follow:

$$\varepsilon_a \triangleq \frac{\varepsilon_r + 1}{2} \tag{4.110}$$

$$\eta = \sqrt{\frac{\mu_0}{\varepsilon_0}} \tag{4.111}$$

$$Z_0 = \eta \frac{1}{\pi\sqrt{\varepsilon_a}}\left[\ln\frac{4a}{b} + \frac{1}{8}\left(\frac{b}{a}\right)^2 - \frac{1}{2}\frac{\varepsilon_r - 1}{\varepsilon_r + 1}\left(\ln\frac{\pi}{2} + \frac{1}{\varepsilon_r}\ln\frac{4}{\pi}\right)\right] \tag{4.112}$$

for $b/a < 0.56$ and

$$Z_0 = \eta\sqrt{\frac{1}{\varepsilon_r}}\left\{\frac{b}{a} + 0.441 + \frac{\varepsilon_r + 1}{2\pi\varepsilon_r}\left[\ln\left(\frac{b}{a} + 0.94\right) + 1.451\right] + \frac{\varepsilon_r - 1}{\varepsilon_r^2}(0.82)\right\}^{-1} \tag{4.113}$$

for $b/a > 0.56$.

Equation (4.112) is Eq. (48) in [4] and Eq. (4.113) is Eq. (31) in [4] except that here b/a represents $2a/2b$ in [4]. The method of images then can be invoked to find the characteristic impedance of the microstrip. This is described more fully in Section 4.8.4. A comparison of Eq. (4.109) with no fringing capacitance, and the formula that includes fringing capacitance, Eqs. (4.112) and (4.113), as well as the microstrip is shown in Fig. 4.17 (*a*, *b*, and *c*) for $\varepsilon_r = 1$, 4, and 10, respectively. A parallel strip line with a given b/a ratio will give a certain value for Z_0. A microstrip line with $w/h = b/(a/2)$ would give a characteristic impedance of $Z_0/2$. The equations used for the microstrip calculation are those in Section 4.8.4, so the factor of 2 is close, but not exact.

4.8.3 Coaxial Transmission Line

A coaxial transmission line comes in the form of rigid, semirigid, and flexible forms. The end view of a coaxial line, which is shown in Fig. 4.18, consists of an inner conductor and the outer conductor, which is normally grounded. The electric field points from the center to the outer conductor, and the longitudinal current on the center conductor produces a magnetic field concentric to the inner conductor. The potential between the two conductors is a solution of the transverse form of Laplace's equation in cylindrical coordinates where there is no potential difference in the longitudinal z direction. The notation for the divergence and curl operators follows that given in [5]:

$$0 = \nabla\nabla_t\Phi$$

$$0 = \frac{1}{r}\frac{\partial}{\partial r}\left(r\frac{\partial\Phi}{\partial r}\right) + \frac{1}{r^2}\frac{\partial^2\Phi}{\partial\phi^2} \tag{4.114}$$

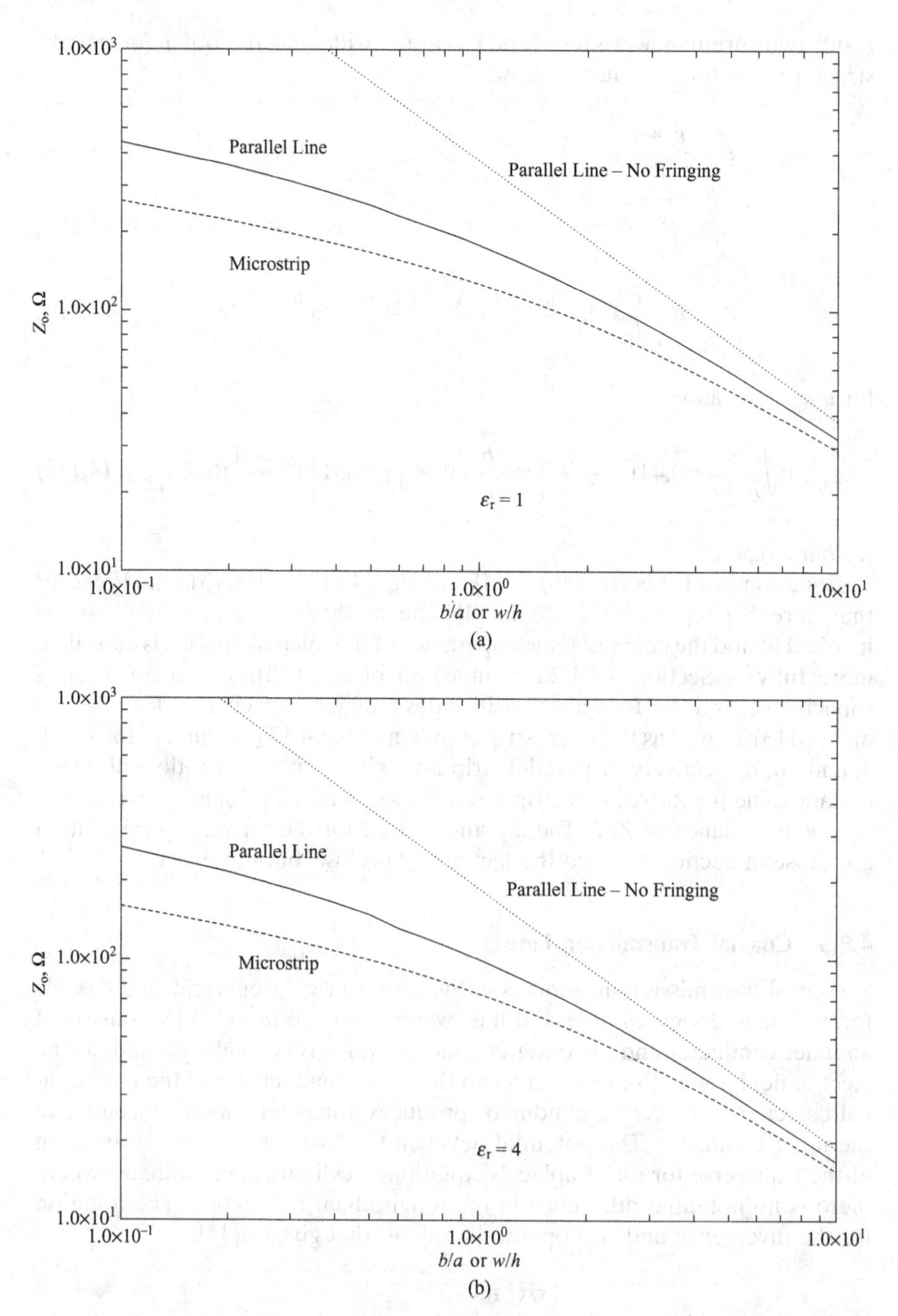

FIGURE 4.18 Characteristic impedance for parallel strip line (with and without fringing) and microstrip for (*a*) $\varepsilon_r = 1$, (*b*) $\varepsilon_r = 4$, and (*c*) $\varepsilon_r = 10$. b/a corresponds to parallel strip line and w/h corresponds to microstrip.

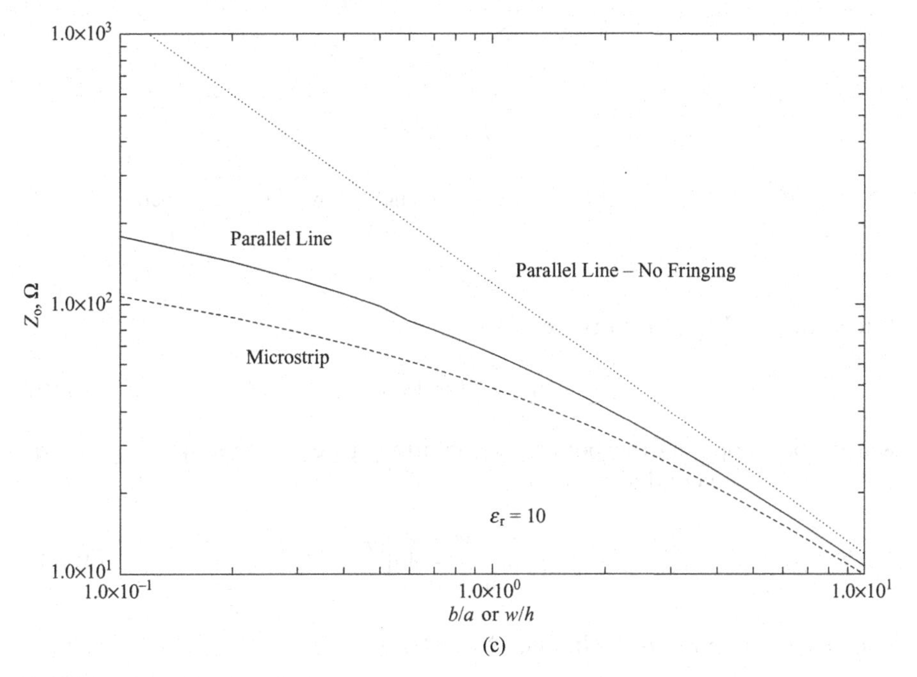

FIGURE 4.18 *Continued*

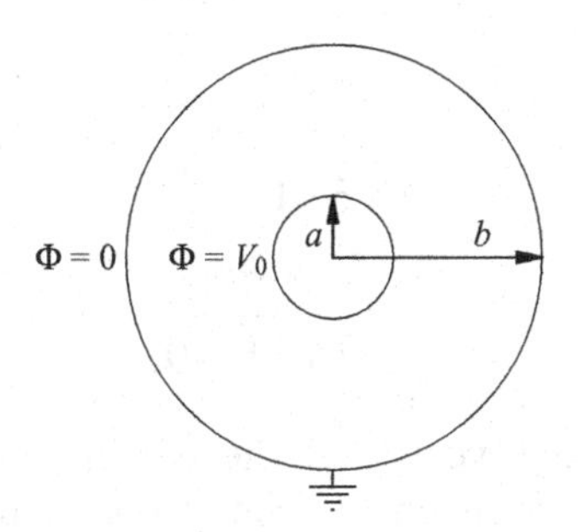

FIGURE 4.19 Coaxial transmission line.

Because there is no potential variation in the z direction, the z derivative of Φ is zero. Because of symmetry, there is no variation of Φ in the ϕ direction either. Thus, Eq. (4.114) simplifies to an ordinary second-order differential equation subject to the boundary conditions that $\Phi = 0$ on the outer conductor and $\Phi = V_0$ on the inner conductor:

$$0 = \frac{1}{r}\frac{d}{dr}\left(r\frac{d\Phi}{dr}\right) \tag{4.115}$$

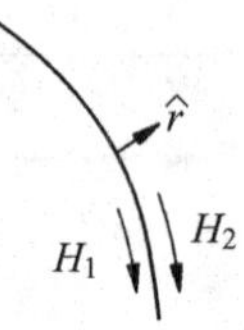

FIGURE 4.20 Continuity of magnetic field along center conductor.

Integration of Eq. (4.115) twice gives

$$\Phi = C_1 \ln r + C_2 \tag{4.116}$$

which upon applying the boundary conditions gives the potential anywhere between the two conductors:

$$\Phi(r) = \frac{V_0}{\ln(a/b)} \ln\left(\frac{r}{b}\right) \tag{4.117}$$

The electric field is easily obtained by differentiation.

$$\mathbf{E} = -\nabla_t \Phi = \frac{V_0 e^{-j\beta z}}{\ln(b/a)} \frac{\hat{r}}{r} \tag{4.118}$$

The magnetic field is then

$$\begin{aligned} \mathbf{H} &= \hat{z} \times \mathbf{E} \\ &= \frac{V_0 e^{-j\beta z}}{r\eta \ln(b/a)} \hat{\phi} \end{aligned} \tag{4.119}$$

The outward normal unit vector of the center conductor, $\hat{r}$, is shown in Fig. 4.20. The surface current on the center conductor is determined by the boundary condition for the tangential magnetic field:

$$\mathbf{J}_S = \hat{r} \times (\mathbf{H}_2 - \mathbf{H}_1) = \hat{r} \times \mathbf{H}_2 \tag{4.120}$$

The later result occurs because the magnetic field is zero inside the conductor. The total current flowing in the center conductor is

$$\begin{aligned} I_0 \hat{\mathbf{z}} &= \int_0^{2\pi} \frac{\hat{z} V_0}{a\eta \ln(b/a)} a \, d\phi \\ &= \frac{\hat{z} 2\pi V_0}{a\eta \ln(b/a)} \end{aligned} \tag{4.121}$$

so that

$$Z_0 = \frac{V_0}{I_0} = \frac{\eta}{2\pi} \ln\left(\frac{b}{a}\right) \tag{4.122}$$

Coaxial Dielectric Loss The differential form of Ampère's law relates the magnetic field to both the conduction current and the displacement current. In the absence of a conductor

$$\nabla \mathbf{H} = \mathrm{J} + \frac{\partial \mathbf{D}}{\partial t} \tag{4.123}$$

$$\approx j\omega\varepsilon\mathbf{E} \tag{4.124}$$

By taking the curl of Eq. (4.124), the Helmholtz wave equation for **H** can be found. A solution of the wave equation would give the propagation constant, γ:

$$\gamma = j\omega\sqrt{\mu\varepsilon} \triangleq jk_0\sqrt{\varepsilon_r} \tag{4.125}$$

where ε_r is the relative dielectric constant and k_0 is the propagation constant in free space. A lossy dielectric is typically represented as the sum of the lossless (real) and lossy (imaginary) parts:

$$\varepsilon_r = \varepsilon_r' - j\varepsilon_r'' \tag{4.126}$$

The revised propagation constant is found by substituting this into Eq. (4.125). The result can be simplified by taking the first two terms of the Taylor series expansion since $\varepsilon_r'' << \varepsilon_r'$.

$$\gamma = \alpha + j\beta = jk_0\sqrt{\varepsilon_r'}\left(1 - j\frac{\varepsilon_r''}{2\varepsilon_r'}\right) \tag{4.127}$$

so that

$$\alpha_d = \frac{k_0\varepsilon_r''}{2\sqrt{\varepsilon_r'}} \tag{4.128}$$

$$\beta = k_0\sqrt{\varepsilon_r'} \tag{4.129}$$

The power loss is proportional to $\exp(-2\alpha z)$.

Coaxial Conductor Loss The power loss per unit length, P_ℓ, is obtained by taking the derivative of the power at a given point along a transmission line:

$$P = P_0 e^{-2\alpha_c z} \tag{4.130}$$

$$P_\ell \triangleq -\frac{dP}{dz} = 2\alpha_c P \tag{4.131}$$

For a low loss conductor where the dielectric losses are negligible, Eq. (4.123) becomes with the help of Ohm's law,

$$\nabla \mathbf{H} = \sigma \mathrm{E} \tag{4.132}$$

where σ is the metal conductivity. This would be the same equation as Eq. (4.124) if

$$\varepsilon \Rightarrow \frac{\sigma}{j\omega} \tag{4.133}$$

With this substitution the wave impedance becomes the metal surface impedance:

$$\sqrt{\frac{\mu}{\varepsilon}} \Rightarrow Z_m = (1+j)\sqrt{\frac{\omega\mu_0}{2\sigma}} \tag{4.134}$$

At the surface there will be a longitudinal electric field of $Z_m J_s$ directed in the $\hat{z}$ direction. Thus, in a lossy line, the fields will no longer be strictly TEM. This longitudinal electric field produces energy flow into the conductor proportional to $E\hat{z} \times H\hat{\phi}$. This energy is dissipated in the center and outer conductors. The power loss per unit length is found in the following way:

$$\begin{aligned} P_\ell &= \frac{R_m}{2} \oint \mathbf{J}_s \cdot \mathbf{J}_s^* \, d\ell \\ &= \frac{R_m}{2} \oint (\hat{\mathbf{r}} \times \mathbf{H})(\hat{\mathbf{r}} \times \mathbf{H}^*) \, d\ell \\ &= \frac{R_m}{2} \oint \mathbf{H} \cdot \mathbf{H}^* \, d\ell \\ &= \frac{R_m V_0^2 \pi}{\eta^2 (\ln b/a)^2} \left(\frac{1}{a} + \frac{1}{b} \right) \end{aligned} \tag{4.135}$$

The power, P, transmitted down the line is found by the Poynting theorem:

$$\begin{aligned} P &= \frac{1}{2} \Re\{Z_m\} \int_a^b \int_0^{2\pi} \mathbf{E} \times \mathbf{H}^* \cdot \hat{\mathbf{z}} r \, dr \, d\phi \\ &= \frac{\pi V_0^2}{\eta \ln(b/a)} \end{aligned} \tag{4.136}$$

The attenuation constant associated with conductor loss is found from Eq. (4.131):

$$\alpha_c = \frac{P_\ell}{2P} = \frac{R_m}{2\eta \ln(b/a)} \frac{a+b}{ab} \tag{4.137}$$

while the dielectric loss found earlier is

$$\alpha_d = \frac{k_0 \varepsilon_r''}{2\sqrt{\varepsilon_r'}} \tag{4.138}$$

The total loss is found from Eq. (4.99).

4.8.4 Microstrip Transmission Line

Microstrip has been a popular form of transmission line for RF and microwave frequencies for some time. The microstrip line shown in Fig. 4.21 consists of a conductor strip of width w on a dielectric of thickness h above a ground plane. Part of the electric field between the strip and the ground plane is in the dielectric and part in the air. The field is more concentrated in the dielectric than in the air. Consequently, the effective dielectric constant, ε_{eff}, is somewhere between ε_r and 1, but closer to ε_r than 1. A variety of methods have been used to find ε_{eff}. However, rather than provide a proof, a simple empirically based procedure for synthesizing a microstrip line will be given. A microstrip line is not strictly a TEM type of transmission line and does have some frequency dispersion. Unless microstrip is being used in a wide bandwidth application, the TEM approximation should be adequate. Determining the characteristic impedance could be done by using the parallel strip line described earlier and the method of images. The microstrip ground plane reflects the image of the narrow strip so that it would appear to be two strips with dielectric between the strips. This is exactly what was done in [6]. However, since the height of the microstrip line is half the height of the parallel strips, the electric field between the two conductors in the microstrip will be twice as strong, the capacitance would be twice as large, so Z_0 is half as large as the eqivalent paralle strip line. The synthesis problem occurs when the desired characteristic impedance, Z_0,

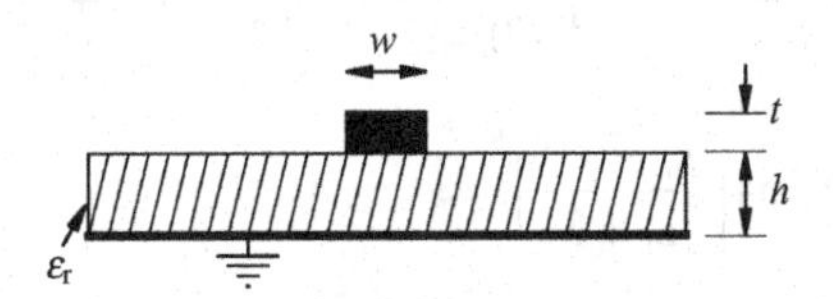

FIGURE 4.21 Microstrip transmission line.

and the dielectric constant of the substrate, ε_r, are known and the geometrical quantity w/h is to be found. The synthesis equations, given by [6] are simple, and give an approximate solution to the microstrip problem.

$$\frac{w}{h}=\frac{8\sqrt{\dfrac{A}{11}\left(7+\dfrac{4}{\varepsilon_r}\right)+\dfrac{1}{0.81}\left(1+\dfrac{1}{\varepsilon_r}\right)}}{A} \tag{4.139}$$

$$A=\exp\frac{Z_0\sqrt{\varepsilon_r+1}}{42.4}-1 \tag{4.140}$$

The analysis equations given below by [7] are more accurate than the synthesis equations. The value given by Eqs. (4.139) and (4.140) provides anititial value for w/h that can be used in an iterative procedure to successfully solve the synthesis problem. This process depends on knowing ε_r, and the conductor thickness, t. The solution results in the effective dielectric constant, ε_{eff}, needed to determine electrical line lengths and Z_0. The procedure for the analysis procedure follows:

$$\Delta u_a=\frac{t/h}{\pi}\ln\left[1+\frac{4\exp(1)}{(t/h)\coth^2\sqrt{6.517w/h}}\right] \tag{4.141}$$

$$\Delta u_r=\frac{1}{2}\left[1+\frac{1}{\cosh\sqrt{\varepsilon_r-1}}\right]\Delta u_a \tag{4.142}$$

$$u_a=\frac{w}{h}+\Delta u_a \tag{4.143}$$

$$u_r=\frac{w}{h}+\Delta u_r \tag{4.144}$$

$$Z_{0a}(x)=\frac{\eta}{2\pi}\ln\left[\frac{f(x)}{x}+\sqrt{1+\left(\frac{2}{x}\right)^2}\right] \tag{4.145}$$

$$f(x)=6+(2\pi-6)\exp\left[-(30.666/x)^{0.7528}\right] \tag{4.146}$$

$$\varepsilon_e(x,\varepsilon_r)=\frac{\varepsilon_r+1}{2}+\frac{\varepsilon_r-1}{2}\left(1+\frac{10}{x}\right)^{-a(x)b(\varepsilon_r)} \tag{4.147}$$

$$a(x)=1+\frac{1}{49}\ln\left[\frac{x^4+(x/52)^2}{x^4+0.432}\right]+\frac{1}{18.7}\ln\left[1+\left(\frac{x}{18.1}\right)^3\right] \tag{4.148}$$

$$b(\varepsilon_r)=0.564\left[\frac{\varepsilon_r-0.9}{\varepsilon_r+3}\right]^{0.053} \tag{4.149}$$

From the given value of t and trial solutions of w/h, Eqs. (4.141) to (4.144) give unique values for u_a and u_r. In Eqs. (4.145) through (4.148), x is replaced by u_a or u_r as required by the calculation for Z_0 and ε_{eff} in Eqs (4.150) and (4.151):

$$Z_0\left(\frac{w}{h}, t, \varepsilon_r\right) = \frac{Z_{0a}(u_r)}{\sqrt{\varepsilon_e(u_r, \varepsilon_r)}} \tag{4.150}$$

$$\varepsilon_{eff}\left(\frac{w}{h}, t, \varepsilon_r\right) = \varepsilon_e(u_r, \varepsilon_r)\left[\frac{Z_{0a}(u_a)}{Z_{0a}(u_r)}\right]^2 \tag{4.151}$$

Since w/h increases when Z_0 decreases and vice versa, one very simple and effective method for finding the new approximation for w/h is done by using the following ratio:

$$\left(\frac{w}{h}\right)_{i+1} = \left(\frac{w}{h}\right)_i \frac{\text{calculated } Z_0 \text{ from Eq.}(4.150)}{\text{desired } Z_0} \tag{4.152}$$

For example, if w/h for trial i is too small, the resulting Z_0 will increase the value for w/h in trial $i + 1$. This procedure has been coded in the program MICSTP, which will converge to a value for Z_0 with acceptable error within five iterations when given an initial value from Eq. (4.139).

The effect of dielectric and conductor loss is summarized here [8–10]:

$$\alpha_c = \begin{cases} 0.159A'\dfrac{R_m\left[32 - u_r^2\right]}{hZ_0\left[32 + u_r^2\right]}, & \dfrac{w}{h} < 1 \\ 7.02 \times 10^{-6}\ A'\dfrac{R_m Z_0 \varepsilon_{eff}}{h}\left[u_r + \dfrac{0.667u_r}{u_r + 1.444}\right], & \dfrac{w}{h} \geq 1 \end{cases} \tag{4.153}$$

$$A' = 1 + u_r\left[1 + \frac{1}{\pi}\ln\left(\frac{2B}{t}\right)\right] \tag{4.154}$$

$$B = \begin{cases} h, & \dfrac{w}{h} \geq \dfrac{\pi}{2} \\ 2\pi w, & \dfrac{w}{h} < \dfrac{\pi}{2} \end{cases} \tag{4.155}$$

The value for the loss caused by the dielectric is

$$\alpha_d = \frac{\varepsilon_r}{2\sqrt{\varepsilon_{eff}}}\frac{\varepsilon_{eff} - 1}{\varepsilon_r - 1}\left(\frac{k_0\varepsilon_r''}{\varepsilon_r'}\right) \tag{4.156}$$

where the wave number in vacuum is k_0. The definition for R_m, ε_r', and ε_r'' were given previously as the real part of Eq. (4.134) and the complex parts of Eq. (4.126), respectively.

While actual microstrip transmission lines are somewhat dispersive, the above analysis presumes that the effective dielectric constant is independent of frequency. Design of broadband circuits will sometimes require knowledge of how ε_{eff} varies with frequency. An accurate formula suitable for computer-aided design is given by [11]. In these expressions, frequency f is in gigahertz, h, and w are in centimeters:

$$\varepsilon_{\text{eff}} = \varepsilon_{\text{r}} - \frac{\varepsilon_{\text{r}} - \varepsilon_{\text{eff}}(f=0)}{1+P(f)} \tag{4.157}$$

where

$$P(f) = P_1 P_2 [(0.1844 + P_3 P_4) 10 fh]^{1.5763} \tag{4.158}$$

$$P_1 = 0.27488 + \left[0.6315 + \frac{0.525}{(1+0.157 fh)^{20}}\right]\frac{w}{h} \tag{4.159}$$

$$P_2 = 0.33622[1 - \exp(-0.03442\varepsilon_{\text{r}})] \tag{4.160}$$

$$P_3 = 0.363 \exp\left(-4.6\frac{w}{h}\right) \times \left\{1 - \exp\left[-\left(\frac{fh}{3.87}\right)^{4.97}\right]\right\} \tag{4.161}$$

$$P_4 = 1 + 2.571\left\{1 - \exp\left[-\left(\frac{\varepsilon_{\text{r}}}{15.916}\right)^{8}\right]\right\} \tag{4.162}$$

These equations for Z_0 and ε_{eff} are widely accepted as the most accurate available for the quasi-TEM microstrip transmission line. Typical silicon integrated circuit transmission lines have a more complicated structure. The "dielectric" under the metal line typically consists of a thin layer of SiO_2 approximately 0.5 to 1 μm thick and a Si substrate that could be approximately 200 μm thick (Fig. 4.22). This results in three distinct propagation modes: (1) the dielectric quasi-TEM mode, (2) the skin effect mode, and (3) the slow wave mode [12]. The analysis of these modes is an approximation based on the transverse resonance technique for parallel plate waveguide. The SiO_2 layer is low loss and thin. The Si layer is thick and has a conductivity, σ, dependent on its doping.

The dielectric quasi-TEM mode occurs when ω/σ is large so that the Si substrate layer is low loss and acts as a dielectric. If the signal wavelength is

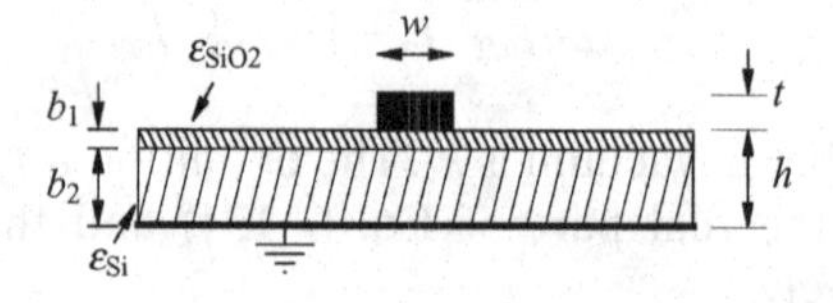

FIGURE 4.22 Microstrip transmission line in Si-integrated circuits.

much greater than the double-layer thickness, h, then propagation will be almost TEM. For this case the dielectric constant is

$$\varepsilon_{\text{eff}} \propto \left[\frac{1}{h}\left(\frac{b_1}{\varepsilon_{\text{SiO}_2}} + \frac{b_2}{\varepsilon_{\text{Si}}}\right)\right]^{-1} \tag{4.163}$$

The skin effect mode occurs when $\omega\sigma$ is large so that the Si substrate acts as a lossy conductor wall. The skin depth, $\delta = \sqrt{2/\sigma\omega\mu_0}$, is large enough to allow signficant penetration of the fields into the Si. Propagation in this mode is highly dispersive. In this case

$$\varepsilon_{\text{eff}} \propto \varepsilon_{\text{SiO2}} \frac{h}{b_1} \tag{4.164}$$

$$\mu_{\text{eff}} \propto \mu_0 \frac{1}{h}\left(b_1 + \frac{\delta}{2}\right) \tag{4.165}$$

The slow wave mode occurs when ω is not too high and σ is moderate. The dielectric constant is similar to that found for the skin effect mode in Eq. (4.164), but the permeability is back to μ_0. The effective dielectric constant is multipled by the h/b_1 ratio, thereby producing slow wave propagation. With the right conductivity, this slow wave mode can extend into the low gigahertz range. It normally pays to stay out of this range in RF circuits. When $\omega\sigma$ is sufficiently small, the field can penetrate to the ground plane.

In multiple layer structures, it is possible to have two metals separted by Si where one metal would act as the ground. Depending on the metal width, it could be used as either microstrip with possibly an over layer or parallel plate line. The t/h ratio for integrated circuit lines will typically be much larger than that used in printed circuit boards. A large metal thickness will then decrease the required metal width for a given Z_0.

There are a variety of other transmission line geometries that could be studied, but these examples should provide information on the most widely used forms. There are cases where one line is located close enough to a neighboring line that there is some electromagnetic coupling between them. There are cases where interactions between discontinuities or nearby structures would preclude an analytic solution. In such cases, solutions can be found from 2.5- and 3-dimensional numerical Maxwell equation solvers.

4.9 SCATTERING PARAMETERS

This chapter began with a discussion of five ways of describing a two-port circuit in terms of its terminal voltages and currents. In principle any one of these is sufficent. One of these, the h parameters, was popularized during the

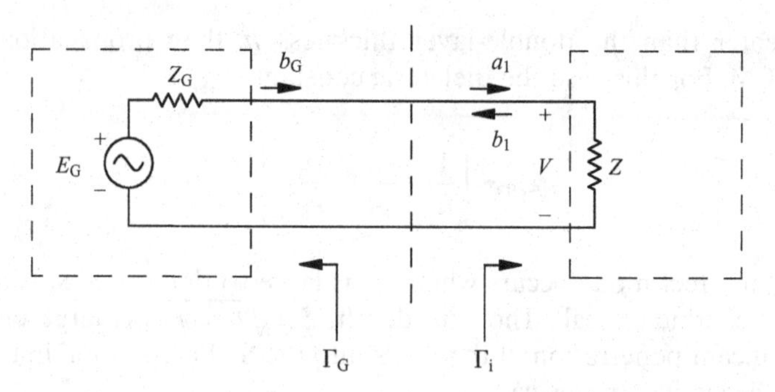

FIGURE 4.23 Wave reflections from unmatched generator source.

early days of the bipolar transistor since they could be directly measured for a transistor. For a similar reason, scattering parameters, or S parameters, have been found convenient by RF and microwave engineers because circuits can be directly measured in terms of them at these frequencies. Scattering parameters represent reflection and transmission coefficients of waves, a quantity that can be measured directly at RF and microwave frequencies. However, these wave quantities can be directly related to the terminal voltages and currents, so that there is a relationship between the scattering paramters and the z, y, h, g, and $ABCD$ parameters.

Consider a one-port circuit excited with a voltage source E_G with an internal generator impedance, Z_G, as shown in Fig. 4.23. Quantity a represents the wave entering into the port. Quantity b represents the wave leaving the port. Both of these quantities are complex and can be related to the terminal voltage and current. The generator and the load are characterized by a reflection coefficient, Γ_G and Γ_i, respectively. The wave b_G from the generator undergoes multiple reflections until finally the reflected wave from the load, b_1, is obtained:

$$\begin{aligned} b_1 &= \Gamma_i b_G + (\Gamma_i b_G)\Gamma_G\Gamma_i + (\Gamma_i^2\Gamma_G b_G) + \cdots \\ &= b_G\Gamma_i\left[1 + \Gamma_i\Gamma_G + (\Gamma_i\Gamma_G)^2 + \cdots\right] \\ &= \frac{b_G\Gamma_i}{1 - \Gamma_i\Gamma_G} \end{aligned} \tag{4.166}$$

This last expression is the sum of a geometric series. Since $\Gamma_i = b_1/a_1$, an expression for the wave entering into the load can be found:

$$a_1 = \frac{b_G}{1 - \Gamma_G\Gamma_i} \tag{4.167}$$

The power actually delivered to the load is then

$$P_1 = \tfrac{1}{2}\left(|a_1|^2 - |b_1|^2\right) \tag{4.168}$$

From the definition of Γ_i and Eq. (4.167), the delivered power can be found:

$$P_1 = \frac{1}{2}\left(\frac{|b_G|^2}{|1-\Gamma_G\Gamma_i|^2}\right)\left(1-|\Gamma_i|^2\right) \tag{4.169}$$

and when matched so that $\Gamma_i = \Gamma_G^*$:

$$P_1 = P_a = \frac{1}{2}\frac{|b_G|^2}{1-|\Gamma_G|^2} \tag{4.170}$$

The latter is the available power from the source. Similar expressions could be found for an n-port circuit.

The wave values will be related to the terminal voltages and currents. With reference to Fig. 4.23, Ohm's law gives

$$E_G = Z_G I + V \tag{4.171}$$

A forward-going voltage wave, V^+, can be related to the forward-going current by

$$V^+ = Z_G^* I^+ \tag{4.172}$$

The reason for this is that if $Z = Z_G^*$, $V = V^+$, because $V^- = 0$. Similarly,

$$V^- = Z_G I^- \tag{4.173}$$

The forward-going voltage and current can be expressed by means of Ohm's law as follows:

$$V^+ = \frac{Z_G^* E_G}{Z_G + Z_G^*} = \frac{Z_G^* E_G}{2\Re\{Z_G\}} \tag{4.174}$$

$$I^+ = \frac{E_G}{2\Re\{Z_G\}} \tag{4.175}$$

These voltages and currents represent root mean square (rms) values, so the incident power is

$$P_{\text{inc}} = \Re\{V^+ I^{+*}\} = \frac{|E_G|^2}{4\Re\{Z_G\}}$$

$$= \frac{|V^+|^2 \cdot \Re\{Z_G\}}{|Z_G|^2} \tag{4.176}$$

The incident power, P_{inc}, is proportional to $|a|^2$ and the reflected power, P_{ref}, to $|b|^2$. Taking the square root of a number to get a complex quantity is strictly speaking not possible mathematically unless a choice is made regarding the phase angle of the complex quantity. This choice is related to choosing Z_G^* and V^+ for a and Z_G and V^- for b:

$$a = \sqrt{P_{\text{inc}}}$$

$$= \frac{V^+\sqrt{\Re\{Z_G\}}}{Z_G^*} \tag{4.177}$$

$$= \frac{I^+ Z_G^* \sqrt{\Re\{Z_G\}}}{Z_G^*} \tag{4.178}$$

and for b,

$$b = \sqrt{P_{\text{ref}}}$$

$$= \frac{V^-\sqrt{\Re\{Z_G\}}}{Z_G} \tag{4.179}$$

$$= \frac{I^- Z_G \sqrt{\Re\{Z_G\}}}{Z_G} \tag{4.180}$$

From Eqs. (4.177) through (4.180) the forward- and reverse-going voltages and currents can be found in terms of the waves a and b. The total voltage and total current are then found.

$$V = V^+ + V^- = \frac{aZ_G^* + bZ_G}{\sqrt{\Re\{Z_G\}}} \tag{4.181}$$

$$I = I^+ - I^- = \frac{a - b}{\sqrt{\Re\{Z_G\}}} \tag{4.182}$$

These are now in the form of two equations where a and b can be solved in terms of V and I. Multiplying Eq. (4.182) by Z_G and adding to Eq. (4.181) gives

$$V + Z_G I = \frac{2a\Re\{Z_G\}}{\sqrt{\Re\{Z_G\}}}$$

which can be solved for a. In a similar fashion b is found:

$$a = \frac{1}{2\sqrt{\Re\{Z_G\}}}(V + Z_G I) \tag{4.183}$$

$$b = \frac{1}{2\sqrt{\Re\{Z_G\}}}(V - Z_G^* I) \tag{4.184}$$

Ordinarily, the generator impedance is equal to the characteristic impedance of a transmission line to which it is connected. The common way then to write Eqs. (4.183) and (4.184) is in terms of Z_0, which is assumed lossless.

$$a = \frac{1}{2\sqrt{Z_0}}(V + Z_0 I) \tag{4.185}$$

$$b = \frac{1}{2\sqrt{Z_0}}(V - Z_0 I) \tag{4.186}$$

The ratio of Eqs. (4.186) and (4.185) is

$$\frac{b}{a} = \frac{V/I - Z_0}{V/I + Z_0} = \Gamma$$

where Γ is the reflection coefficient of the wave. The transmission coefficient is defined as the voltage across the load V due to the incident voltage V^+:

$$T = \frac{V}{V^+} = 1 + \frac{b}{a} = 1 + \Gamma \tag{4.187}$$

This is to be contrasted with the conservation of power represented by $|T|^2 + |\Gamma|^2 = 1$.

For the two-port circuit shown in Fig. 4.24 there are two sets of ingoing and outgoing waves. These four quantities are related together by the scattering matrix.

$$\begin{bmatrix} b_1 \\ b_2 \end{bmatrix} = \begin{bmatrix} S_{11} & S_{12} \\ S_{21} & S_{22} \end{bmatrix} \begin{bmatrix} a_1 \\ a_2 \end{bmatrix} \tag{4.188}$$

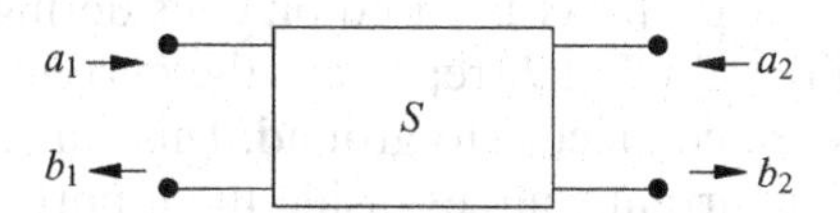

FIGURE 4.24 Two-port circuit described with scattering parameters.

The individual S parameters are found by setting one of the independent variables to zero:

$$S_{11} = \left.\frac{b_1}{a_1}\right|_{a_2=0}$$

$$S_{12} = \left.\frac{b_1}{a_2}\right|_{a_1=0}$$

$$S_{21} = \left.\frac{b_2}{a_1}\right|_{a_2=0}$$

$$S_{22} = \left.\frac{b_2}{a_2}\right|_{a_1=0}$$

Thus, S_{11} is the reflection coefficient at port 1 when port 2 is terminated with a matched load. The value, S_{12}, is the reverse transmission coefficient when port 1 is terminated with a matched load. Similarly, S_{21} is the forward transmission coefficient, and S_{22} is the port 2 reflection coefficient when port 1 is matched.

The formulas for converting between the scattering parameters and the volt–ampere relations given in Section 4.1 are given in Appendix D. In each of these formulas there is a Z_0 because a reflection or transmission coefficient is always relative to a reference impedance, which in this case is Z_0. This is further corroborated by Eqs. (4.185) and (4.186) where the wave values are related to a voltage, current, and Z_0.

4.10 INDEFINITE ADMITTANCE MATRIX

Typically, a certain node in a circuit is designated as being the ground node. Similarly, in an n-port network, at least one of the terminals is considered to be the ground node. In an n-port circuit in which none of the terminals is considered the reference node, it can be described by the indefinite admittance matrix. This can be used, for example, when converting the y parameters of a common emitter transistor to the y parameters of a common base transistor. The indefinite admittance matrix has the property that the sum of the rows = 0 and the sum of the columns = 0. Thus, the indefinite admittance matrix can be easily obtained from the usual definite y matrix, which has been defined with at least one terminal connected to ground.

The derivation of this property is based on considering the currents in the indefinite circuit of Fig. 4.25. Let J_k represent the current going into terminal k when all terminals are connected to ground. This current, J_k, is therefore a result of indepenedent current sources inside the n port or currents resulting from initial conditions. The currents going into each terminal are

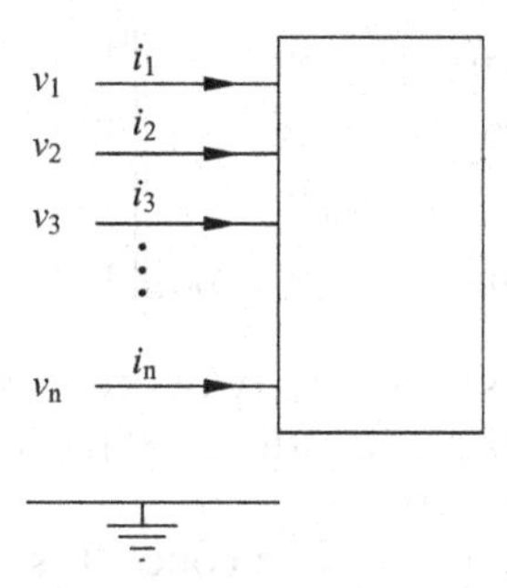

FIGURE 4.25 An n-port indefinite circuit.

$$\begin{bmatrix} i_1 \\ i_2 \\ \vdots \\ i_n \end{bmatrix} = \begin{bmatrix} y_{11} & y_{12} & \cdots & y_{1n} \\ y_{21} & y_{22} & \cdots & y_{2n} \\ \vdots & \vdots & \ddots & \vdots \\ y_{n1} & y_{n2} & \cdots & y_{nn} \end{bmatrix} \begin{bmatrix} v_1 \\ v_2 \\ \vdots \\ v_n \end{bmatrix} + \begin{bmatrix} J_1 \\ J_2 \\ \vdots \\ J_n \end{bmatrix} \tag{4.189}$$

The sum of all the equations represented by Eq. (4.189) gives the total current going into a node, which by Kirchhoff's law must be zero:

$$\sum_{i=1}^{n}\sum_{j=1}^{n} y_{ji}v_i = \sum_{k=1}^{n} i_k - \sum_{k=1}^{n} J_k = 0 \tag{4.190}$$

All the terminal voltages except the jth are set to 0 by connecting them to the external ground. Then since $v_k \neq 0$, the only nonzero term on the left-hand side of Eq. (4.190) would be

$$v_k \sum_{j=1}^{n} y_{jk} = 0 \tag{4.191}$$

Thus, the sum of the columns of the indefinite admittance matrix is 0.

The sum of the rows can also be shown to be 0. If the same voltage v_0 is added to each of the terminal voltages, the terminal currents would remain unchanged:

$$\begin{bmatrix} i_1 \\ i_2 \\ \vdots \\ i_n \end{bmatrix} = \begin{bmatrix} y_{11} & y_{12} & \cdots & y_{1n} \\ y_{21} & y_{22} & \cdots & y_{2n} \\ \vdots & \vdots & \ddots & \vdots \\ y_{n1} & y_{n2} & \cdots & y_{nn} \end{bmatrix} \begin{bmatrix} v_1 + v_0 \\ v_2 + v_0 \\ \vdots \\ v_n + v_0 \end{bmatrix} + \begin{bmatrix} J_1 \\ J_2 \\ \vdots \\ J_n \end{bmatrix} \tag{4.192}$$

Comparison of this with Eq. (4.189) shows that

$$\begin{bmatrix} y_{11} & y_{12} & \cdots & y_{1n} \\ y_{21} & y_{22} & \cdots & y_{2n} \\ \vdots & \vdots & \ddots & \vdots \\ y_{n1} & y_{n2} & \cdots & y_{nn} \end{bmatrix} \begin{bmatrix} v_0 \\ v_0 \\ \vdots \\ v_0 \end{bmatrix} = 0 \tag{4.193}$$

Thus, the sum of the rows is 0. A variety of other important properties of the indefinite admittance matrix are described in [1, Chapter 2] to which reference should be made for further details.

One of the useful properties of this concept is illustrated by the problem of converting common source hybrid parameters of an FET to common gate hybrid parameters. This might be useful in designing a common gate oscillator with a transistor characterized as a common source device. The first step in the process is to convert the hybrid parameters to the equivalent definite admittance matrix (which contains 2 rows and 2 columns) by using the formulas in Appendix D. The definite admittance matrix, which has a defined ground, can be changed to the corresponding 3×3 indefinite admittance matrix by adding a column and row such that Σ rows = 0 and the Σ columns = 0. If the y_{11} corresponds to the gate and y_{22} corresponds to the drain, then y_{33} would correspond to the source.

$$[Y] = \begin{matrix} \\ g \\ d \\ s \end{matrix} \begin{matrix} \begin{matrix} g & d & s \end{matrix} \\ \begin{pmatrix} y_{11} & y_{12} & y_{13} \\ y_{21} & y_{22} & y_{23} \\ y_{31} & y_{32} & y_{33} \end{pmatrix} \end{matrix} \tag{4.194}$$

The common gate parameters are found by forcing the gate voltage to be 0. Consequently, the first column may be removed since it is multiplied by the zero gate voltage anyway. At this point the first row represents a redundant equation and can be removed. Row 1 and column 1 have been deleted and a new common gate definite admittance matrix is formed. This can then be converted to the equivalent common gate hybrid matrix.

4.11 INDEFINITE SCATTERING MATRIX

A similar property can be determined for the scattering matrix. The indefinite scattering matrix has the property that the sum of the rows = 1 and the sum of the colmuns = 1. For the first property, the three-port shown in Fig. 4.26 is excited at all three terminals by the same voltage value. The output wave is

$$b_j = S_{j1}a_1 + S_{j2}a_2 + S_{j3}a_3 \qquad j = 1, 2, 3 \tag{4.195}$$

Under this excitation, all the input waves, a_j, have the same amplitude, so Eq. (4.195) becomes

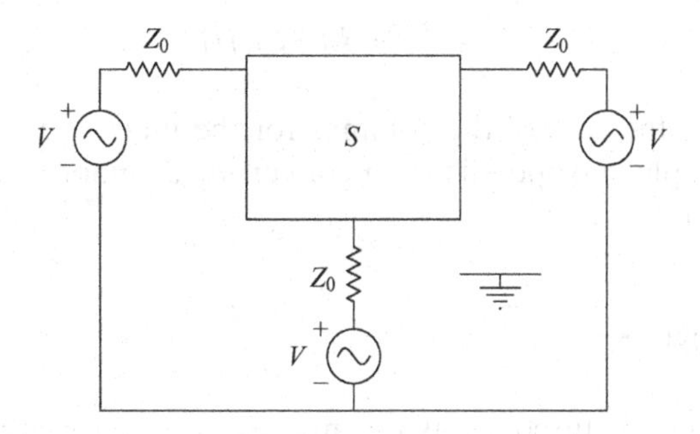

FIGURE 4.26 Indefinite scattering parameter circuit.

$$b_j = (S_{j1} + S_{j2} + S_{j3})a_1 \qquad j = 1, 2, 3 \tag{4.196}$$

Since from Eqs. (4.178) and (4.180) $a_k = \sqrt{Z_0} I_k^+$ and $b_k = \sqrt{Z_0} I_k^-$, Eq. (4.196) can be written in terms of the incident and reflected currents:

$$I_j^- = (S_{j1} + S_{j2} + S_{j3})I_1^+ \tag{4.197}$$

When all the terminal voltages are set equal, then all the terminal currents must be zero, since there can be no voltage difference between any two ports. Thus, $I_j^- = I_1^+$, which means that

$$S_{j1} + S_{j2} + S_{j3} = 1 \tag{4.198}$$

proving the sum of the rows = 1.

To show that the sum of the columns = 1, only port 1 is excited with a voltage source. Thus, $a_1 \neq 0$ and $a_2 = a_3 = 0$. By Kirchhoff's currnet law the sum of the currents into the three terminal circuit is zero:

$$0 = I_1 + I_2 + I_3 \tag{4.199}$$

$$= (I_1^+ - I_1^-) + (I_2^+ - I_2^-) + (I_3^+ - I_3^-) \tag{4.200}$$

Now since $I_2^+ = I_3^+ = 0$ because of a_2, a_3

$$I_1^+ = I_1^- + I_2^- + I_3^- \tag{4.201}$$

In addition

$$b_k = S_{k1}a_1$$

$$I_k^- = S_{k1}I_1^+ \tag{4.202}$$

so that

$$I_1^+ = (S_{11} + S_{21} + S_{31}) I_1^+ \tag{4.203}$$

which shows that the sum of the columns for the indefinite scattering matrix is 1. See the example in Appendix E for converting common source to common gate S parameters.

4.12 CONCLUSIONS

This chpater began with concepts essential for two-port circuit design. The *ABCD* transmission parameters were used to descirbe a lumped-element analog to the transmission line circuit that pervades all of RF circuit design practice. Various types of transmission lines were described in this chapter: two-wire line, parallel strip line, coaxial line, and microstrip. There are other widely used transmission line structures that are used such as slotline, coplanar line, and assorted types of coupled lines. Finally, scattering parameters were described and used in an indefinite matrix approach similar to that used in lumped-circuit analysis.

PROBLEMS

4.1. Determine the image impedance for the two-port circuit shown in Fig. 4.27, which is terminated by a 5-Ω resistor on each end.

4.2. If the load side of the circuit in Fig. 4.28 is terminated with its matching impedance, what is the impedance that will match the input side? In this circuit the reactance $X = \omega L = 2\,\Omega$, $B_1 = \omega C_1 = \frac{1}{8}$ S, and $B_2 = \omega C_2 = \frac{7}{8}$ S.

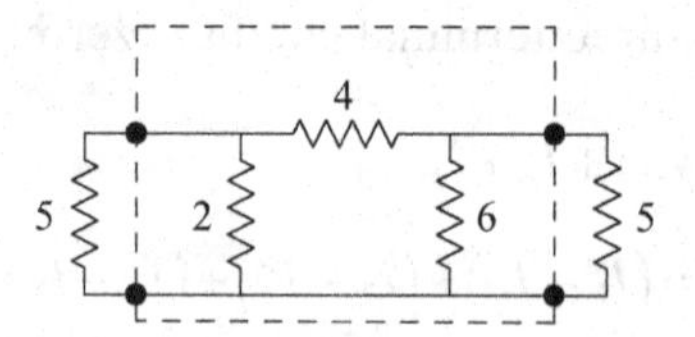

FIGURE 4.27 Image impedance for Problem 4.1.

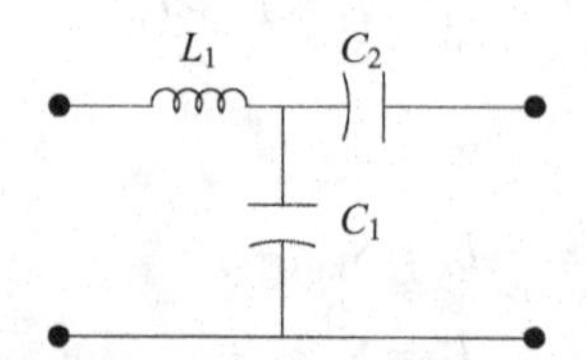

FIGURE 4.28 Circuit to be matched in Problem 4.2.

4.3. A reciprocal circuit is described in terms of the *ABCD* parameters, where $A = 2$, $B = 7$, and $C = 3$.

a. What is the value for D?

b. What is the value for the image propagation constant?

c. What is the value for Z_{I1} and Z_{I2}?

d. If the output were connected to an impedance $Z_L = \sqrt{77}$, what would be the input impedance?

4.4. Convert the following scattering parameters (related to $50\,\Omega$) to *ABCD* parameters:

$\|S_{11}\|$	$\angle S_{11}$	$\|S_{21}\|$	$\angle S_{21}$	$\|S_{12}\|$	$\angle S_{12}$	$\|S_{22}\|$	$\angle S_{22}$
0.49	–29	3.25	85	0.10	65	0.65	–33

4.5. Given the S parameters, derive the z parameters as given in Table D.1 in Appendix D.

4.6. Two transmission lines are cascaded together as shown in Fig. 4.29. What is the input impedance at the left-hand side of the line?

4.7. The transmission line circuit of length ℓ and characteristic impedance Z_0 is terminated by a resistance R_L. Determine the Q for this circuit at the first appropriate nonzero frequency.

4.8. The impedance of a circuit is given by

$$Z_{in} = R_1\left(\frac{\omega}{\omega_0}\right) + j\left[\frac{L}{\omega}(\omega^2 - \omega_0^2) + Z_0 \tan\frac{\omega\ell}{c}\right]$$

where $\ell = \pi c/\omega_0$ and c is the velocity of light in a vacuum. What is the Q of this circuit in terms of ω_0? What is the Q of this circuit if $\ell = 2\pi c/\omega_0$?

4.9. In the shunt transmission line circuit in Fig. 4.30, determine the value for Z_{02} that would produce a real value for Z_{in}. What is the value for Z_{in}?

4.10. Determine the S_{11} scattering parameter for the circuit in Fig. 4.31. Express your answer in the form of a real and imaginary part.

4.11. In the four parts below, design the transmission line when $Z_0 = 40\,\Omega$ and $\varepsilon_r = 2.4$.

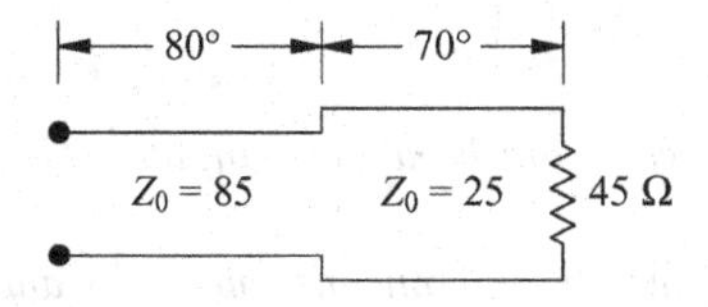

FIGURE 4.29 Transmission line circuit for Problem 4.6.

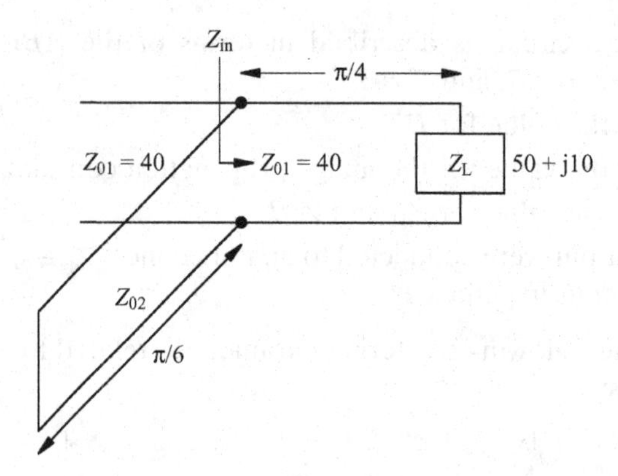

FIGURE 4.30 Shunt transmission line circuit for Problem 4.9.

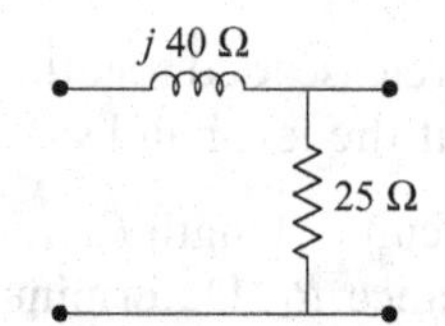

FIGURE 4.31 Circuit for Problem 4.10.

a. Determine the wire spacing of a two-wire transmission line that uses wires with a diameter of 80.1 mils (AWG#12) and that are surrounded with a dielectric layer.

b. Design a parallel plate transmission line using a double-sided copper-clad printed circuit board of thickness 0.25 in.

c. Determine the center conductor diameter of a dielectric-filled coaxial line where a standard outer diameter of 0.502 in. is used.

d. Design a microstrip line using a 1-oz copper-clad dielectric board that is 25 mils thick. The 1-oz specification refers to the thickness of the copper, where 1 oz ≈ 1.4 mils. Find the conductor width and effective dielectric constant.

REFERENCES

1. W.-K. Chen, *Active Network and Feedback Amplifier Theory*, New York: McGraw-Hill, 1980.
2. S. R. Seshadri, *Fundamentals of Transmission Lines and Electromagnetic Fields*, Reading MA: Addison Wesley, pp. 335–350, 1971.

3. R. E. Collin, *Foundations for Microwave Engineering*, 2nd ed., New York: McGraw-Hill, 1992.
4. H. A. Wheeler, "Transmission-Line Properties of Parallel Strips Separated by a Dielectric Sheet," *IEEE Trans. Microwave Theory Tech.*, **MTT-13**, pp. 172–185, Mar. 1965.
5. C. T. Tai, *Generalized Vector and Dyadic Analysis*, Piscataway, NJ: IEEE Press, 1997.
6. H. A. Wheeler, "Transmission-Line Properties of a Strip on a Dielectric Sheet on a Plane," *IEEE Trans. Microwave Theory Tech.*, **MTT-25**, pp. 631–647, Aug. 1977.
7. E. Hammerstad and O. Jensen, "Accurate Models for Microstrip Computer-Aided Design," *1980 IEEE MTT-S Int. Microwave Symposium Digest*, pp. 407–409, May 1980.
8. K. C. Gupta, R. Garg, and R. Chadha, *Computer Aided Design of Microwave Circuits*, Dedham, MA: Artech House, 1981.
9. R. A. Pucel, D. J. Massé, and C. P. Hartwig, "Losses in Microstrip," *IEEE Trans. Microwave Theory Tech.*, **MTT-16**, pp. 342–350, Jun. 1968.
10. R. A. Pucel, D. J. Massé, and C. P. Hartwig, "Corrections to 'Losses in Microstrip,'" *IEEE Trans. Microwave Theory Tech.*, **MTT-16**, p. 1064, Dec. 1968.
11. M. Kirschning and R. H. Jansen, "Accurate Model for Effective Dielectric Constant of Microstrip with Validity up to Millimetre-Wave Frequencies," *Elect. Lett.*, **18**(6), pp. 272–273, Mar. 1982.
12. H. Hasegawa, M. Furukawa, and H. Yanai, "Properties of Microstrip Line on Si-SiO_2 System," *IEEE Trans. Microwave Theory Tech.*, **MTT-19**, pp. 869–881, Nov. 1971.

CHAPTER FIVE

Filter Design and Approximation

5.1 INTRODUCTION

The development of filters has been extensively studied and refined in the twentieth century by engineers and scientists in Germany and the United States. This work was associated with the development of telephony and used simple mathematics. In the 1930s and 1940s major progress was made in filter theory and optimum filter designs. To work through that body of knowledge would be beyond the present purpose of reviewing filter design and filter synthesis techniques. New types of analog filters emerged in the 1950s and 1960s called active filters, which for the most part used operational amplifiers. At this point in time, operational amplifiers have not made a significant inroad to the RF range. Active filter design discussions can be found in Chen [1].

The present chapter concentrates on basic filter types and techniques. The solutions to the filter approximation problem and implementations that have arisen because of special requirements are treated in this chapter. The treatment is further limited to filters that perform certain tasks on continuous time-varying analog signals by way of linear time-invariant circuit elements.

5.2 IDEAL AND APPROXIMATE FILTER TYPES

The function of a filter is to separate different frequency components of the input signal that passes through the filter network. The characteristics of the network are specified by a transfer function, $H(j\omega)$ or $H(s)$, where $s = +j\omega$

Radio Frequency Circuit Design, Second Edition, by W. Alan Davis

represents the complex frequency defined for the Laplace transform. The transfer function is the ratio of the output signal to the input signal:

$$H(j\omega) = \frac{V_{\text{out}}}{V_{\text{in}}} \tag{5.1}$$

$$= |H(j\omega)| e^{j\phi(\omega)} \tag{5.2}$$

The transfer phase function, $\phi(\omega)$, is related to the transfer group delay through a derivative with respect to frequency as follows:

$$\tau_{\text{d}}(\omega) = -\frac{d\phi(\omega)}{d\omega} \tag{5.3}$$

For constant group delay, the phase function must be linear with frequency. In most filters only the magnitude of the transfer function is of interest. However, with modern-day systems using signals with complex modulation schemes, phase and group delay functions are also important.

A filter network passes some of the input signal frequencies and stops others, and being a linear circuit this function is performed without adding or generating new frequency components. The frequency band that passes, ideally without losses (0-dB insertion loss) defines the pass band and the band that stops the frequencies, ideally with infinite loss, is called the stop band. Figure 5.1 shows this loss representation of the ideal low-pass filter. It is low pass because it passes all low-frequency signals from dc to some given frequency, ω_c, and stops all signals above ω_c. The frequency, ω_c, is called the cutoff frequency of the filter. An ideal low-pass filter is physically not realizable since this requires a circuit with an infinite number of elements due to an abrupt change from pass band to stop band. Such a change is not practical.

This raises a practical issue of how does one specify the filters? The concept of the transition band, ω_p to ω_s, is the frequency range that separates the pass

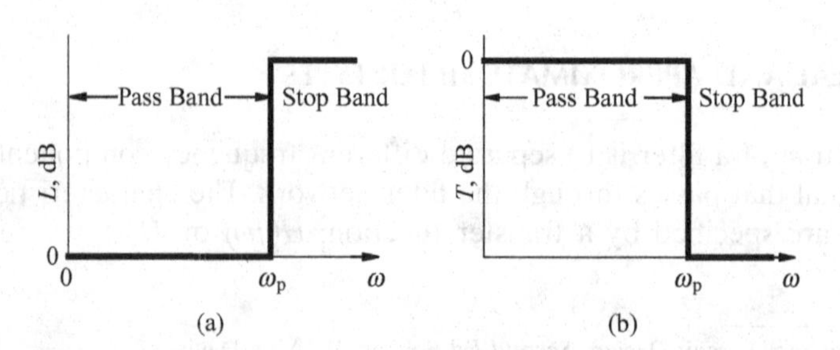

FIGURE 5.1 Loss or attenuation characteristics of ideal low-pass filter.

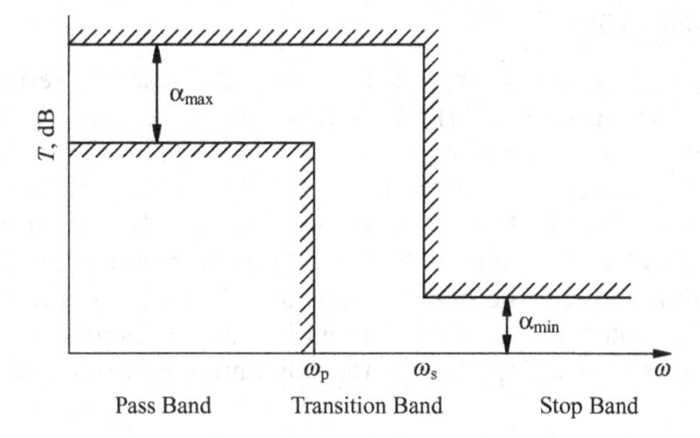

FIGURE 5.2 Method of specifying practical low-pass filter.

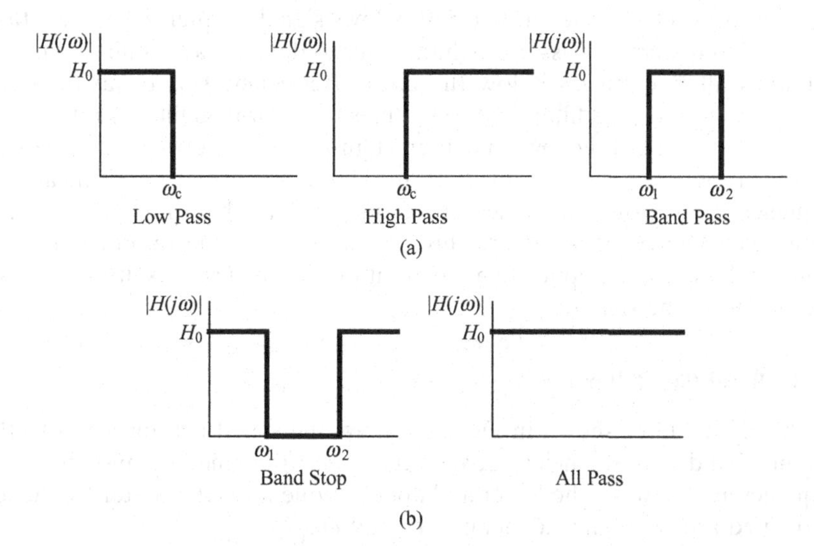

FIGURE 5.3 Five major filter types using transfer function amplitude response.

band and stop band where the loss transitions from a minimum to a maximum value. This is shown in Fig. 5.2. The ratio of ω_s/ω_p is sometimes referred to as filter selectivity, ratio, or filter steepness. The closer the selectivity is to one the more complex and costly the filter circuit is. Similar considerations can be applied in the design of filters using phase linearity and/or group delay flatness.

The concepts of pass band, stop band, and transition band permit specifications of five major types of filters: (a) low pass, (b) high pass, (c) band pass, (d) band stop, and (e) all pass. The transmission behavior of these filters is shown in Fig. 5.3.

5.2.1 Low-Pass Filter

Low-pass filter networks are realized by using a cascade of series inductors and shunt capacitors (Fig. 5.3*a*). The number of these elements determines the steepness of the edge of the pass band, with the larger number resulting in a more complex and steeper filter. The penalties are the complexity of the filter, the in-band loss, higher cost, and larger size. At low frequencies, series inductances produce low impedance, and shunt capacitors produce a high impedance, thus allowing the signal to appear at the output of the filter. Above the cutoff frequency, the series inductors behave as large impedances and shunt capacitors as low impedances, thereby impeding the signal transfer to the load.

5.2.2 High-Pass Filter

The high-pass filter shown in Fig. 5.3*b* allows signal frequencies higher than the cutoff frequency to pass through the filter to the load with a minimum loss and stops all frequencies below the cutoff frequency. This behavior is the reverse of the low-pass filter, and sometimes the high-pass filter is referred to as the complement of the low-pass filter. High-pass filter networks are realized by using a cascade of series capacitors and shunt inductors. Capacitors at high frequencies have low impedance and inductors have high impedance. Thus, the high-frequency signal passes through the filter to the output load with minimum loss. Just the opposite happens at low frequencies resulting in a high attenuation of the signal.

5.2.3 Band-Pass Filter

The band-pass filter shown in Fig. 5.3*c* shows the signal is transferred to the load in a band of frequencies between the lower frequency, ω_1, and the upper frequency, ω_2. Between the lower and upper frequency is the center frequency, ω_0, defined as the geometric mean of ω_1 and ω_2.

5.2.4 Band-Stop Filter

The band-stop filter is the complement of the band-pass filter and is shown in Fig. 5.3*d*. The signal in a band-stop filter is transferred to the load in two frequency bands, one from a low frequency to the lower stop-band frequency, ω_1, and the other from the upper stop band frequency, ω_2, to infinite frequency. The signal experiences high loss between ω_1 to ω_2, hence the name band-stop or band-reject.

5.2.5 All-Pass Filter

The all-pass filter allows the signal amplitude for all frequencies to pass through the network without any significant loss (Fig. 5.3*e*). This network has no frequency-selective pass band or stop band. The transmitted signal ideally experiences a linear phase shift or constant group delay with frequency. Unfortunately, minimum phase networks do not have constant group delay: Rather there are peaks near the corner frequency. All passive ladder networks, that have frequency selectivity, are minimum phase. In the design process there is a trade-off between flat group delay and filter selectivity. However, a network that is nonminimum phase can be cascaded with a minimum phase network to achieve both flat group delay and selectivity. All-pass networks with a nonminimum phase characteristic are used as group delay compensation devices.

5.3 TRANSFER FUNCTION AND BASIC FILTER CONCEPTS

Before proceeding with the design of filters, it is important to understand the transfer function in the complex frequency domain, $s = \sigma + j\omega$. As described in Section 5.2, the filter transfer function is the ratio of the output signal voltage to the input signal voltage. One could also easily select the ratio of currents. The transfer function in general can be written as a ratio of two polynomials:

$$H(s) = \frac{P(s)}{Q(s)} = \frac{a_0 + a_1 s + a_2 s^2 + \cdots + a_m s^m}{b_0 + b_1 s + b_2 s^2 + \cdots + b_n s^n} \tag{5.4}$$

where polynomials $P(s)$ and $Q(s)$ in general are of order m and n. The order of polynomial $Q(s)$ is the order of the filter. Polynomials $P(s)$ and $Q(s)$ can be factored and rewritten in the form

$$H(s) = \frac{(s - z_1)(s - z_2)(s - z_3)\ldots(s - z_m)}{(s - p_1)(s - p_2)(s - p_3)\ldots(s - p_n)} \tag{5.5}$$

The values $z_1, z_2, z_3, \ldots, z_m$ are called the zeros of the transfer function or simply transmission zeros. The roots of $Q(s), p_1, p_2, p_3, \ldots, p_n$ are the poles of the transfer function. The poles and zeros can be real or complex, but complex poles and zeros must occur in conjugate pairs. That is, if $-2 + j3$ is a pole, than $-2 - j3$ must be a pole as well. The magnitude plot of a voltage transfer function represents the loss or attenuation of the filter circuit, and in decibels is given by

$$L_{\text{dB}} = 20\log|H(s)| \tag{5.6}$$

Poles and zeros of realizable passive networks must follow certain rules:

1. All poles of a transfer function occur in the left half s-plane. The left half s-plane includes the imaginary $j\omega$ axis.

2. The transfer function be realized as a lossless circuit terminated in a resistor (Darlington's Theorem).

5.4 LADDER NETWORK FILTERS

The class of minimum phase filters are those filters in which the zeros of the impedance function are chosen to be in the left half plane. This means that the topology of a low-pass filter is a ladder network. The filter is assumed to be lossless and terminated on each side by a real resistance. As indicated in Fig. 5.4, there are four possible choices for the first and last reactive elements. The final reactive element depends on whether the number of elements is even or odd. The basic procedure is to develop a design for a low-pass filter whose terminating resistors are 1 Ω and whose cutoff frequency is $\omega_c = 1$ rad/s. Once this normalized low-pass filter is designed, the impedance level is adjusted to the desired value, the cutoff frequency is adjusted, and the circuit topology is transformed to a high-pass, band-pass, or band-stop filter as desired.

The notation for the low-pass filter prototype filter shown in Fig. 5.4 is widely used. The g values are the prototype capacitances, inductances, and terminating resistances. For the normalized filter, $g_0 = 1$. For some important filter types, the values for the g_k can be found from recursion formulas.

5.4.1 Butterworth Filter

The transducer power gain for a two-port circuit is

$$G_T = \frac{\text{power delivered to the load}}{\text{power avialable from the source}} \tag{5.7}$$

which for a passive filter is a quantity ≤1. A filter with many reactive elements would be expected to more closely approximate an ideal filter with rectangular shape (infinitely steep band edge skirts and flat pass band) than one with few

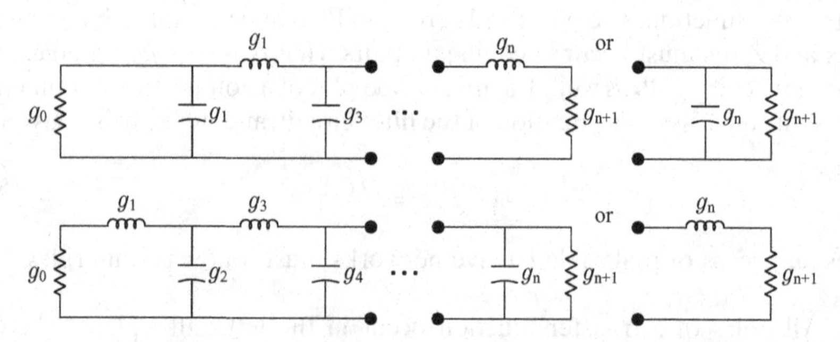

FIGURE 5.4 Lumped-element prototype low-pass filter.

reactive elements. For a filter with n poles (n reactive elements), the low-pass Butterworth approximation provides the maximum flatness in its pass band near $\omega = 0$. The gain function for this type of filter is given by

$$|H(j\omega)|^2 = G_T = \frac{H_0}{1+(\omega/\omega_c)^{2n}} \tag{5.8}$$

where $H_0 \le 1$. The first $(2n - 1)$ derivatives of the denominator of this function are all zero at $\omega = 0$ leading to it being maximally flat. The poles of this function all have a magnitude of 1 and are separated from one another on the unit circle by π/n radians. Furthermore, there are no poles on the $j\omega$ axis. At the edge of the pass band, the filter attenuates the power by ½ or −3 dB. A recursion formula for the filter elements that would produce this response can be found in a variety of references, one of which is [1]:

$$g_0 = g_{n+1} = 1 \tag{5.9}$$

$$g_k = 2\sin\left[\frac{(2k-1)\pi}{2n}\right] \qquad k = 1, 2, 3, \ldots, n \tag{5.10}$$

One set of filter values worthy of remembering is the three-pole Butterworth filter, where g_1, g_2, g_3 = 1, 2, 1. The low-pass prototype filter starting with a shunt C has C_1 = 1 F, L_2 = 2 H, and C_3 = 1 F.

Often minimum requirements are placed on the shape of the pass band. In this instance the minimum number of poles needed to produce a desired specification is

$$n = \frac{\log\left(10^{\alpha_{min}/10} - 1\right) - \log\left(10^{\alpha_{max}/10} - 1\right)}{2\log(\omega_s/\omega_p)} \tag{5.11}$$

In this expression the maximum attenuation in the pass band $0 \le \omega \le \omega_p$ is α_{max}, and the minimum attenuation in the stop band $\omega_s \le \omega < \infty$, is α_{min}. When the pass band is defined as the same as the cutoff frequency, then $\omega_p = \omega_c$ and α_{max} = +3 dB. In this case,

$$n = \frac{\log\left(10^{\alpha_{min}/10} - 1\right)}{2\log(\omega_s/\omega_c)} \tag{5.12}$$

5.4.2 Chebyshev Filter

The slope of the pass-band skirts for a given number of poles of a filter can be improved by allowing small ripples in the pass band. In antenna theory, the Dolph–Chebyshev amplitude weighting of the array elements provides the minimum beam width for a specified side lobe level. Similarly, in filter design, the Chebyshev function provides the maximum possible bandwidth for a given pass-band ripple or the minimum possible pass-band ripple for a given bandwidth. The Chebyshev (equal ripple) low-pass filter transducer gain function is

$$|H(j\omega)|^2 = G_T = \frac{H_0}{1+\varepsilon^2 T_n^2(\omega/\omega_c)} \tag{5.13}$$

where ω_c is the low-pass cutoff frequency. The value ε is a number <1 and is a measure of the pass-band ripple. The Chebyshev function, $T_n(x)$, oscillates between +1 and −1 when its argument is less than 1. The poles of this transfer function lie on an ellipse with no $j\omega$ axis poles.

For $x > 1$, $T_n(x)$ rapidly becomes large. The Chebyshev function can be written in a form that clearly shows this characteristic:

$$T_n(x) = \cos[n \arccos(x)] \qquad 0 \le x \le 1 \tag{5.14}$$

$$T_n(x) = \cosh[n \operatorname{arccosh}(x)] \qquad x > 1 \tag{5.15}$$

Since $|T_n(x)| < 1$ in the pass band, the pass-band transfer function is

$$\frac{1}{1+\varepsilon^2} \le |H(j\omega)|^2 \le 1$$

Outside the pass band, $T_n(x)$ increases approximately exponentially.

The Chebyshev functions can be found in terms of a polynomial of its argument from a recursion formula:

$$T_{n+1}(x) = 2xT_n(x) - T_{n-1}(x) \tag{5.16}$$

The formula begins by setting $T_0(x) = 1$, and $T_1(x) = x$. Furthermore for n odd,

$$T_n(0) = 0 \quad \text{and} \quad T_n(\pm 1) = \pm 1 \tag{5.17}$$

while for n even

$$T_n(0) = (-1)^{n/2} \quad \text{and} \quad T_n(\pm 1) = 1 \tag{5.18}$$

The next few Chebyshev functions are shown below:

$$T_2(x) = 2x^2 - 1 \tag{5.19}$$

$$T_3(x) = 4x^3 - 3x \tag{5.20}$$

$$T_4(x) = 8x^4 - 8x^2 + 1 \tag{5.21}$$

$$T_5(x) = 16x^5 - 20x^3 + 5x \tag{5.22}$$

$$T_6(x) = 32x^6 - 48x^4 + 18x^2 - 1 \tag{5.23}$$

If the maximum pass-band frequency is ω_c and the minimum stop-band frequency beyond which the attenuation is always greater than α_{min} is ω_s, then the number of poles required in the function is n [1]:

$$n = \frac{\operatorname{arccosh}\left[1/\varepsilon\left(10^{\alpha_{min}/10} - 1\right)^{1/2}\right]}{\operatorname{arccosh}(\omega_s/\omega_c)} \tag{5.24}$$

where α_{min} is the minimum attenuation in the stop band.

Just as in the Butterworth approximation, there is a set of recursion formulas for the low-pass prototype Chebyshev filter. Finding expressions for the g values for the filter requires first expanding the Chebyshev functions by its own set of recursion formulas. The low-pass prototype filter structure (for a given number n of reactive elements) is then equated to the nth-order filter function so that a correlation is made between the circuit and the function. Fortunately, the hard work has been done in network synthesis books (e.g. [1, 2]). The final recursion formulas are given below:

$$g_0 = 1 \tag{5.25}$$

$$g_{n+1} = \begin{cases} 1 & n \text{ odd} \\ \tanh^2(\beta/4) & n \text{ even} \end{cases} \tag{5.26}$$

$$g_1 = \frac{2a_1}{\gamma} \tag{5.27}$$

$$g_k = \frac{4a_{k-1}a_k}{b_{k-1}g_{k-1}} \qquad k = 2, 3, \ldots, n \tag{5.28}$$

$$a_k = \sin\left[\frac{(2k-1)\pi}{2_n}\right] \qquad k = 1, 2, \ldots, n \tag{5.29}$$

$$b_k = \gamma^2 + \sin^2\left(\frac{k\pi}{n}\right) \qquad k = 1, 2, \ldots, n \tag{5.30}$$

$$\beta = \ln\left[\coth\left(\frac{A_m}{17.32}\right)\right] \tag{5.31}$$

$$A_m = 10\log(\varepsilon^2 + 1) \tag{5.32}$$

$$\gamma = \sinh\left(\frac{\beta}{2n}\right) \tag{5.33}$$

One important difference between the Butterworth and Chebyshev approximations is the value for g_{n+1}. The unequal impedance levels for the even-order Chebyshev termination impedances is often avoided by simply restricting the choices of n for the Chebyshev function to odd values. The circuit element values for these two filter functions were found by using network synthesis techniques after determining the poles of the transfer function. Other filter functions are available and are briefly mentioned here.

5.4.3 Inverse Chebyshev Filter

In this filter, equal ripples are found in the stop band while the pass band is smooth. The inverse Chebyshev function is useful when it is necessary to control the attenuation over the entire stop band. The transfer function and the minimum number of poles needed to guarantee a maximum attenuation in the pass band for the inverse Chebyshev function is given below:

$$|H(j\omega)|^2 = \frac{H_0 \varepsilon^2 T_n^2(\omega_c/\omega)}{1+\varepsilon^2 T_n^2(\omega_c/\omega)} \tag{5.34}$$

$$n = \frac{\operatorname{arccosh}\left[1/\varepsilon\left(10^{\alpha_{\max}/10}-1\right)^{-1/2}\right]}{\operatorname{arccosh}(\omega_c/\omega_p)} \tag{5.35}$$

The actual circuit is generated using the Cauer network synthesis techniques described in Section 5.6.1. To do this the polynomial expansion of the Chebyshev functions will be needed.

5.4.4 Bessel–Thompson Filter

The previous functions were designed to provide a specific magnitude in the transfer response, while the phase was left uncontrolled. The Bessel–Thompson filter is designed to provide a maximally flat time delay response. The ideal Bessel–Thompson response would have a flat magnitude response, $|H(j\omega)| = 1$, and a phase response proportional to frequency, $\arg[H(j\omega)] = -\omega T$. The normalized time delay D is

$$D(\omega T) = \frac{d[\arg H(j\omega T)]}{d\omega T} \tag{5.36}$$

where T is the filter time delay. The transfer function designed to provide maximally flat time delay is

$$D(sT) = \frac{B_n(0)}{B_n(sT)} \tag{5.37}$$

Letting $y = sT$, the recursion formula for the Bessel polynomials is

$$B_n(y) = (2n-1)B_{n-1}(y) + y^2 B_{n-2}(y) \tag{5.38}$$

where $B_0(y) = 1$ and $B_1(y) = y + 1$. The first few polynomials are as follows.

$$B_2(y) = y^2 + 3y + 3 \tag{5.39}$$

$$B_3(y) = y^3 + 6y^2 + 15y + 15 \tag{5.40}$$

$$B_4(y) = y^4 + 10y^3 + 45y^2 + 105y + 105 \tag{5.41}$$

Once the polynomials are known, network synthesis techniques can be used to derive a low-pass prototype filter [1–3].

5.5 ELLIPTIC FILTER

The low-pass filter can be characterized as having a pass band from $\omega = 0$ to $\omega = \omega_p$ with an attenuation no greater than H_0 plus a small ripple. In addition it is characterized as having a stop band from ω_s to ∞ with an insertion loss no less than some high value, α_{min} (Fig. 5.2). In the Chebyshev filter, the passband ripple is fixed to a certain maximum, but small, value while the attenuation in the stop band increases monotonically with ω. The inverse Chebyshev filter produces an equal ripple in the stop band and a monotonically decreasing insertion loss for ω going from ω_p toward $\omega = 0$. The elliptic function filter produces an equal ripple response in both the pass band and in the stop band. The name, elliptic, comes from its association with the Jacobian elliptic functions. This design provides a way of not throwing away excess stop-band attenuation at high frequencies, but allows redistribution of the attenuation over the whole stop band. As a consequence, the rate of cutoff may be increased by putting some of the transmission zeros near the pass band. The cost for having equal ripple response in both the pass band and in the stop band is a slightly more complicated circuit topology for the elliptic filter (Fig 5.5).

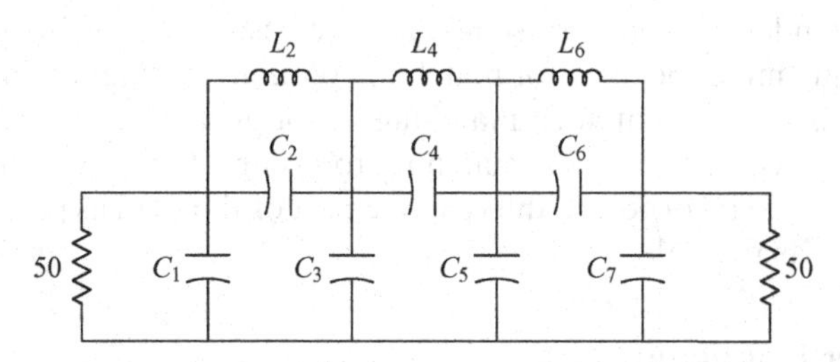

FIGURE 5.5 Seven-pole low-pass elliptic filter topology. When $f_p = 0.8\,\text{GHz}$ and $f_s = 1\,\text{GHz}$, $C_1 = 3.285\,\text{pF}$, $C_2 = 0.547\,\text{pF}$, $L_2 = 12.653\,\text{nH}$, $C_3 = 5.459\,\text{pF}$, $C_4 = 2.682\,\text{pF}$, $L_4 = 9.947\,\text{nH}$, $C_5 = 4.846\,\text{pF}$, $C_6 = 2.040\,\text{pF}$, $L_6 = 8.963\,\text{nH}$, and $C_7 = 2.231\,\text{pF}$.

There is no simple recursion formula for the design of elliptic function filters. Typically, tables of values are derived numerically [4, 5] and are used for the low-pass prototype filter. These tabulated values have been incorporated in a program called ELLIPTIC. In this program the desired maximum attenuation level in the pass band, minimum attenuation in the stop band, the frequencies where the pass band ends and the stop band begins, and finally the number of poles, n, are balanced against each other to provide an elliptic filter design. If so desired, the program will produce a SPICE net list that can be used to analyze the design. In the SPICE program, the voltage is plotted using V(21) or VDB(21) to display the insertion loss on a linear or log scale, respectively.

5.6 MATCHING BETWEEN UNEQUAL RESISTANCE LEVELS

For a low-pass filter, a perfect match cannot in principle be achieved when impedance matching is used. In the preceding Butterworth and Chebyshev functions, the constant, H_0 is ≤ 1 since a passive filter cannot produce gain greater than 1. When the input and output resistance levels are equal, then H_0 is 1. The ratio of the load-to-generator resistances introduces a constraint on H_0. For Butterworth filters this constraint is

$$\frac{R_L}{R_G} = \left(\frac{1+\sqrt{1-H_0}}{1-\sqrt{1-H_0}} \right)^{\pm 1} \tag{5.42}$$

while for Chebyshev filters this constraint is [1]:

$$\frac{R_L}{R_G} = \left(\frac{1+\sqrt{1-H_0}}{1-\sqrt{1-H_0}} \right)^{\pm 1} \quad n \text{ odd} \tag{5.43}$$

$$\frac{R_L}{R_G} = \left(\frac{\sqrt{1+\varepsilon^2}+\sqrt{1+\varepsilon^2-H_0}}{\sqrt{1+\varepsilon^2}-\sqrt{1+\varepsilon^2-H_0}} \right)^{\pm 1} \quad n \text{ even} \tag{5.44}$$

One might wonder if the generator and load consisted of complex impedances, what technique might be used for matching. Without getting too involved with that issue, it is known that such matching is not always possible. The impedances must be "compatible" for matching to occur. One thing a designer can do though is try to incorporate the reactive part of the load as part of the filter itself as much as possible.

5.6.1 Cauer Procedure

A doubly terminated filter can be designed for any physically realizable transfer function. A variety of different circuit realizations may be possible, but

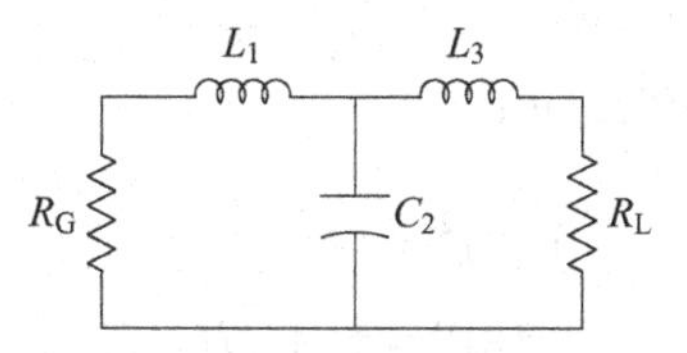

FIGURE 5.6 Butterworth low-pass filter with unequal resistance terminations. When $R_G = 20\,\Omega$ and $R_L = 80\,\Omega$, $L_1 = 127.75\,H$, $C_2 = 0.01804\,F$, and $L_3 = 43.400\,H$.

only one will be described. However, this particular realization method is widely used and provides practical filter design. Approximation theory determines the transfer function $|H(j\omega)|^2$ that will approximate the ideal filter characteristics. In a lossless, low-pass circuit with possibly unequal termination resistances (Fig. 5.6), the reflected power $|\Gamma(j\omega)|^2$ can be found:

$$1-|H(j\omega)|^2 = |\Gamma(j\omega)|^2 = \Gamma(j\omega)\Gamma(-j\omega) \tag{5.45}$$

The final expression results from the magnitude being the product of the reflection coefficient and its complex conjugate. This can be generalized by replacing $j\omega$ with the complex frequency s:

$$\Gamma(s)\Gamma(-s) = 1-|H(s)|^2 \tag{5.46}$$

The right-hand side is a known function that is given in the form of a ratio of polynomials in s. A requirement for realizability of an impedance or reflection coefficient is that it be positive real. All the poles of the function must lie in the left half side of the complex plane in order to avoid unrealizable growing exponentials. Half of the poles of $|\Gamma(s)|^2$ lie in the left half side and half in the right half side of the complex frequency plane. The function $\Gamma(s)$ can be extracted from $\Gamma(s)\Gamma(-s)$ by choosing only those poles in the left half side. The choice of which zeros to choose is more arbitrary since there is not the same realizability restrictions on the zeros. If the choice is made to use only the left half plane zeros, the resulting reflection coefficient and the corresponding driving point impedance is the minimum phase function. This leads to a synthesis of a ladder network. The $j\omega$ axis zeros are even multiples of complex-conjugate pairs and are divided equally between $\Gamma(s)$ and $\Gamma(-s)$.

The problem of actually finding the poles and zeros requires finding the roots of the denominator and numerator polynomials. While these roots can be found analytically for the Butterworth and Chebyshev filters, the roots for other functions such as the Bessel–Thompson filter function must be found numerically. The transfer function is in the form of

$$|H(s)|^2 = \frac{H_0}{F(s)} \tag{5.47}$$

where $F(0) = 1$.

The dc reflection coefficient is

$$|\Gamma(0)|^2 = 1 - H_0 \tag{5.48}$$

$$\Gamma(0) = \pm\sqrt{1 - H_0} \tag{5.49}$$

Since at dc in a low-pass circuit the series reactive elements are short circuits and the shunt reactive elements are open circuits, the reflection coefficient is

$$\Gamma(0) = \frac{R_L - R_G}{R_L + R_G} \tag{5.50}$$

Consequently,

$$H_0 = \frac{4\,R_L/R_G}{(R_L/R_G + 1)^2} \tag{5.51}$$

which, of course, equals 1 when both sides of the filter have equal terminations.

Once the reflection coefficient is determined, the Cauer synthesis procedure is used to obtain the circuit elements. The input impedance to the filter at any frequency is given in terms of the reflection coefficient:

$$Z_{in} = R_G \frac{1 + \Gamma(s)}{1 - \Gamma(s)} \tag{5.52}$$

The Cauer extraction technique for a ladder network can now be used. The polynomials in the numerator and denominator are arranged in descending powers of s. It will always be the case for a lossless transfer function that the highest power of the numerator and denominator will differ by at least 1. If the numerator is the higher order polynomial, then an impedance pole at $s = \infty$ (i.e., a series inductor) can be extracted from the impedance function. This is done by synthetic division. The fractional remainder is now inverted and synthetic division is again carried out to extract an admittance pole at $s = \infty$ (i.e., a shunt capacitor). The process continues until only the load resistance or conductance remains.

As an example, consider a three-pole Butterworth with a 3-dB cutoff frequency at 1 rad./s. The input resistance is $R_G = 20\,\Omega$, and the output resistance is $R_L = 80\,\Omega$. The Butterworth transfer function is therefore

$$|H(\omega)|^2 = \frac{H_0}{1 + \omega^6} \tag{5.53}$$

where from Eq. (5.51),

$$H_0 = \frac{16}{25} \tag{5.54}$$

and

$$|\Gamma(\omega)|^2 = 1 - |H(\omega)|^2 = \frac{1+\omega^6 - H_0}{1+\omega^6} \tag{5.55}$$

Now replace ω with $-js$, factor the denominator into the six roots of 1, and recombine into two cubic factors where one factor contains the left half plane roots and the other the right half plane roots. This is the standard Butterworth polynomial:

$$\Gamma(s)\Gamma(-s) = \frac{9/25 - s^6}{\left(-s^3 + 2s^2 - 2s + 1\right)\left(s^3 + 2s^2 + 2s + 1\right)} \tag{5.56}$$

In this case the denominator is readily factored analytically, but the roots of the numerator when $H_0 \neq 1$ must be found numerically. The program POLY can provide the complex roots of a polynomial with complex coefficients. In this example, all values are calculated using double-precision arithmetic, though for clarity only three or four significant figures are shown.

The reflection coefficient containing only left half plane poles and zeros is

$$\Gamma(s) = \frac{s^3 + 1.687s^2 + 1.423s + 0.599}{s^3 + 2s^2 + 2s + 1} \tag{5.57}$$

The input impedance is found from Eq. (5.52):

$$Z_{\text{in}} = 20\frac{2s^3 + 3.687s^2 + 3.423s + 1.599}{0s^3 + 0.313s^2 + 0.577s + 0.400} \tag{5.58}$$

Extraction of the impedance pole at $s = \infty$ is done by synthetic division:

$$\begin{array}{r|l} & 6.387s \\ 0.313s^2 + 0.577s + 0.400 & 2s^3 + 3.3687s^2 + 3.423s + 1.599 \\ & 2s^3 + 3.3687s^2 + 2.555s \\ \hline & 0s^3 + 0s^2 + 0.868s + 1.599 \end{array}$$

The remainder is inverted and an admittance pole at $s = \infty$ is extracted:

$$\begin{array}{r|l} & 0.361s \\ 0.868s + 1.599 & 0.313s^2 + 0.577s + 0.400 \\ & 0.313s^2 + 0.577s \\ \hline & 0s^3 \quad + 0s \quad + 0.400 \end{array}$$

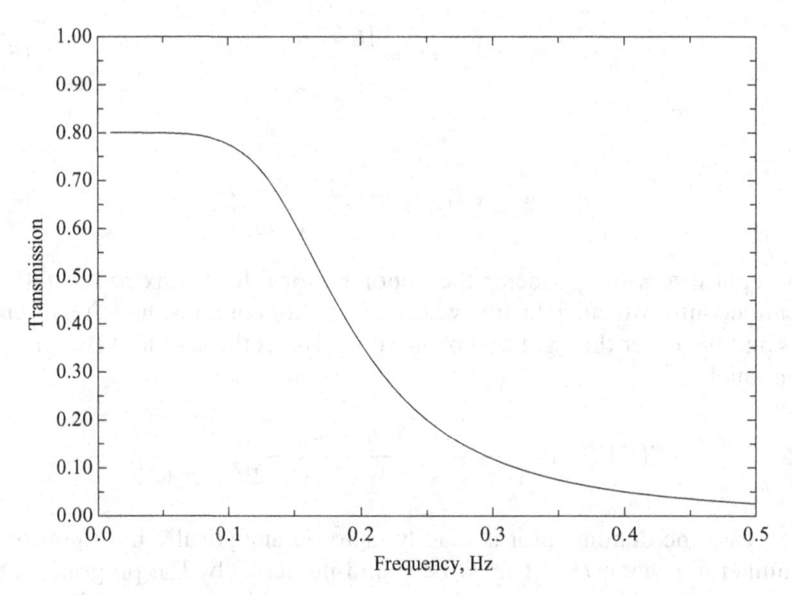

FIGURE 5.7 Frequency response of Butterworth low-pass filter.

By inversion again and performing synthetic division once more, another impedance pole at $s = \infty$ is removed:

$$0.400 \overline{\smash{)}\,0.868s + 1.599} \quad \text{quotient: } 2.170s$$

The final remainder, 1.599/0.400 = 4.000, represents the normalized load resistance, which is the expected value. Hence, $L_1' = 6.387$ H, $C_2' = 0.361$ F, and $L_3' = 2.170$ H. The impedance level of the circuit is now adjusted from 1Ω to $R_G = 20\Omega$ by multiplying all the inductances and dividing all capacitances by 20Ω. Thus, L_1' becomes $L_1 = 127.75$ H, C_2' becomes $C_2 = 0.01804$ F, and L_3' becomes $L_3 = 43.400$ H. The final circuit is shown in Fig. 5.6. Verification of this circuit is shown by a SPICE analysis found in Fig. 5.7. Near zero frequency the insertion loss is 0.8 or −1.938 dB and at 1 rad/s (0.159 Hz) the loss has increased by 3 dB.

Easier analytical methods are available for the Chebyshev filter, and these are in fact used in the Chebyshev impedance transforming circuit described in Section 5.6.4. The Cauer method shown here can be used where a closed-form solution for the roots is not available.

5.6.2 Filter-Type Transformation

Filter design is based on the design of a low-pass prototype circuit whose impedance level is 1Ω and whose low-pass cutoff frequency is $\omega_c = 1$ rad/s. If the desired impedance level is to be changed from 1 to R_G, then all inductors and resistors should be multiplied by R_G, and all capacitors should be divided by R_G as was done in the previous example. If the circuit elements of the low-pass prototype are denoted by a p subscript, then the new adjusted values can be found:

$$L = R_G L_p \tag{5.59}$$

$$C = \frac{C_p}{R_G} \tag{5.60}$$

$$R = R_G R_p \tag{5.61}$$

To adjust the cutoff frequency from 1 rad/s to ω_c, the low-pass circuit elements are further modified in the following way:

$$L' = \frac{L}{\omega_c} \tag{5.62}$$

$$C' = \frac{C}{\omega_c} \tag{5.63}$$

$$R' = R \tag{5.64}$$

Transformation of the low-pass filter to a high-pass filter can be accomplished by another frequency transformation. The normalized complex frequency variable for the low-pass prototype circuit is s_n. On the $j\omega$ axis the pass band of the low-pass filter occurs between $\omega = -1$ and $+1$. If the cutoff frequency for the high-pass filter is ω_c, then the high-pass frequency variable is

$$s = \frac{\omega_c}{s_n} \tag{5.65}$$

Applying this transformation will transform the pass-band frequencies of the low-pass filter to the pass band of the high-pass filter. This is illustrated in Fig. 5.8. The reactance of an inductor, L, in the low-pass filter becomes a capacitance, C_h, in the high-pass filter:

$$Ls_n = \frac{L\omega_c}{s} = \frac{1}{C_h s} \tag{5.66}$$

or

$$C_h = \frac{1}{L\omega_c} \tag{5.67}$$

Similarly, application of the frequency transformation Eq. (5.65) will convert a capacitor in the low-pass filter to an inductor in the high-pass filter:

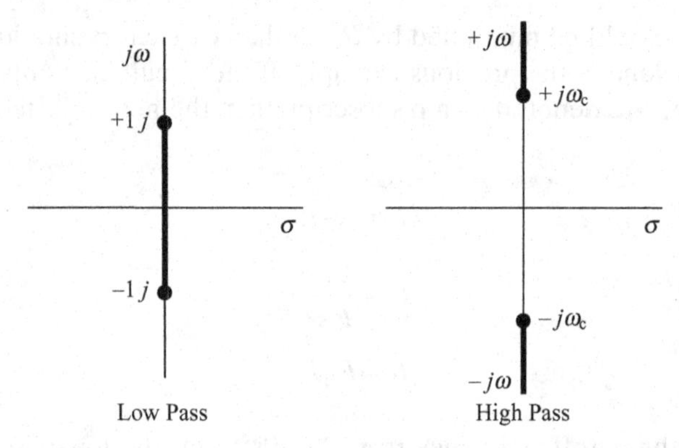

FIGURE 5.8 Low-pass to high-pass transformation.

$$L_h = \frac{1}{C\omega_c} \tag{5.68}$$

A band-pass filter is specified to have a pass band from ω_1 to ω_2. The "center" of the pass band is the geometric mean of the band edge frequencies, $\omega_0 = \sqrt{\omega_1\omega_2}$. The fractional bandwidth is $w = (\omega_2 - \omega_1)/\omega_0$. A band-pass circuit can be formed from the low-pass prototype by using a frequency transformation that will map the pass band of the low-pass filter to the pass band of the band-pass filter. The desired frequency transformation is

$$s_n = \frac{1}{w}\left(\frac{s}{\omega_0} + \frac{\omega_0}{s}\right) \tag{5.69}$$

where s is the frequency variable for the band-pass circuit. To verify this expression for the $j\omega$ axis, Eq. (5.69) is rewritten as

$$\omega_n = \frac{1}{w}\left(\frac{\omega}{\omega_0} - \frac{\omega_0}{\omega}\right) \tag{5.70}$$

A short table of specific values for the normalized low-pass prototype circuit and the corresponding band-pass frequencies are shown in Table 5.1.

A graphic illustration of the frequency transformation is shown in Fig. 5.9. A consequence of this transformation is that an inductor L in the low-pass prototype filter becomes a series LC circuit in the band-pass circuit:

$$Ls_n = \frac{Ls}{w\omega_0} + \frac{L\omega_0}{ws} \tag{5.71}$$

TABLE 5.1 Low-Pass to Band-Pass Mapping

Band-Pass ω	Low-Pass ω_n
ω_2	+1
ω_0	0
ω_1	−1
$-\omega_1$	+1
$-\omega_0$	0
$-\omega_2$	−1

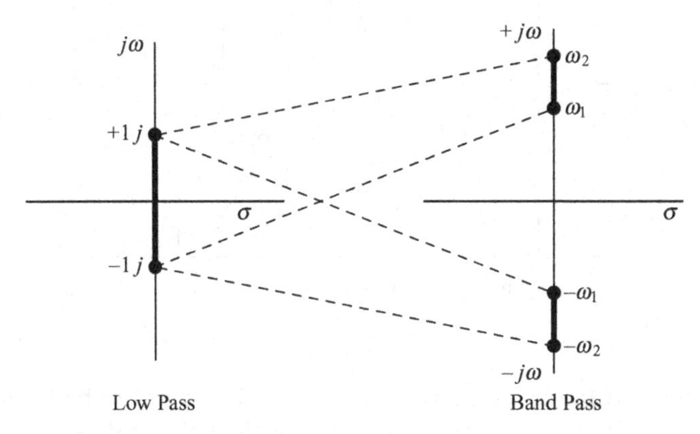

FIGURE 5.9 Low-pass to band-pass transformation.

Similarly, a capacitance in the low-pass filter is transformed to a parallel *LC* circuit:

$$Cs_n = \frac{Cs}{w\omega_0} + \frac{C\omega_0}{ws} \tag{5.72}$$

Finally, the low-pass to band-stop filter frequency transformation is the reciprocal of Eq. (5.69):

$$s_n = w\left(\frac{s}{\omega_0} + \frac{\omega_0}{s}\right)^{-1} \tag{5.73}$$

All these transformations from the low-pass prototype filter are summarized in Fig. 5.10.

5.6.3 Number of Poles for a Band-Pass Filter

The functions given by Eqs. (5.11) and (5.24) are used to find the number of poles required for a low-pass Butterworth or Chebyshev filter. The number

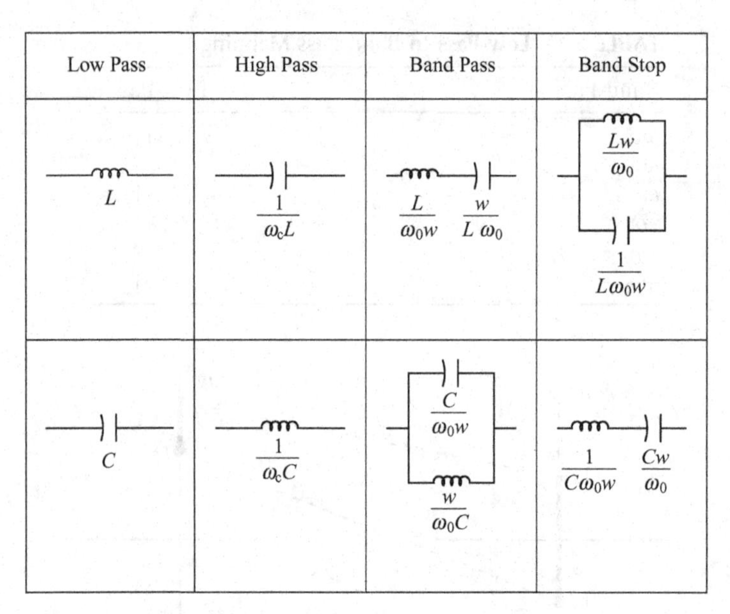

FIGURE 5.10 Filter conversion chart.

of poles required for a band-pass filter that gives a prescribed attenuation above the upper band edge frequency or below the lower band edge frequency can also be determined. The ratio of stop-band to cutoff frequency for the low-pass circuit is

$$F_c \triangleq \frac{\omega_s}{\omega_c} \tag{5.74}$$

A similar ratio can be defined for a band-pass filter:

$$F_{c1\text{-bp}} = \frac{\omega_1}{\omega_{s1}} \tag{5.75}$$

or

$$F_{c2\text{-bp}} = \frac{\omega_{s2}}{\omega_2} \tag{5.76}$$

When meeting the required level of attenuation at the upper stop frequency, $F_{c2\text{-bp}}$ is more difficult to achieve than the lower stop frequency. Using $F_{c2\text{-bp}}$ provides the more conservative design. However, either pass-band edge may be specified. The design specifications for the band-pass filter determines the value for $F_{c2\text{-bp}}$. It is now necessary to find F_c for the equivalent low-pass circuit and apply the result to Eq. (5.11) or (5.24) to determine the number of poles for the filter.

The relationship between the normalized low-pass prototype circuit frequency and the band-pass circuit frequency is given by Eq. (5.70). This needs to be solved for the equivalent stop to the upper pass-band edge frequency ratio for the band-pass circuit. Solving Eq. (5.70) for the band-pass frequency, ω, gives

$$w\omega\omega_0\omega_n = \omega^2 - \omega_0^2 \tag{5.77}$$

$$0 = \omega^2 - \omega(\omega_0\omega_n w) - \omega_0^2 \tag{5.78}$$

$$\omega = \frac{\omega_0\omega_n w \pm \sqrt{(\omega_0\omega_n w)^2 + 4\omega_0^2}}{2} \tag{5.79}$$

$$= \frac{\omega_0}{2}\left[\omega_n w + \sqrt{(\omega_n w)^2 + 4}\right] \tag{5.80}$$

The frequency, ω, is the variable corresponding to the band-pass circuit while ω_n corresponds to the low-pass prototype frequency variable. The cutoff frequency for the low-pass prototype circuit is $\omega_n = \omega_c = 1\,\text{rad/s}$, and this corresponds to the edge of the pass band, ω_2, for the band-pass circuit. The low-pass prototype stop frequency is $\omega_s = \omega_c \cdot F_c$. The corresponding stop-to-band edge frequency ratio for the band-pass circuit is presumed to be a design specification along with the fractional bandwidth, w. These frequency points for the band-pass circuit are illustrated in Fig. 5.11. It is necessary to find the corresponding value, F_c, of the low-pass circuit in order to determine n. The band-pass ratio, $F_{c2\text{-bp}}$, is evaluated by using Eq. (5.80) to provide expressions for ω_{s2} and ω_2:

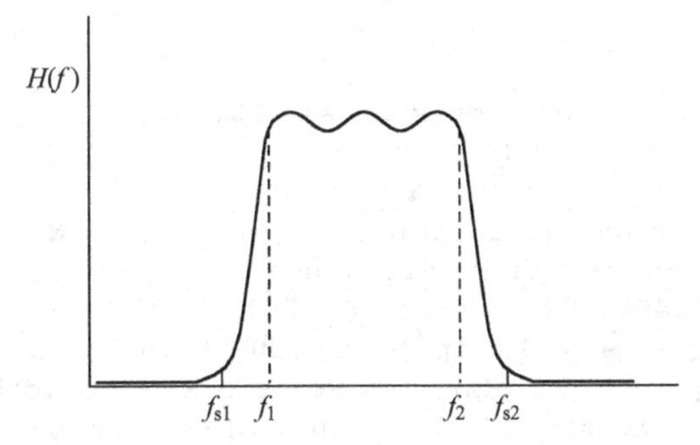

FIGURE 5.11 Upper pass-band frequency and stop-band frequency for a band-pass filter.

$$F_{c2-bp} = \frac{\omega_{s2}}{\omega_2} = \frac{wF_c + \sqrt{(wF_c)^2 + 4}}{w + \sqrt{w^2 + 4}} \tag{5.81}$$

$$\triangleq \frac{a(F_c)}{w + \sqrt{(w)^2 + 4}} \tag{5.82}$$

The value for $a(F_c)$ is determined by the specified values, F_{c2-bp} and w, in Eq. (5.82). This equation can be solved for F_c in terms of a known a by equating Eqs. (5.81) and (5.82):

$$F_c = (a^2 - 4)\frac{1}{2aw} \tag{5.83}$$

Once F_c is known for the low-pass circuit, the number of poles required for a given stop-band attenuation can be found from Eq. (5.11) or (5.24) for the Butterworth and Chebyshev filters, respectively. For example, if a band-pass Butterworth filter is to have 15 dB attenuation at a point $1.1 \times f_2$, and have a fractional bandwidth of 20%, then $\alpha_{min} = 15$ dB, and $\alpha_{max} = 3$ dB. The above equations give $a(F_c) = 2.431$, $F_c = 1.964$ and from Eq. (5.12) $n = 2.535$. So the number of required poles would be three.

5.6.4 Chebyshev Band-Pass Filter Example

The analytical design technique for a Chebyshev filter with two unequal resistances has been implemented in the program called CHEBY. As an example of its use consider the design of a Chebyshev filter that matches a 15-Ω to a 50-Ω load resistance. It will have $n = 3$ poles, a center frequency of 1.9 GHz, and a fractional bandwidth $w = (f_2 - f_1)/f_0 = 20\%$. The program, CHEBY, is used to find the filter circuit elements. The program could have used the Cauer procedure described in Section 5.6.1, but instead it used simpler analytical formulas [1]. The following is a sample run of the program:

```
Generator  AND  Load  resistances  15.,50.
Pass   band  ripple  (dB)  0.2
Band-pass Filter? <Y/N>  Y
Specify  stopband attenuation  OR  n,   <A/N>  N
Number  of  transmission  poles n  =  3
L( 1)  =  .62405E+02  C(  2)  =  .25125E-01 L( 3) = .36000E+02
Number  of  poles  = 3  Ripple =  .20000E+00   dB
Center Frequency, Fo (Hz), AND Fractional Bandwidth, w 1.9E9,.2
Through  series LC. L1(  1) = .261370E-07 C1(  1) = .268458E-12
Shunt  parallel LC. L2(  2) =  .666805E-09 C2(  2) = .105229E-10
Through  series LC. L3(  3)  =  .150777E-07  C3(  3) = .465370E-12
```

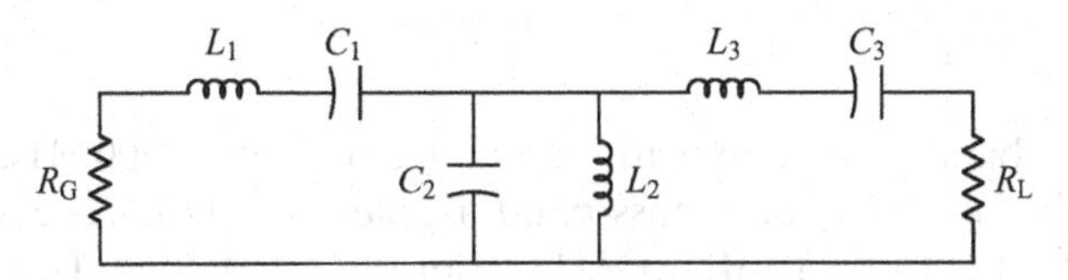

FIGURE 5.12 A 15:50-Ω Chebyshev band-pass filter where $L_1 = 26.14\,\text{nH}$, $C_1 = 0.2685\,\text{pF}$, $L_2 = 0.6668\,\text{nH}$, $C_1 = 10.52\,\text{pF}$, $L_3 = 15.08\,\text{nH}$, $C_3 = 0.4654\,\text{pF}$.

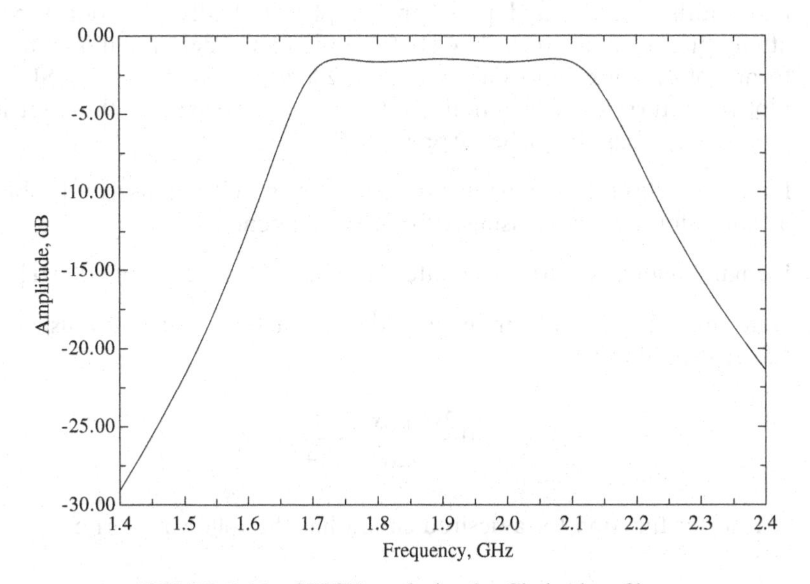

FIGURE 5.13 SPICE analysis of a Chebyshev filter.

The resulting circuit shown in Fig. 5.12 can be analyzed using the SPICE template described in Appendix G. The results in Fig. 5.13 show that the minimum loss in the pass band is −1.487 dB which corresponds to $\sqrt{H_0} \le 0.7101$.

5.7 CONCLUSIONS

The filter designs described in this chapter can be used in the transceiver for blocking unwanted frequencies from the antenna or mixer and providing critical impedance matching in the amplifier. The following chapter describes impedance transformers that are unique to the radio frequency range.

PROBLEMS

5.1. Design a band-pass filter with center frequency of 500 MHz, fractional bandwidth $w = 5\%$, and pass-band ripple of 0.1 dB. The out-of-band attenuation is to be 10 dB 75 MHz from the band edge. The terminating impedances are each 50 Ω. Using SPICE, plot the return loss (reflection coefficient in decibels) and the insertion loss over the pass band. See Appendix G.

5.2. Design a band-pass filter with center frequency of 500 MHz, fractional bandwidth $w = 5\%$, and pass-band ripple of 0.1 dB. The out-of-band attenuation is to be 10 dB 75 MHz from the band edge, and it is to transform a 50-Ω source impedance to a 75-Ω load impedance. Using SPICE, plot the return loss (reflection coefficient in decibels) and the insertion loss over the pass band. See Appendix G.

5.3. Design an elliptic function filter with the same specifications as in Problem 5.1 and plot the results using SPICE. See Appendix G.

5.4. Design a high-pass three-pole filter with cutoff frequency of 900 MHz.

5.5. Determine the circuit from the impedance function given below using the Cauer procedure:

$$Z(s) = 20\frac{2s^3 + 8s^2 + 11s + 12}{0.5s^2 + s2 + 2}$$

5.6. A transfer function for a desired circuit has the following form:

$$|H(s)|^2 = \frac{1}{1+s^4} = \frac{1}{D(s)}$$

where

$$D(s) = \left(s - \frac{\sqrt{2}}{2} - j\frac{\sqrt{2}}{2}\right)\left(s - \frac{\sqrt{2}}{2} + h\frac{\sqrt{2}}{2}\right)\left(s + \frac{\sqrt{2}}{2} + j\frac{\sqrt{2}}{2}\right)\left(s + \frac{\sqrt{2}}{2} - j\frac{\sqrt{2}}{2}\right)$$

Determine the expression for the required input impedance for this circuit.

5.7. Design a Chebyshev 50:50 Ω filter with a pass-band ripple of 0.1 dB from 2.0 to 3.0 GHz. The attenuation should be at least 15 dB at $f <$ 1 GHz and $f >$ 4.0 GHz. Determine the actual filter attenuation from 1.0 to 4.0 GHz using SPICE or other simulator to plot the insertion loss.

REFERENCES

1. Wai-Kai Chen, *Passive and Active Filters*, New York: Wiley, 1986.
2. E. A. Guillemin, *Synthesis of Passive Networks*, New York: Wiley, 1957.
3. F. F. Kuo, *Network Analysis and Synthesis*, New York: Wiley, 1962.
4. A. Zverev, *Handbook of Filter Synthesis*, New York: Wiley, 1967.
5. H. Howe, *Stripline Circuit Design*, Norwood MA: Artech, 1974.

CHAPTER SIX

Transmission Line Transformers

6.1 INTRODUCTION

An RF transceiver often requires impedance transformation, power splitting, or transformation from a balanced to an unbalanced (balun) transmission line. Such circuits appropriate to the RF range are described in this chapter. The subject matter of Chapter 3 was impedance transformation. This subject is taken up here again, but now with more careful attention given to the special problems and solutions required for RF designs. The discrete-element designs described previously can be used in RF designs with the understanding that element values will change as frequency changes. The alternative to discrete-element circuits are transmission line circuits. The classical microwave quarter-wavelength transformer can be used up to hundreds of gigahertz in the appropriate transmission line medium. However, at 1 GHz, a three-section quarter-wavelength transformer would be a little less than a meter long! The solution lies in finding a transformation structure that may not work at 100 GHz but will be practical at 1 GHz.

The conventional transformer consists of two windings on a high-permeability iron core. The flux, ϕ, is induced onto the core by the primary winding. By Faraday's law, the secondary voltage is proportional to $d\phi/dt$. For low-loss materials, the primary and secondary voltages will be in phase. Ideal transformers have perfect coupling and no losses. The primary-to-secondary voltage ratio is equal to the turns ratio, n, between the primary and secondary windings, namely $V_p/V_s = n$. The ratio of the primary-to-secondary current ratio is $I_p/I_s = 1/n$. This implies conservation of power, $V_pI_p = V_sI_s$. As a consequence, the impedance seen by the generator or primary side in terms of the load impedance is

Radio Frequency Circuit Design, Second Edition, by W. Alan Davis

$$Z_G = n^2 Z_L \tag{6.1}$$

When the secondary side of the ideal transformer is an open circuit, the input impedance of the transformer on the primary side is infinity.

In a physical transformer the ratio of the leakage inductances on primary and secondary sides is $L_p/L_s = n$. For the ideal transformer, L_p and L_s approach ∞, but their ratio remains finite at n. The physical transformer has an associated mutual inductance, $M = k\sqrt{L_p/L_s}$, where k is the coupling coefficient. The leakage inductance together with the interwire capacitances limits the high-frequency response. The transmission line transformer avoids these frequency limitations.

6.2 IDEAL TRANSMISSION LINE TRANSFORMERS

It was found in Chapter 2 that inductive coils always come with stray capacitance. It was this capacitance that restricted the frequency range for a standard coupled-coil transformer. The transmission line transformer can be thought of as simply tipping the coupled-coil transformer on its side. The coil inductance and stray capacitance now form the components for an artificial transmission line whose characteristic impedance is

$$Z_0 = \sqrt{\frac{L}{C}} \tag{6.2}$$

The artificial transmission line can be used, in principle, up to very high frequencies because the shunt capacitance forms part of the transmission line characteristic impedance. The transmission line transformer can be made from a variety of forms of transmission lines such as two parallel lines, a twisted pair of lines, a coaxial cable, or a pair of wires on a ferrite core. The transmission line transformer can be defined as having the following characteristics:

1. The transmission line transformer is made up of interconnected lines whose characteristic impedance is a function of such mechanical things as wire diameter, wire spacing, and insulation dielectric constant.
2. The transmission lines are designed to suppress even-mode currents and allow only odd-mode currents to flow (Fig. 6.1).

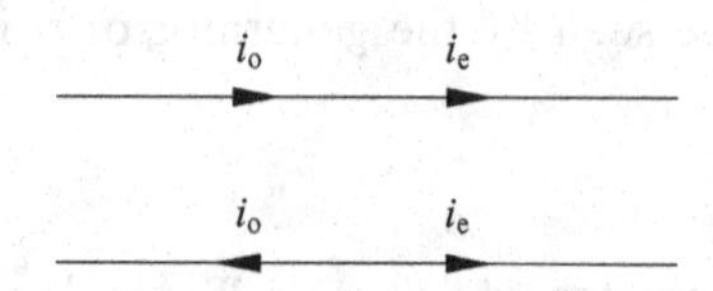

FIGURE 6.1 Two-wire transmission line showing odd- and even-mode currents.

3. The transmission lines carry their own "ground" so that transmission lines relative to true ground are unintentional.
4. All transmission lines are of equal length and typically $<\lambda/8$.
5. The transmission lines are connected at their ends only.
6. Two different transmission lines are not coupled together either by capacitance or inductance.
7. For a short transmission line, the voltage difference between the terminals at the input port is the same as the voltage difference at the output port.

Some explanation of these points is needed to clarify the characteristics of the transmission line transformer. In property 2, for a standard transmission line, the current going to the right in one conductor must be equal to the current going to the left in the other in order to preserve current continuity (Fig. 6.1). Since only odd-mode currents are allowed, the external magnetic fields are negligible. The net current driving the magnetic field outside of the transmission line is low. The third point is implied by the second. The transmission line is isolated from other lines as well as the ground. The equality of the odd-mode currents in the two lines of the transmission line as well as the equivalence of the voltages across each end of the transmission line is dependent on the transmission line being electrically short in length. The analysis of transmission line transformers will be based on the given assumptions above.

As an example, consider the transmission line transformer consisting of one transmission line with two conductors connected as shown in Fig. 6.2. The transformation ratio will be found for this connection. Assume first that v_1 is the voltage across R_G and i_1 is the current leaving the generator resistance:

1. Current i_1 passes through the upper conductor of the transmission line.
2. The odd-mode current i_1 flows in the opposite direction in the lower conductor of the transmission line.
3. The sum of the two transmission line currents at the output node is $2i_1$.

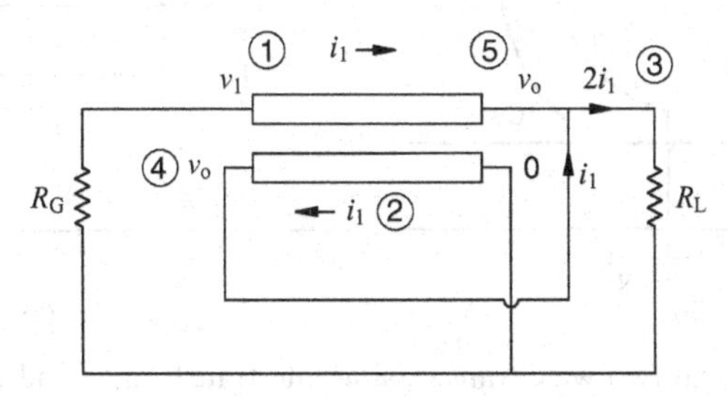

FIGURE 6.2 Analysis steps for transmission line transformer.

4. The voltage at the output node is assumed to be v_o. Consequently, the voltage at the left side of the lower conductor in the transmission line is v_o above ground.
5. On the left-hand side, the voltage difference between the two conductors is $v_1 - v_o$. This is the same voltage difference on the right-hand side. Consequently,

$$v_o - 0 = v_1 - v_o$$
$$v_o = v_1/2$$

If $R_G = v_1/i_1$, then

$$R_L = \frac{v_o}{2i_1} = \frac{v_1/2}{2i_1} = \frac{R_G}{4} \tag{6.3}$$

This 4:1 circuit steps down the impedance level by a factor of 4.

A physical connection for this transformer is shown in Fig. 6.3, where the transmission line is represented as a pair of lines. In this diagram the nodes in the physical representation are matched to the corresponding nodes of the schematic representation. The transmission line is bent around to make the *B–C* distance a short length. The transmission line, shown here as a two-wire line, can take a variety of forms such as a coupled line around a ferromagnetic core, flexible microstrip line, or coaxial line. If the transformer is rotated about a vertical axis at the center, the circuit shown in Fig. 6.4 results. Obviously, this results in a 1:4 transformer where $R_L = 4R_G$. Similar analysis to that given above verifies this result. In addition multiple two-wire transmission line trans-

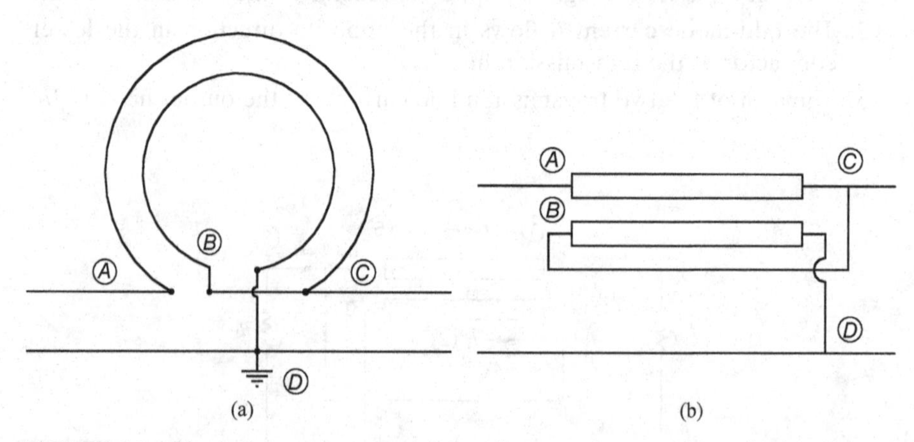

FIGURE 6.3 Physical two-wire transmission line transformer and equivalent formal representation.

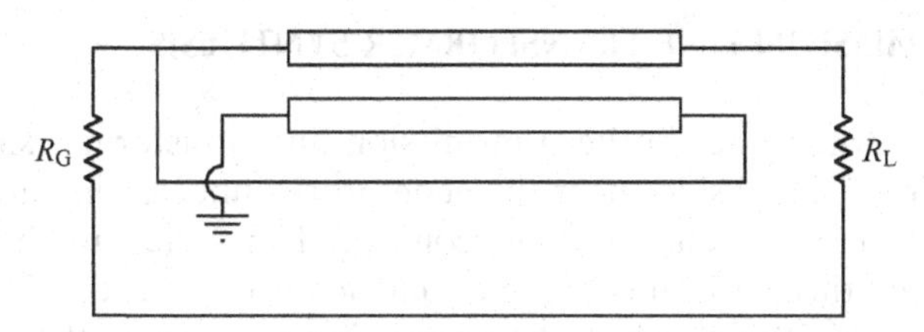

FIGURE 6.4 Alternate transmission line transformer connection.

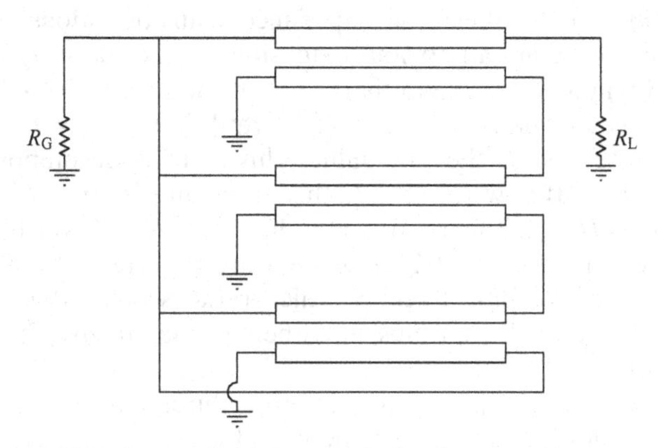

FIGURE 6.5 A 16:1 transmission line transformer.

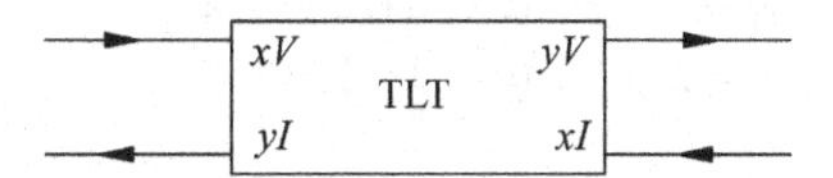

FIGURE 6.6 Symbol for general transmission line transformer.

formers may be tied together to achieve a variety of different transformation ratios. An example of three sections connected together is shown in Fig. 6.5. In this circuit the current from the generator splits into four currents going into the transmission lines. Because of the equivalence of the odd-mode currents in each line, these four currents are all equal. The voltages on the load side of each line pair build up from ground to 4× the input voltage. As a result, for match to occur, $R_L = 16R_G$.

The voltages and currents for a transmission line transformer (TLT) having a wide variety of different interconnections and numbers of transmission lines can be represented by the simple diagram in Fig. 6.6, where x and y are integers. The impedance ratios, $R_G = (x/y)^2 R_L$, range from 1:1 for a 1-transmission line circuit to 1:25 for a 4-transmission line circuit with a total of 16 different transformation ratios [1]. A variety of transmission line transformer circuits are found in [1] and [2].

6.3 TRANSMISSION LINE TRANSFORMER SYNTHESIS

All the transmission lines in the transmission line transformer shown in Fig. 6.5 have their left-hand sides near the generator connected in parallel and all their right-hand sides near the load connected in series. In this particular circuit, there are three transmission lines, and analysis shows that $V_{in}:V_{out} = 1:4$, and $R_G:R_L = 1:16$. The number of transmission lines, m, is the order of the transformer, so that when all the transmission lines on the generator side are connected in shunt and on the load side in series, the voltage ratio is $V_{in}:V_{out} = 1:(m+1)$. Synthesis of impedance transformations of 1:4, 1:9, 1:16, 1:25, and so on are all obvious extensions of the transformer shown in Fig. 6.5. To obtain a voltage ratio that is not of the form $1:(m+1)$ there is a simple synthesis technique [3]. The voltage ratio is $V_{in}:V_{out} = H:L$, where H is the high value and L the low value. This ratio is decomposed into an $V_{in}:V_{out} = H - L:L$. If now $H - L < L$, this procedure is repeated where now $H' = L$ and $L' = H - L$. This ratio is now $V_{out}:V_{in} = H':L'$, which in turn can be decomposed into $H' - L':L'$. These steps are repeated until a 1:1 ratio is achieved, all along keeping track of which ratio is being done, $V_{in}:V_{out}$ or $V_{out}:V_{in}$. The allowed voltage ratios, upon being squared, give the impedance ratios as shown in Table 6.1.

An example given in [3] illustrates the procedure. If an impedance ratio of $R_G:R_L = 9:25$ is desired, the corresponding voltage ratio is $V_{in}:V_{out} = 3:5$:

Step 1 $H:L = V_{out}:V_{in} = 5:3 \rightarrow (5-3):3 = 2:3$
Step 2 $H:L = V_{in}:V_{out} = 3:2 \rightarrow (3-2):2 = 1:2$
Step 3 $H:L = V_{out}:V_{in} = 2:1 \rightarrow (2-1):1 = 1:1$

Now working backward from step 3, a $V_{in}:V_{out} = 1:2$ transmission line transformer is made by connecting two transmission lines in shunt on the input side and in series on the output side (Fig. 6.7*a*). From step 2, the V_{out} is already 2, so another transmission line is attached to the first pair in shunt on the output side and series on the input side (Fig. 6.7*b*). Finally from step 1, $V_{in} = 3$ already,

TABLE 6.1 Voltage Ratios for Transmission Line Transformers

Number of Lines	1	2	3	4
	1:1	2:3	3:4	4:5
	1:2	1:2	3:5	5:7
	—	1:3	2:5	5:8
	—	—	1:4	4:7
	—	—	—	3:7
	—	—	—	3:8
	—	—	—	2:7
	—	—	—	1:5

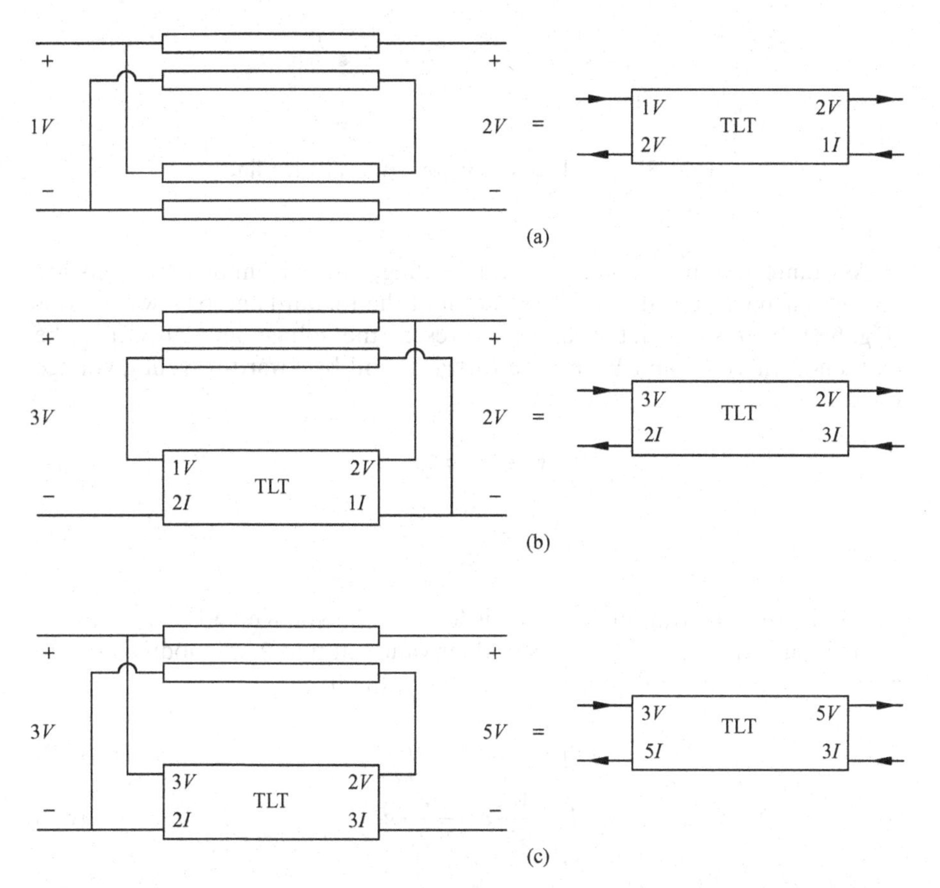

FIGURE 6.7 Step-by-step procedure for synthesis for desired impedance ratio.

so the input is connected in shunt with another added transmission line and the outputs connected in series (Fig. 6.7*c*). The final design has $V_{in}:V_{out} = 3:5$ as desired.

6.4 ELECTRICALLY LONG TRANSMISSION LINE TRANSFORMERS

One of the assumptions given in the previous section was that the electrical length of the transmission lines was short. Because of this the voltages and currents at each end of an individual line could be said to be equal. However, as the line becomes electrically longer (or the frequency increases), this assumption ceases to be accurate. It is the point of this section to provide a means of determining the amount of error in this assumption. Individual design goals would dictate whether a full frequency-domain analysis is needed.

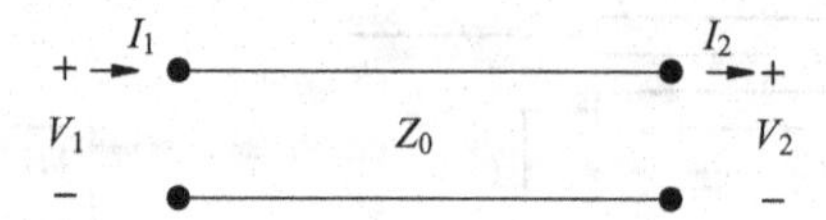

FIGURE 6.8 Electrically long transmission line.

As pointed out in Chapter 4, the total voltage and current on a transmission line are each expressed as a combination of the forward and backward terms (Fig. 6.8). In this case let V_2 and I_2 represent the voltage and current at the load end, where V^+ and V^- are the forward- and backward-traveling voltage waves:

$$V_2 = V^+ + V^- \tag{6.4}$$

$$I_2 = \frac{V^+}{Z_0} - \frac{V^-}{Z_0} \tag{6.5}$$

Assuming that the transmission line is lossless, the voltage and current waves at the input side, 1, given in terms of their values at port 2 are modified by the phase associated with the electrical length of the line:

$$V_1 = V^+ e^{j\theta} + V^- e^{-j\theta} \tag{6.6}$$

$$I_1 = \frac{V^+}{Z_0} e^{j\theta} - \frac{V^-}{Z_0} e^{-j\theta} \tag{6.7}$$

The sign associated with the phase angle, $+\theta$, for V^+ is used because the reference is at port 2 while a positive phase is associated with traveling from left to right. The Euler formula is used in converting the exponentials to sines and cosines. The voltage at the input, V_1, is found in terms of V_2 and I_2 with the help of Eqs. (6.4) and (6.5):

$$V_1 = V_2 \cos\theta + jZ_0 I_2 \sin\theta \tag{6.8}$$

Similarly, I_1 can be expressed in terms of the voltage and current at port 2:

$$I_1 = I_2 \cos\theta + j\frac{V_2}{Z_0} \sin\theta \tag{6.9}$$

The 1:4 transmission line transformer shown in Fig. 6.4 is now reconsidered in Fig. 6.9 to determine its frequency response. The generator voltage can be expressed in terms of the transmission line voltages and currents:

$$V_G = (I_1 + I_2) R_G + V_1 \tag{6.10}$$

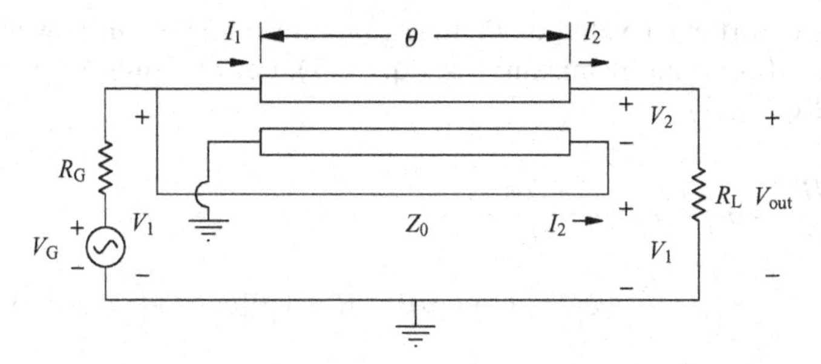

FIGURE 6.9 Electrically long 1:4 transmission line transformer.

The nontransmission line connections are electrically short. Therefore, the output voltage across R_L is $V_o = V_1 + V_2$, and

$$V_G = (I_1 + I_2)R_G + I_2 R_L - V_2 \tag{6.11}$$

In Eqs. (6.11), (6.8), and (6.9), V_1 is replaced by $I_2R_L - V_2$ to give three equations with three unknowns I_1, I_2, and V_2:

$$V_G = I_1 R_G + I_2(R_G + R_L) - V_2 \tag{6.12}$$

$$0 = 0 + I_2(jZ_0 \sin\theta - R_L) + V_2(1+\cos\theta) \tag{6.13}$$

$$0 = -I_1 + I_2 \cos\theta + j\frac{V_2}{Z_0}\sin\theta \tag{6.14}$$

The determinate of this set of equations is

$$\Delta = -2R_G(1+\cos\theta) - R_L \cos\theta + j\sin\theta\left(\frac{-R_G R_L}{Z_0} - Z_0\right) \tag{6.15}$$

and the current I_2 is

$$I_2 = \frac{-V_G(1+\cos\theta)}{\Delta} \tag{6.16}$$

Consequently, the power delivered to the load from the source voltage is

$$\begin{aligned} P_o &= \tfrac{1}{2}|I^2|^2 R_L \\ &= \frac{1}{2}\frac{|V_G|^2(1+\cos\theta)^2 R_L}{[2R_G(1+\cos\theta) + R_L(\cos\theta)]^2 + [(R_G R_L + Z_0^2)/Z_0]^2 \sin^2\theta} \end{aligned} \tag{6.17}$$

Now the particular value of R_L that guarantees maximum power transfer into the load is found by maximizing Eq. (6.17). Let D represent the denominator in Eq. (6.17):

$$\frac{dP_o}{dR_L} = 0 = \frac{1}{2}|V_G|^2 \frac{(1+\cos\theta)^2}{D} \times \left(1 - \frac{R_L}{D}\{2[2R_G(1+\cos\theta) + R_L\cos\theta]\cos\theta + [\cdots]\sin^2\theta\}\right) \quad (6.18)$$

In the low-frequency limit where $\theta \to 0$, the coefficient of $\sin^2\theta$ in Eq. (6.18) will be multiplied by zero. Furthermore, $D(\theta = 0) = (4R_G + R_L)^2$ so that Eq. (6.18) requires that $R_L = 4R_G$. The optimum characteristic impedance is found by maximizing P_o with respect to Z_0, while this time keeping the line length $\neq 0$. The result is not surprising, as it is the geometric mean between the generator and load resistance:

$$Z_0 = \sqrt{R_L/R_G} = 2R_G \quad (6.19)$$

From Eq. (6.17) the output power when $Z_0 = 2R_G$ and $R_L = 4R_G$ is

$$P_o = \frac{1}{2}\frac{|V_G|^2(1+\cos\theta)^2}{R_G(1+3\cos\theta)^2 + 4R_G\sin^2\theta} \quad (6.20)$$

This reduces to the usual form for the available power when $\theta \to 0$.

More complicated transmission line transformers might benefit from using SPICE to analyze the circuit. The analysis above gives a clue to how the values of Z_0 and the relative values of R_G and R_L might be chosen with the help of a low-frequency analysis.

As an example, consider the circuit in Fig. 6.9 again where $R_G = 50\,\Omega$ so that $R_L = 200\,\Omega$, $Z_0 = \sqrt{50 \times 200} = 100\,\Omega$, the length of the transformer is 4 cm, and the frequency is 1.0 GHz. The return loss (= 20 log of the reflection coefficient) in Fig. 6.10 shows that in principle a good match is obtained even at 1 GHz.

The SPICE net list used to analyze this circuit makes use of the conversion of voltages to S parameters:

```
Analysis of a circuit for S11 and S21
*
* R01 and R02 are input and output resistance levels.
* RL is the load resistance.  The load may be supplemented
* with additional elements.
* Lines beginning with ** may be used for PSPICE instead
**.PARAM R01=50, R02=50.   RLOAD=50.     IN1=-1/R01
**.PARAM R01=50, R02=200.  RLOAD=200.    IIN=-1/R01
**.FUNC N(R01,R02)  SQRT(R02/R01)
**R01   1    0    {R01}
r01     1    0    50
```

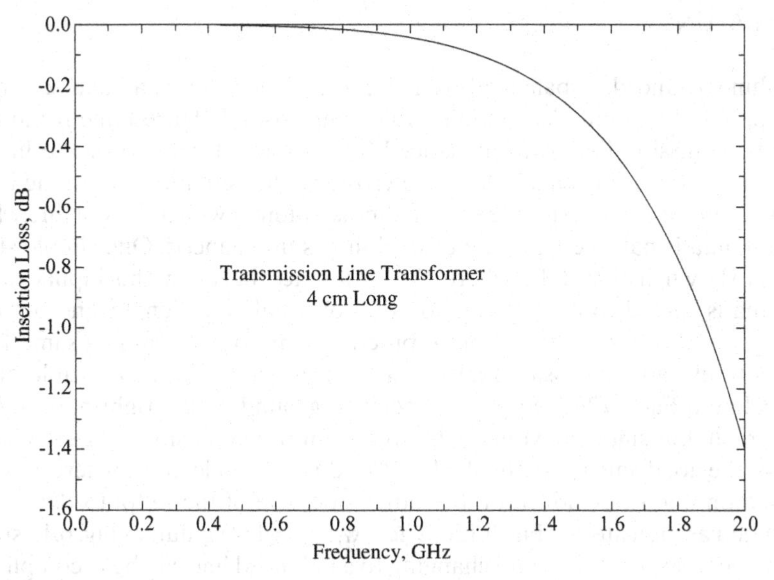

FIGURE 6.10 Return loss for frequency-dependent transmission line transformer of Fig. 6.9.

```
vin      10   11    ac   1
**GI1    1    0     VALUE={-V(10,11)/R01}
*gi1     1    0     10     11     "-1/R01"
gi1      1    0     10     11     -.02
e11      10   0     1      0    2
r11      11   0     1
xcircuit      1     2      tltckt
**RL     2    0     {RLOAD}
rl       2    0     200
**E21    21   0     VALUE={V(2)*2/N(R01,R02)}
*    n = SQRT(R02/R01)
*e21     21   0   2   0   "2/n"
e21      21   0   2   0   1
r21      21   0   1
*
.subckt  tltckt  1   4
* Input side
* 4  cm = 0.1333 wavelength  at  1  GHz
TLT4    1    0    4   1    Z0=100  F=1GHZ   NL=.1333
* Output  side
.ends     tltckt
* Code  for  S11 and S21
*.AC DEC  "num"  "f1"    "f2"
.ac   lin  301  .1meg    2ghz
**.PROBE V(11)  V(21)
.end
```

6.5 BALUNS

A balun (balanced–unbalanced) is a circuit that transforms a balanced transmission line to an unbalanced one. An example of a balanced line is the two-wire transmission line. An unbalanced line is one where one of the lines is grounded, such as in coaxial line or microstrip. One situation where the balun plays an important role is in feeding a dipole antenna with a coaxial line where the antenna is balanced and the coaxial line is unbalanced. One simple structure is shown in Fig. 6.11 where the difference between the inputs of the antenna is forced to be 180° by addition of a half wavelength line between them. At radio frequencies, a more practical way to perform this same function is to use a transmission line transformer as shown in the example of the 1:1 balun in Fig. 6.12*a*. There is no specified ground on the right-hand side of this circuit, but since the voltage difference on the input side is V, the voltage across the load must also be V. For the dipole application, where a $+V$ is needed on one side and $-V$ on the other side, one of the output sides can be grounded as indicated in Fig. 6.12*b*. The (R_G:R_L = 1:4) balun in Fig. 6.13 shows that impedance matching and changing to a balanced line can be accomplished with a balun. Analysis of this circuit may be aided by assuming some voltage, V_x, at the bottom side of R_L. When the voltage at the top side of R_L is found, it also contains V_x. The difference between the bottom and top sides of R_L removes the V_x.

6.6 DIVIDERS AND COMBINERS

Transmission lines can be used to design power dividers and power combiners. These are particularly important in design of high-power solid-state RF amplifiers where the input can be split between several amplifiers or where the outputs of several amplifiers may be effectively combined into one load. A very simple two-way power divider is shown in Fig. 6.14. In this circuit $R_L = 2R_G$ and the transmission line characteristic impedance should be $Z_0 = \sqrt{2}R_G$. The current in R_n ordinarily would be 0 because of the equal voltages on either

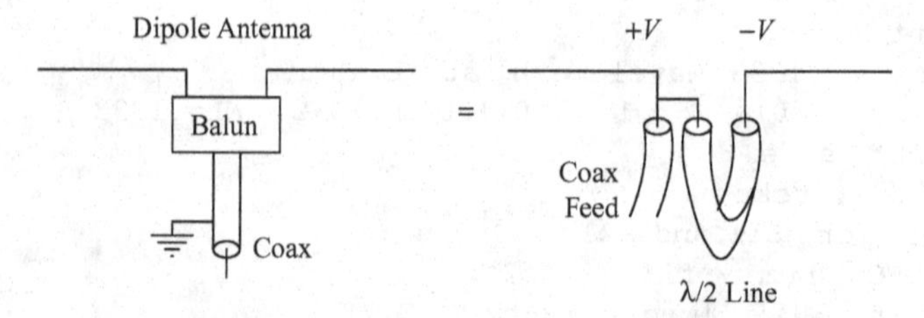

FIGURE 6.11 Balun example used for dipole antenna.

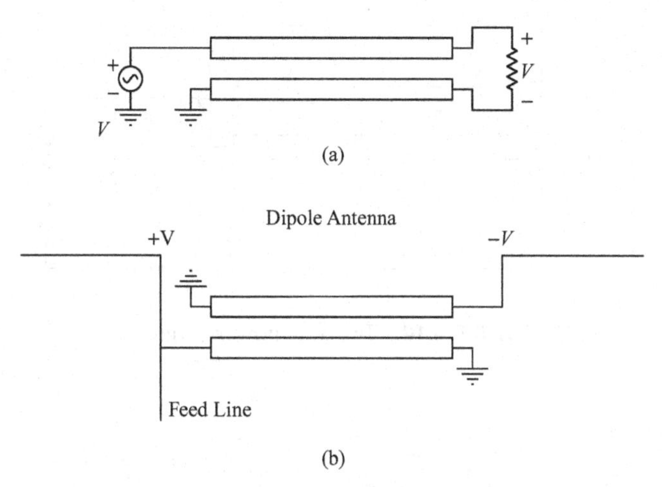

FIGURE 6.12 (*a*) Transmission line transformer implementation of a (1:1) balun, and (*b*) grounding one side gives a +V and −V to two sides of dipole antenna.

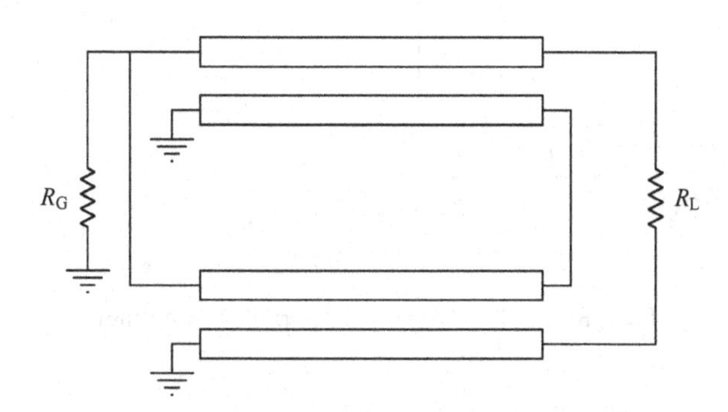

FIGURE 6.13 Balun with $R_G:R_L = 1:4$ impedance ratio.

side of that resistance. Under unbalanced load conditions, R_n can absorb some of the unbalanced power and thus protect the load. The two loads are both $2R_G$. The input voltage is V_1 on the top conductor, and the voltage on the bottom conductor is V_x (Fig. 6.14). On the right-hand side of the transmission line, the bottom conductor is V_1 and so the top conductor must be $2V_1 - V_x$ to ensure that both sides of the transmission line have the same voltage across the terminals, that is, $V_1 - V_x$. Since the current flowing through the top load resistor and the bottom load resistor must be the same, the voltage on either side of R_n is the same. Consequently, $2V_1 - V_x = V_x$ or $V_x = V_1$, so the voltage-to-current ratio at the load is

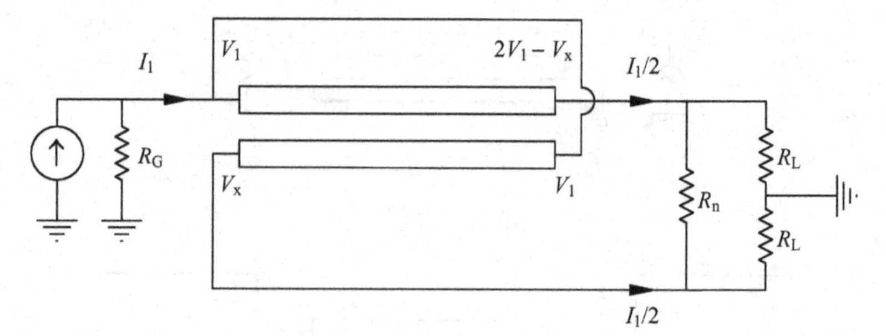

FIGURE 6.14 Two-way power divider.

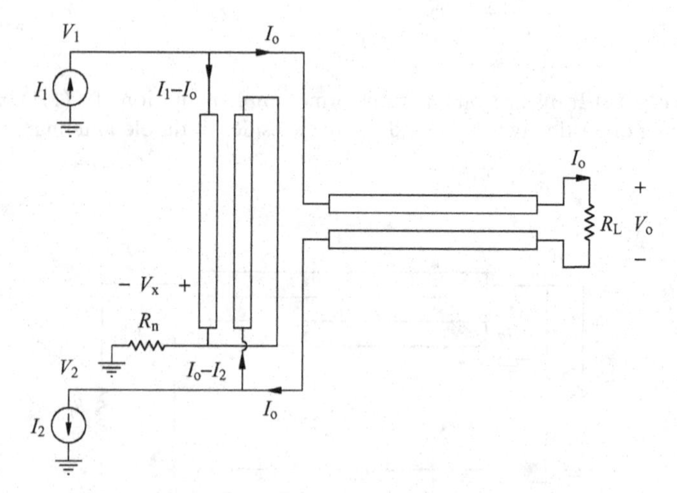

FIGURE 6.15 Two-way 180° power combiner.

$$R_L = \frac{V_1}{I_1/2} = 2R_G \tag{6.21}$$

A two-way 180° power combiner shown in Fig. 6.15 makes use of a hybrid coupler and a balun. The resistor, R_n, is used to dissipate power when the two inputs are not exactly equal amplitude or exactly 180° out of phase so that matched loading for the two sources is maintained. For example, consider when $I_1 = I_2$ as shown in Fig. 6.15 so that I_1 is entering the circuit and I_2 is leaving the circuit. The current flowing through the load, R_L, is I_o. The current flowing into the hybrid transmission line from the top is $I_1 - I_o$ while the current flowing into the bottom is $I_o - I_2$. The odd-mode current in the transmission line forces

$$I_1 - I_o = I_o - I_2$$

or

$$I_o = I_1 \tag{6.22}$$

All the current goes through the balun and no current flows through the hybrid. The current through R_n is therefore 0, leading to $V_x = 0$. The voltage difference between the two ends of the transmission lines of the hybrid is the same and implies that

$$V_1 - V_x = V_x - V_2$$

or

$$V_1 = -V_2 \tag{6.23}$$

and

$$V_o = V_1 - V_2 = 2V_1 \tag{6.24}$$

The matching load resistance is then

$$\frac{V_o}{I_o} = R_L = \frac{2V_2}{I_1} = 2R_G \tag{6.25}$$

When I_1 and I_2 are both entering the circuit so that $I_1 = -I_2$, and $V_1 = V_2$, then voltages across the top and bottom of the transmission line in the hybrid circuit of Fig. 6.15 are

$$V_1 - V_x = V_x - V_2$$

or

$$V_x = V_1 \tag{6.26}$$

The voltage across the load is $V_o = 0$ and $I_o = 0$. The current in the hybrid transmission line is I_1, so the current flowing through R_n is $2I_1$:

$$R_n = \frac{V_x}{2I_1} = \frac{V_1}{2I_1} = \frac{R_G}{2} \tag{6.27}$$

The choices for R_L and R_n assure impedance matching for an arbitrary phase relationship between I_1 and I_2. Optimum performance would be expected if the characteristic impedances of the transmission lines were

$$Z_{0-\text{balun}} = \sqrt{2}R_G \tag{6.28}$$

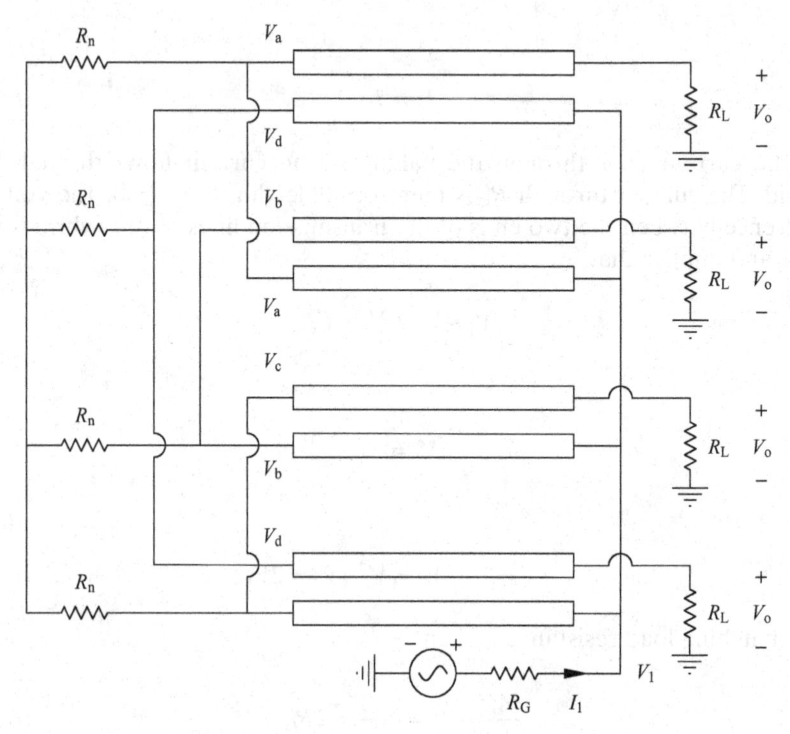

FIGURE 6.16 Four-way power divider.

$$Z_{0\text{-hybrid}} = R_G/\sqrt{2} \tag{6.29}$$

The four-way power divider illustrated in Fig. 6.16 has some similarities with the Wilkinson power divider used at microwave frequencies. In the Wilkinson divider, matching impedances between the input and output is done by choosing the quarter-wavelength transmission lines to have a characteristic impedance $Z_0 = \sqrt{N}R_G$ where N is the power division ratio, and $R_n = R_G$. In the present circuit, impedance matching is done using an impedance transformer at the voltage source (not shown in Fig. 6.16). If it is desired that all the output loads and voltages be equal to one another, then it follows that the currents in the R_n resistors is 0. This can be shown easily. The voltage difference between the conductors on the right-hand side in each of the transmission lines is $V_o - V_1$. Then for the left-hand side,

$$V_o - V_1 = V_a - V_d = V_b - V_a = V_c - V_b = V_d - V_c \tag{6.30}$$

Combining the second and third expressions, then the third and fourth expressions, and so on leads to the following:

$$2V_a = V_b + V_d \tag{6.31}$$

$$2V_b = V_c + V_a \tag{6.32}$$

$$2V_c = V_b + V_d \tag{6.33}$$

Equations (6.31) and (6.33) clearly show that $V_a = V_c$ and Eq. (6.32) shows $V_b = V_a$ and finally $V_d = V_a$. This means there is no current flowing in the R_n resistors and that on the right-hand side, $V_o = V_1$. The current entering each transmission line must then be $I_1/4$ where I_1 is the input current from the source. The load currents are also $I_1/4$ so the impedance transformation at the input requires $R_G = R_L/4$.

6.7 THE 90° COUPLER

The 90° coupler is commonly used to do power division and combining. This is a four-port lossless circuit in which power entering one port will divide between two output ports. The two output signals are 90° out of phase with one another. The fourth port is isolated from the input. The typical branch line or rat race coupler used at microwave frequencies use quarter-wavelength transmission lines. Even techniques such as capacitive loading or folding of transmission lines would still produce a cumbersome design in the lower RF range. A compact design using lumped capacitances and coupled inductors is given in [4, 5]. The coupled inductors are essentially an iron core transformer turned on its side. The four terminals become four ports when they are all referenced with respect to a ground plane. An alternative design that does not require coupled inductors is shown in Fig. 6.17 [6].

The circuit is excited with a voltage wave amplitude of ½ for the odd mode and ½ for the even mode. The superposition of inputs will give an input amplitude of 1 at port 1 and zero at port 4. The outputs at ports 2 and 3 are b_2 and b_3. The symmetry across a central horizontal line enables use of this odd- and even-mode analysis [7]. When ports 1 and 4 are excited in the odd mode,

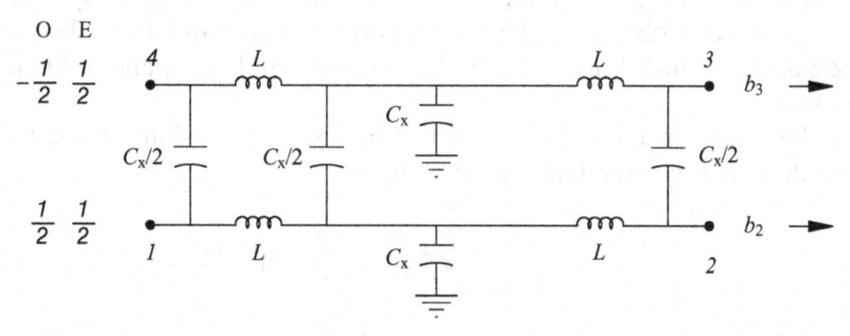

FIGURE 6.17 Four-port 90° lumped-element coupler design with odd- and even-mode excitation.

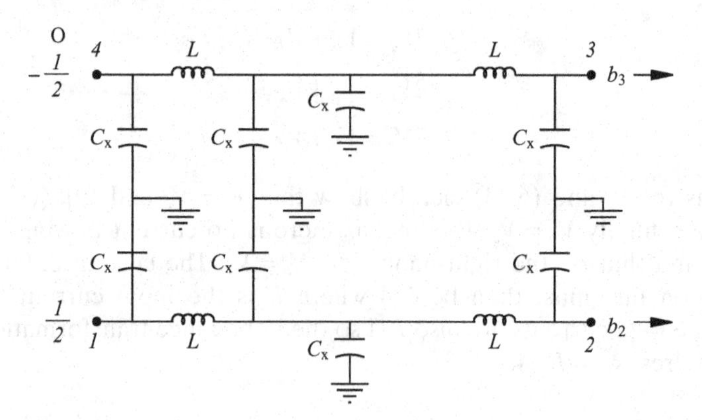

FIGURE 6.18 Half circuit that results from odd-mode excitation.

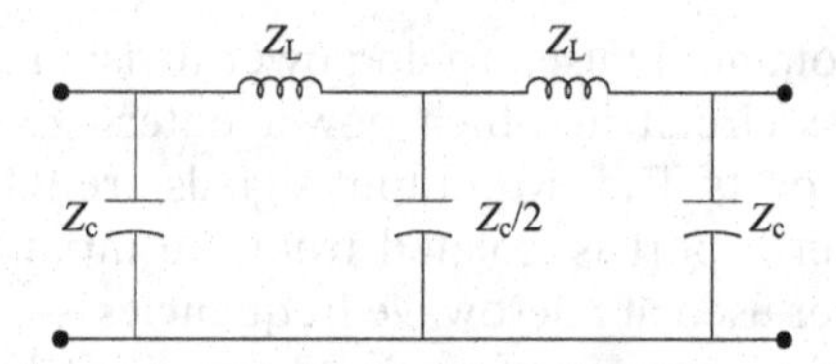

FIGURE 6.19 Odd-mode two-port half circuit.

voltages on the horizontal centerline are zero, so the voltages and currents in the circuit would not be affected if these points were all grounded. When ports 1 and 4 are excited in the even mode, no currents flow across the central line of symmetry, and these points can be open circuited without changing any of the internal voltages or currents. The odd-mode circuit is shown in Fig. 6.18.

The two halves of the circuit are each two-port circuits that can be analyzed separately, one of which is shown in Fig. 6.19 where $Z_L = sL$ and $Z_C = 1/sC_x$. The analysis process proceeds by (1) determine the *ABCD* parameters of the circuit, (2) find the reflection and transmission coefficients of the four-port circuit, (3) repeat the process for the even-mode excitation, (4) specify that $b_1 = 0$, thereby enforcing a condition for impedance matching, (5) specify that $b_4 = 0$ to enforce isolation of the fourth port, (6) determine the values of Z_L and Z_C, and (7) show that b_2 and b_3 have equal amplitudes and a 90° phase difference.

For the circuit in Fig. 6.19, the *ABCD* parameters are determined for the odd-mode circuit by standard circuit analysis:

$$A = \left.\frac{V_1}{V_2}\right|_{I_2=0} = \frac{2Z_L^2}{Z_C^2} + \frac{4Z_L}{Z_C} + 1 \tag{6.34}$$

$$B = \left.\frac{V_1}{-I_2}\right|_{V_2=0} = 2Z_L\left(\frac{Z_L}{Z_C} + 1\right) \tag{6.35}$$

$$C = \left.\frac{I_1}{V_2}\right|_{I_2=0} = \frac{2}{Z_C}\left(\frac{Z_L}{Z_C}+2\right)\left(\frac{Z_L}{Z_C}+1\right) \tag{6.36}$$

$$D = \left.\frac{I_1}{-I_2}\right|_{V_2=0} = \frac{2Z_L^2}{Z_C^2}+\frac{4Z_L}{Z_C}+1 \tag{6.37}$$

The impedance, Z_0, is the impedance level to which the coupler is attached.

The reflection and transmission coefficients for the two-port circuit may be found from Table D.1 in Appendix D as $S_{11} = \Gamma$ and $S_{21} = T$. It should be noted that $AD - BC = 1$ for a linear reciprocal network. For the odd mode:

$$\Gamma_o = \frac{A + B/Z_0 - CZ_0 - D}{A + B/Z_0 + CZ_0 + D} \tag{6.38}$$

$$T_o = \frac{2}{A + B/Z_0 + CZ_0 + D} \tag{6.39}$$

It will be shown later that if $b_1 = b_4 = 0$, then $\Gamma_o = \Gamma_e = 0$. The latter is the even-mode reflection coefficient that is yet to be found. Since $A = D$ from Eqs. (6.34) and (6.37), then for Γ_o to be zero:

$$\frac{1}{Z_0^2} = \frac{C}{B} \tag{6.40}$$

The impedances in Fig. 6.19 are really reactances, which are $Z_L = jX_L$ and $Z_C = -jX_C$. Filling in these values for C/B gives

$$\left(\frac{X_L}{Z_0}\right)^2 = \frac{X_L}{X_C}\left(2 - \frac{X_L}{X_C}\right) \tag{6.41}$$

Equation (6.41) was obtained by requiring a match at the input port so that $\Gamma_o = 0$. Since $A = D$ and $B/Z_0 = CZ_0$, the transmission coefficient is

$$T_o = \frac{1}{A + B/Z_0} \tag{6.42}$$

The denominator is complex, so after multiplying numerator and denominator by its complex conjugate and substituting the requirement for match, Eq. (6.41), the resulting real denominator can be shown to be equal to 1 and

$$T_o = \left(1 - 2\frac{X_L^2}{Z_0^2}\right) - j2\frac{X_L}{Z_0}\left(1 - \frac{X_L}{X_C}\right) \tag{6.43}$$

This process for finding the odd-mode transmission coefficient for Fig. 6.19 must now be repeated for the even-mode circuit in Fig. 6.20. The even-mode

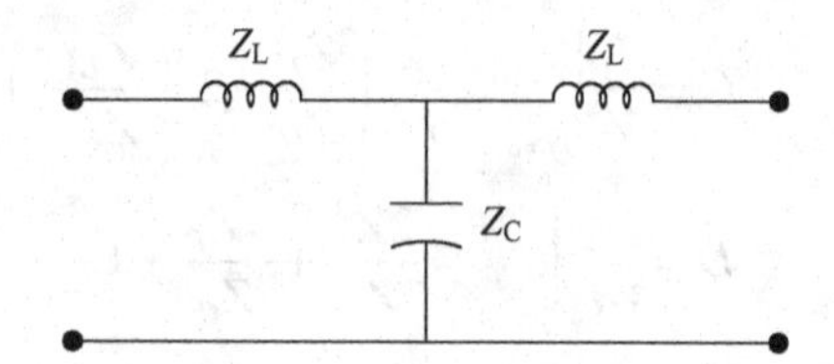

FIGURE 6.20 Even-mode two-port half-circuit.

circuit is found by putting an open circuit at the line of symmetry rather than a short circuit to ground. The *ABCD* parameters are found again by circuit analysis:

$$A = \left.\frac{V_1}{V_2}\right|_{I_2=0} = \frac{Z_L}{Z_C} + 1 \tag{6.44}$$

$$B = \left.\frac{V_1}{-I_2}\right|_{V_2=0} = Z_L\left(\frac{Z_L}{Z_C} + 2\right) \tag{6.45}$$

$$C = \left.\frac{I_1}{V_2}\right|_{I_2=0} = \frac{1}{Z_C} \tag{6.46}$$

$$D = \left.\frac{I_1}{-I_2}\right|_{V_2=0} = \frac{Z_L}{Z_C} + 1 \tag{6.47}$$

The requirement for match at port 1 and isolation at port 4 requires $B/Z_0 = CZ_0$ just as required for the odd mode. For the even-mode circuit:

$$\frac{Z_0^2}{X_C^2} = \frac{X_L}{X_C}\left(2 - \frac{X_L}{X_C}\right) \tag{6.48}$$

Then

$$T_e = \frac{1}{A + CZ_0} \tag{6.49}$$

$$T_e = 1 - \frac{X_L}{X_C} - j\frac{Z_0}{X_C} \tag{6.50}$$

Equation (6.50) for T_e is found by a similar process used to find T_o. When the coupler in Fig. 6.17 is excited as shown, the four outputs are

$$b_1 = \tfrac{1}{2}(\Gamma_e + \Gamma_o) = 0 \quad \text{matched} \tag{6.51}$$

$$b_2 = \tfrac{1}{2}(T_e + T_o) \tag{6.52}$$

$$b_3 = \tfrac{1}{2}(T_e - T_o) \tag{6.53}$$

$$b_4 = \tfrac{1}{2}(\Gamma_e - \Gamma_o) = 0 \quad \text{isolation} \tag{6.54}$$

The only way for b_1 and b_4 to be zero is for $\Gamma_o = \Gamma_e = 0$. The $\Gamma_o = 0$ requirement gave Eq. (6.41) and the $\Gamma_e = 0$ requirement gave Eq. (6.48). The variables X_L and X_C can be normalized with respect to Z_0 so that $\bar{X}_L = X_L/Z_0$ and $\bar{X}_C = X_C/Z_0$, From Eq. (6.41),

$$\bar{X}_L = \frac{2\bar{X}_C}{\bar{X}_C^2 + 1} \quad \text{odd mode} \tag{6.55}$$

and from Eq. (6.48)

$$\bar{X}_C = \frac{\bar{X}_L^2 + 1}{2\bar{X}_L} \quad \text{even mode} \tag{6.56}$$

Simultaneous solution gives

$$0 = \left(\bar{X}_L^2 - 1\right)\left(\bar{X}_L^2 + 3\right) \tag{6.57}$$

The only physically meaningful root is $\bar{X}_L = 1$, which implies $\bar{X}_C = 1$. From Eqs. (6.52) and (6.53),

$$b_{2,3} = \frac{1}{2}\left\{\left[1 - \frac{X_L}{X_C} - j\frac{Z_0}{X_C}\right] \pm \left[\left(1 - \frac{2X_L^2}{Z_0^2}\right) - j\frac{2X_L}{Z_0}\left(1 - \frac{X_L}{X_C}\right)\right]\right\} \tag{6.58}$$

This is evaluated for the known values of X_L and X_C giving:

$$b_2 = \frac{\sqrt{2}}{2}\angle -135° \tag{6.59}$$

$$b_3 = \frac{\sqrt{2}}{2}\angle -45° \tag{6.60}$$

Thus, the outputs are each 3 dB down from the input and are in phase quadrature.

This somewhat tedious process has yielded some fruit. It has shown how a compact 90° coupler might be designed for radio frequencies too low for quarter-wavelength transmission lines. It has also demonstrated how symmetry may be used for analysis of four-port networks.

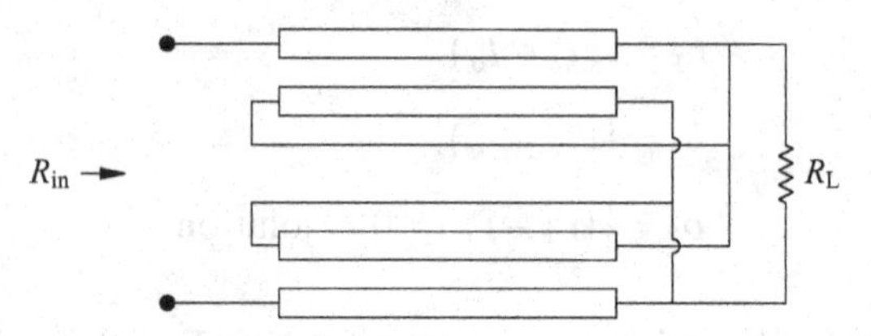

FIGURE 6.21 Transmission line transformer for Problem 6.1.

PROBLEMS

6.1. Indicate the direction of the currents in the transmission line transformer shown in Fig. 6.21. Determine the value of R_{in} in terms of R_L.

6.2. Design a transmission line transformer that matches a 250-Ω generator impedance to a 10-Ω load impedance. What characteristic impedance would you use for the transmission lines? Verify that your design gives the desired result.

6.3. For the 4:1 transformer shown in Fig. 6.2, find the output power, $P_o = |I_2|^2 R_L/2$, where the frequency dependence of the transmission lines is used. You will have three equations in the three unknowns I_1, I_2, and V_2. The final answer is similar to Eq. (6.17).

6.4. Design a transmission line transformer that matches a 200-Ω load to a 50-Ω source impedance. The transmission lines are to be 4 cm long, but the transmission line characteristic impedance can be chosen to give an acceptable match by not deviating from 50 Ω by more than 25 Ω to at least 2.5 GHz. Using SPICE, plot the return loss at the input side as a function of frequency. What is the return loss at 1 GHz?

6.5. Repeat Problem 6.4 for a transmission line transformer that matches 800 Ω to 50 Ω. The SPICE analysis should again show the return loss versus frequency. For this circuit, what is the return loss at 1 GHz?

6.6. Synthesize a transmission line transformer using the technique described in Section 6.3 to give a resistance ratio of $R_G : R_L = 25:16$. You may use the TLT symbol given in Fig. 6.6 as long as each symbol is defined by a known two-conductor transmission line.

6.7. Derive Eq. (6.50).

REFERENCES

1. P. L. D. Abrie, *The Design of Impedance-Matching Networks for Radio-Frequency and Microwave Amplifiers*, Dedham, MA: Artech, 1985.

2. H. L. Krauss, C. W. Bostian, and F. H. Raab, *Solid State Radio Engineering*, New York: Wiley, pp. 371–382, 1980.
3. E. Rotholz, "Transmission Line Transformers," *IEEE Trans. Microwave Theory Tech.*, **MTT-29**, pp. 327–331, April 1981.
4. G. F. Avitable, A. Cidronali, C. Salvador, and M. Speciale, "A Compact MMIC 90° Coupler for ISM Applications," *1997 MTT-S Digest*, Piscataway, NJ: IEEE Press, pp. 281–284, 1997.
5. R. W. Vogel, "Analysis and Design of Lumped- and Lumped-Distributed-Element Directional Couplers for MIC and MMIC Applications," *IEEE Trans. Microwave Theory Tech.*, **40**, pp. 253–262., 1992.
6. P. Vizmuller, *RF Design Guide: Systems, Circuits, and Equations*, Boston: Artech, p. 100, 1995.
7. J. Reed and G. J. Wheeler, "A Method of Analysis of Symmetrical Four-Port Networks," *IRE Trans. Microwave Theory Tech.*, **MTT-4**, pp. 246–252, 1956.

CHAPTER SEVEN

Noise in RF Amplifiers

7.1 SOURCES OF NOISE

The dynamic range of a communication transmitter or receiver circuit is usually limited at the high-power point by nonlinearities and at the low-power point by noise. Noise is the random fluctuation of electrical power that interferes with the desired signal. There can be interference with the desired signal by other unwanted deterministic signals, but at this point only the interference caused by random fluctuations will be considered. There are a variety of physical mechanisms that account for noise, but probably the most common source is thermal (also referred to as *Johnson noise* or *Nyquist noise*). This can be illustrated by simply examining the voltage across an open-circuited resistor (Fig. 7.1). The resulting voltage is not zero! The average voltage is zero but not the instantaneous voltage. At any temperature above absolute 0 K the Brownian motion of the electrons will produce random instantaneous currents. These currents will produce random instantaneous voltages, and this leads to noise power.

Noise arising in electron tubes, semiconductor diodes, bipolar transistors, or field-effect transistors come from a variety of mechanisms. For example, for tubes, these include random times of emission of electrons from a cathode (called *shot noise*), random electron velocities in the vacuum, nonuniform emission energy over the surface of the cathode, and secondary emission from the anode. Similarly for diodes, a random emission of electrons and holes produces noise. In a bipolar transistor, in addition to the diode noise there is partition noise. This represents the fluctuation in the path that charge carriers take through the base to the collector after leaving the emitter. There is in addition $1/f$ or *flicker* noise (where f is frequency), which is probably caused

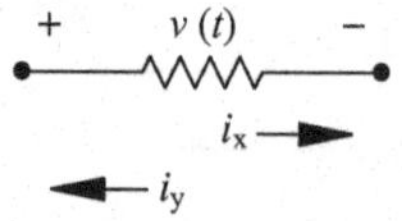

+ $v(t)$ −

i_x

i_y

FIGURE 7.1 Voltage across open-circuit resistor.

by surface recombination of base minority carriers at the base–emitter junction [1]. Clearly, as the frequency approaches dc, the flicker noise increases dramatically. As a consequence, intermediate-frequency amplifier stages in transceivers are designed to operate well above the frequency where $1/f$ noise is a significant contributor to the total noise. Typically, the $1/f$ noise is significant in the frequency ranges from 100 Hz to 10 kHz. In a field-effect transistor, there is thermal noise arising from channel resistance, $1/f$ noise, and a coupling of the channel noise back to the gate where it is, of course, amplified by the transistor gain. Noise also arises from reverse breakdown in the avalanching of electrons in such devices as Zener diodes and IMPATT diodes. At radio frequencies, the two most common noise sources are the thermal noise and the shot noise.

7.2 THERMAL NOISE

The random fluctuation of electrons in a resistance would be expected to rise as the temperature increases since the electron velocities and the number of collisions per second increases. The mean-square noise voltage is expressed as an autocorrelation of the instantaneous voltage over a time period T:

$$\langle v^2 \rangle = \lim_{T\to\infty} \frac{1}{2T} \int_{-T}^{T} v^2(t)\, dt \tag{7.1}$$

The expression for the thermal noise voltage has been derived in a variety of ways. Harry Nyquist first solved the problem based on a transmission line model. Other approaches included using a lumped-element circuit, the random motion of electrons in a metal conductor, or the radiation from a black body. These are all basically thermodynamic models and each method results in the same expression. The black-body method is based on quantum mechanics and therefore provides a solution for noise sources at both cryogenic and room temperatures.

7.2.1 Black-Body Radiation

Classical mechanics is based on the continuity of energy states. When this theory was applied to calculation of the black-body radiation, it was found that the radiation increased without limit. This so-called ultraviolet catastrophe was clearly not physical. However, Planck was able to correct the situation by postulating that energy states are not continuous but are quantized in discrete states. These energy values are obtained by solving the Schrödinger equation for the harmonic oscillator. The actual derivation is found in most introductory texts on quantum mechanics [2]:

$$\varepsilon = \left(n + \frac{1}{2}\right)hf \qquad n = 0, 1, 2 \ldots \tag{7.2}$$

In this equation $h = 6.547 \times 10^{-34}$ J·s is Planck's constant. If energy were continuous, then the average energy could be obtained from the Boltzmann probability distribution function, $P(\varepsilon)$, by the following integral:

$$\bar{\varepsilon} = \frac{\int_0^\infty \varepsilon P(\varepsilon)\, d\varepsilon}{\int_0^\infty P(\varepsilon)\, d\varepsilon} \tag{7.3}$$

where

$$P(\varepsilon) = C \exp(-\beta\varepsilon) \tag{7.4}$$

$$\beta = \frac{1}{kT}$$

and

$$C = \frac{1}{\sum e^{-\beta\varepsilon}}$$

The value, $k = 1.380 \times 10^{-23}$ J/K, is the Boltzmann constant and is essentially the proportionality constant between energy measured in terms of joules and energy measured in terms of absolute temperature. Planck replaced the continuous integrals in Eq. (7.3) with summations of the discrete energy levels [3]:

$$\bar{\varepsilon} = \frac{\sum_{n=0}^{\infty} \varepsilon P(\varepsilon)}{\sum_{n=0}^{\infty} P(\varepsilon)} \tag{7.5}$$

$$= \frac{\sum_{n=0}^{\infty} (n + \tfrac{1}{2}) hf e^{-\beta(n+1/2)hf}}{\sum_{n=0}^{\infty} e^{-\beta(n+1/2)hf}} \tag{7.6}$$

It may be easily verified by differentiation that

$$\frac{d}{d\beta}\ln\sum e^{-\beta(n+1/2)hf}=-\bar{\varepsilon} \tag{7.7}$$

The argument of the logarithm can be evaluated by recognizing it as an infinite geometric series:

$$e^{-\beta hf/2}\sum_{n=0}^{\infty}e^{-n\beta hf}=\frac{e^{-\beta hf/2}}{1-e^{-\beta hf}} \tag{7.8}$$

If Eq. (7.8) is substituted back into Eq. (7.7), the average energy can be found:

$$\begin{aligned}\bar{\varepsilon}&=\frac{d}{d\beta}\left[\ln\left(1-e^{-\beta hf}\right)-\ln e^{-\beta hf/2}\right]\\&=\frac{hfe^{-\beta hf}}{1-e^{-\beta hf}}+\frac{hf}{2}\end{aligned} \tag{7.9}$$

or

$$\bar{\varepsilon}=\frac{hf}{e^{hf/kT}-1}+\frac{hf}{2} \tag{7.10}$$

This will be used as the starting point for finding the noise power.

7.2.2 Nyquist Formula

The thermal noise power in a given bandwidth, Δf, is obtained directly from Eq. (7.10):

$$N_{\mathrm{T}}=\frac{hf\,\Delta f}{e^{hf/kT}-1}+\frac{hf\,\Delta f}{2} \tag{7.11}$$

At room temperature the second term, $hf\,\Delta f/2$, plays no role, but it may be essential in finding the minimum noise figure for cryogenically cooled devices [4]. An approximation for the noise power can be found by expanding Eq. (7.11) into a Taylor series:

$$N_{\mathrm{T}}\approx hf\,\Delta f\left(1+\frac{hf}{kT}-1\right)^{-1}+\frac{hf\,\Delta f}{2}=kT\,\Delta f\left(1+\frac{hf}{2kT}\right) \tag{7.12}$$

At room temperature, $hf/kT \ll 1$, so that this reduces to the usual practical formula for noise power as given by Nyquist [5]:

$$N_T = kT\,\Delta f \tag{7.13}$$

If this is the available power, the corresponding mean-squared voltage is obtained by multiplying this by four times the resistance, R:

$$\begin{aligned}\langle v^2\rangle &= 4RN_T \\ &= 4kTR\,\Delta f\end{aligned} \tag{7.14}$$

The mean-squared noise current is

$$\langle i^2\rangle = 4kTG\,\Delta f \tag{7.15}$$

where G is the associated conductance.

7.3 SHOT NOISE

Shot noise arises from random variations of a direct current, I_0, and is especially associated with current-carrying active devices. Shot noise is most apparent in a current source with zero shunt source admittance. For the purpose of illustration, consider a current source feeding a parallel RLC circuit (Fig. 7.2). The inductor provides a dc path and is open to ac variations of the current. Hence, the resulting noise voltage appears across the resistor (which is presumed free of any thermal noise). If an instrument could measure the current produced by randomly arriving electrons, the instrument would record a series of current impulses for each electron. If n is the average number of electrons emitted by the source in a given time interval Δt, then the dc current is

$$I_0 = \frac{qn}{\Delta t} \tag{7.16}$$

where q is the charge of an electron. Each current pulse provides an energy pulse to the capacitor with the value of

$$\varepsilon = \frac{q^2}{2C} \tag{7.17}$$

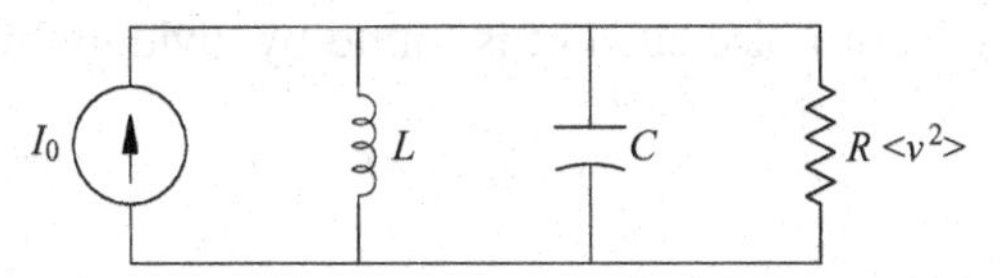

FIGURE 7.2 Equivalent circuit for shot noise and for certain thermal noise calculations.

The average shot noise power delivered to the load is then

$$N_S = \frac{n\varepsilon}{\Delta t}$$

which in the light of Eqs. (7.16) and (7.17) becomes

$$N_S = \frac{nq^2}{2C\,\Delta t}$$

$$= \frac{qI_0}{2C} \tag{7.18}$$

The equipartition theorem as found in thermodynamics books states that the average energy of a system of uniform temperature is equally divided among the degrees of freedom of the system. If there are N degrees of freedom, then

$$\bar{\varepsilon} = \frac{N}{2}kt \tag{7.19}$$

A system with N degrees of freedom can be described uniquely by N variables. The circuit in Fig. 7.2 has two energy storage elements, each containing an average energy of $kT/2$. For the capacitor this average energy is

$$\bar{\varepsilon} = \tfrac{1}{2}C\langle v^2\rangle = \tfrac{1}{2}kT \tag{7.20}$$

But, it was found that the Nyquist noise formula predicted that $\langle v^2\rangle = 4kTR\,\Delta f$. Consequently,

$$C = \frac{1}{4R\,\Delta f} \tag{7.21}$$

Using Eq. (7.21) to replace the value of the capacitance in Eq. (7.18) gives the desired formula for the shot noise power:

$$N_S = 2qRI_0\,\Delta f \tag{7.22}$$

The corresponding shot noise current is found by dividing by R

$$\langle i^2\rangle = 2qI_0\,\Delta f \tag{7.23}$$

The shot noise current is directly proportional to the direct current as has been verified experimentally.

7.4 NOISE CIRCUIT ANALYSIS

When a circuit contains several resistors, the total noise power can be calculated by suitable combination of the resistors. Two resistors in series each produce a mean-squared voltage, $\langle v^2 \rangle$. Since the individual noise voltage sources are uncorrelated, the total $\langle v^2 \rangle$ is the sum of the $\langle v^2 \rangle$ of each of the two resistors. Similarly, two conductances in parallel each produce a mean-squared noise current, $\langle i^2 \rangle$, that may be added when the two conductances are combined since the noise currents are uncorrelated. It should be emphasized that two noise voltages $\langle v \rangle$ cannot be added together; only the mean-square values can be added. The use of an arrow in the symbol for a noise current source is used to emphasize that this is a current source. The use of + and – signs in the symbol for a noise voltage source are used to emphasize that this is a voltage source. They do not imply anything about the phase of the noise sources. When both series and parallel resistors are present as shown in Fig. 7.3, then Thévenin's theorem provides an equivalent circuit and associated noise voltage. In Fig. 7.3 the output resistance is $(R_1 + R_2) \parallel R_3$, and the corresponding noise voltage delivered to the output is

$$\langle v^2 \rangle = 4kT(R_1 + R_2) \| R_3 \tag{7.24}$$

When there is a reactive element in the circuit such as that shown in the simple *RLC* circuit in Fig. 7.4, the output noise voltage would be attenuated by the magnitude of the total admittance. Typically, the the measured noise frequency, f, is approximately a sinusoid, which occurs when $f >> \Delta f$. Then

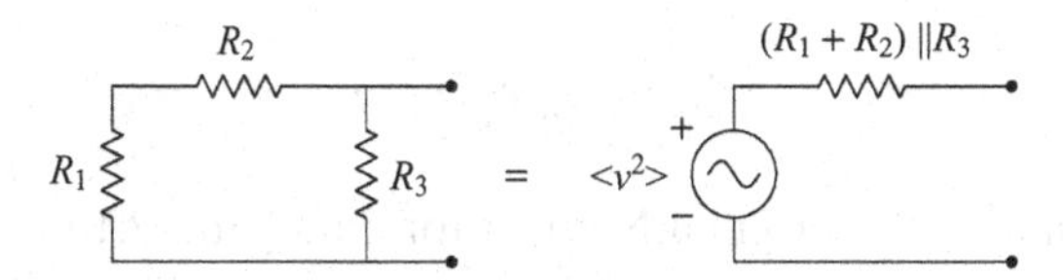

FIGURE 7.3 Noise voltage from series and parallel resistors.

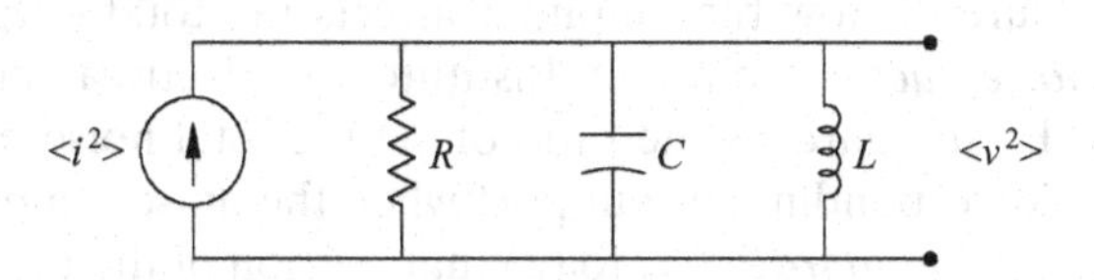

FIGURE 7.4 Noise voltage from *RLC* circuit.

$$\langle v^2 \rangle = \frac{\langle i^2 \rangle}{|G + j(\omega C - 1/\omega L)|^2}$$
$$= \frac{4kT\,\Delta f G}{|Y|^2} \tag{7.25}$$

If the output resistance varies appreciably over the range of the noise bandwidth Δf, then the individual noise "sinusoids" must be summed over the bandwidth, resulting in the following integral:

$$\langle v^2 \rangle = 4kT \int_{\Delta f} \frac{G}{|Y|^2} df \tag{7.26}$$

As a simple example, consider the noise generated from a shunt *RC* circuit that would result from removing the inductance in Fig. 7.4:

$$\langle v^2 \rangle = \int_0^\infty \frac{4kTG\,df}{G^2 + (\omega C)^2} \tag{7.27}$$

$$= \frac{4kTG}{2\pi G^2}\left(\frac{G}{C}\right)\int_0^\infty \frac{d(\omega C/G)}{1 + (\omega C/G)^2}$$

$$= \frac{2kT}{\pi C}\left(\frac{\pi}{2}\right) = \frac{kT}{C} \tag{7.28}$$

This expression does not say that the capacitor is the source of the noise voltage. Indeed experiments have shown that changing the temperature of the resistor is what changes the output noise. When the 3-dB frequency point of the circuit output impedance [$f_{3\text{dB}} = 1/(2\pi RC)$] is considered, the noise voltage in Eq. (7.28) becomes

$$\langle v^2 \rangle = 2\pi f_{3\text{dB}} kTR$$

This looks similar to the original Nyquist formula [Eq. (7.14)] in its form.

7.5 AMPLIFIER NOISE CHARACTERIZATION

One important quality factor of an amplifier is a measure of how much noise it adds to the signal while it amplifies it. The "actual noise factor," F, is a convenient measure of how the amplifier affects the total output noise. The definition of *noise factor* from the Institute of Electrical and Electronic Engineers (IEEE) standards is the ratio of (1) the total noise power per unit bandwidth at a corresponding output port when the noise temperature of the input termination is *standard* 290 K to (2) that portion of the total noise power engendered at the input frequency by the input termination [6]. The standard

$T_0 = 290\,\text{K}$ noise temperature approximates the actual noise temperature of most input terminations:

$$F = \frac{\text{actual noise output power at } T_0}{\text{available noise input power}}\left(\frac{1}{G_T}\right)$$
$$= \frac{N_{T,\text{out}}}{kT_0 G_T\,\Delta f} \tag{7.29}$$

In this expression G_T is the transducer power gain. Noise factor is a measure of the total output noise after it leaves the amplifier divided by the input noise power entering the amplifier and amplified by an ideal noiseless gain, G_T. In an analog amplifier, an amplifier can only add noise so F must always be greater than 1. The noise factor can also be expressed in terms of the signal-to-noise ratio at the input to that at the output. If P represents the input signal power, then

$$F = \frac{P/kT_0\,\Delta f}{G_T P/N_{T,\text{out}}}$$
$$= \frac{S_{\text{in}}/N_{T,\text{in}}}{S_{\text{out}}/N_{T,\text{out}}} \tag{7.30}$$

The signal-to-noise ratio will always be degraded as the signal goes through the amplifier. The expression Eq. (7.30) is strictly true only if the input temperature is 290 K. This is called the spot noise factor.

The portion of the total thermal noise output power contributed by the amplifier itself is

$$N_a = N_{T,\text{out}} - kT_0 G_T\,\Delta f \tag{7.31}$$
$$= (F-1)kT_0 G_T\,\Delta f \tag{7.32}$$

The factor $(F - 1)$ is used in two alternative measures of noise. One of these is noise temperature, which is particularly useful when dealing with very low noise amplifiers where the decibel scale typically used in describing noise figure becomes too compressed to give insight. In this case, the equivalent noise temperature is defined as

$$T_e = T_0(F-1) \tag{7.33}$$

This is the temperature of the source resistance that when connected to the noise-free two-port circuit will give the same output noise as the original noisy circuit.

Another useful parameter for the description of noise is the noise measure [7]:

$$M = \frac{F-1}{1-(1/G)} \tag{7.34}$$

This is particularly useful for optimizing a receiver in which, for example, a trade-off has to be made between a low-gain low-noise amplifier and a high-gain high-noise amplifier.

7.6 NOISE MEASUREMENT

Measurement of noise figure (noise factor measured in decibels) can be accomplished by using a power meter and determining the circuit bandwidth and gain. However, it is inconvenient to determine gain and bandwidth each time a noise measurement is to be taken. The *Y* factor method for determining noise factor is an approach where these two quantities need not be determined explicitly. Actual noise measurements are done over a range of frequencies. The average noise factor over a given bandwidth is [6]

$$\bar{F} = \frac{\int F(f) G_T(f)\, df}{\int G_T(f)\, df} \tag{7.35}$$

This represents a more realistic expression for an actual noise measurement than the spot noise factor such as in Eq. (7.30).

An equivalent noise bandwidth Δf_0 can be defined in terms of the maximum gain over the band as

$$\int G_T(f)\, df = G_0\, \Delta f_0 \tag{7.36}$$

so that

$$\bar{F} = \frac{N_{T,\text{out}}}{kT_0 G_0\, \Delta f_0} \tag{7.37}$$

A measurement system that can be used to measure the noise factor of an amplifier is shown in Fig. 7.5. This excess noise source in this circuit is gated on and off to produce two values of noise measured at the output power detector, N_1 and N_2.

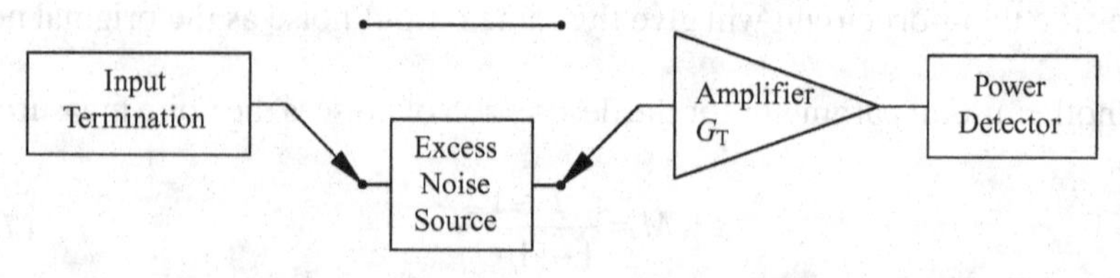

FIGURE 7.5 Noise measurement using *Y* factor method.

N_{ex} = calibrated excess noise source at $T_2 - T_0$
$N_1 = N_{T,out}$ when excess noise source is off
$N_2 = N_{T,out}$ when excess noise source is on
N_{in} = noise from input termination
N_a = noise added by the amplifier itself

The Y factor as the ratio of N_2 to N_1 is easily obtained:

$$\begin{aligned} Y &= \frac{N_2}{N_1} = \frac{G_0 N_{in} + G_0 N_{ex} + N_a}{G_0 N_{in} + N_a} \\ &= \frac{G_0 k T_0 \, \Delta f_0 + G_0 k (T_2 - T_0) \, \Delta f_0 + (\bar{F} - 1) k T_0 G_0 \, \Delta f_0}{G_0 k T_0 \, \Delta f_0 + (\bar{F} - 1) k T_0 G_0 \, \Delta f_0} \\ &= \frac{T_2 - T_0 + \bar{F} T_0}{\bar{F} T_0} \end{aligned} \tag{7.38}$$

When solved for $\bar{F}$

$$\bar{F} = \frac{T_2 - T_0}{T_0 (Y - 1)} \tag{7.39}$$

Since a calibrated noise source is used, $(T_2 - T_0)/T_0$ is known. Also Y is known from the measurement. The amplifier noise factor is then obtained.

7.7 NOISY TWO-PORT CIRCUITS

The noise delivered to the output of a two-port circuit depends on the two-port circuit itself and the impedance of the input excitation source. The noise factor for a two-port circuit is given by the following:

$$F = F_{min} + \frac{R_m}{G_G} \left[(G_G - G_{opt})^2 + (B_G - B_{opt})^2 \right] \tag{7.40}$$

where F_{min} = minimum noise factor
R_n = equivalent noise resistance (often device data are given in terms of a normalized resistance, $r_n = R_n/50$)
$Y_G = G_G + jB_G$ excitation source admittance
$Y_{opt} = G_{opt} + jB_{opt}$ optimum source admittance where the minimum noise factor occurs

While a designer can choose Y_G to minimize the noise factor, such a choice will usually reduce the gain somewhat. Sometimes the noise factor is expressed in terms of the reflection coefficient at the input:

$$\Gamma_G = \frac{Y_0 - Y_G}{Y_0 + Y_G} = \frac{Z_G - Z_0}{Z_G + Z_0} \tag{7.41}$$

where Y_0 and Z_0 are the characteristic admittance and impedance, respectively. Then the noise factor is

$$F = F_{min} + 4r_n \frac{|\Gamma_G - \Gamma_{opt}|^2}{\left(1-|\Gamma_G|^2\right)|1+\Gamma_{opt}|^2} \tag{7.42}$$

The noise factor expression in Eq. (7.40) and its equivalent (7.42) are the basic expressions used to optimize transistor amplifiers for noise figure. The derivation of Eq. (7.40) is the subject of the following section. Readers not wishing to pursue these details at this point may proceed to Section 7.9 without loss of continuity.

7.8 TWO-PORT NOISE FACTOR DERIVATION

The work described here is based on the Institute of Radio Engineers (IRE) standards published between 1956 and 1960 [8, 9]. A noisy resistor can be modeled as a noiseless resistor in series with a voltage noise source. In similar fashion a two-port circuit can be represented as a noiseless two-port and two noise sources. These two noise sources are represented in Fig. 7.6*a* as a voltage v_n and a current i_n. The two-port circuit can be described in terms of its *ABCD* parameters and internal noise sources as

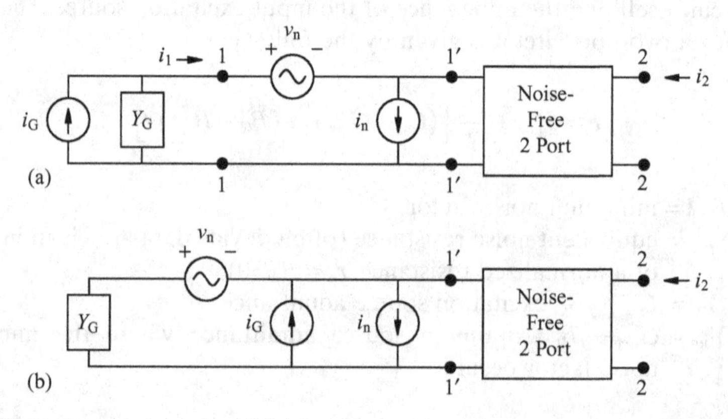

FIGURE 7.6 Equivalent circuit (*a*) for two-port noise calculation and (*b*) equivalent Thévenin circuit.

$$\begin{aligned} v_1 &= Av_2 + Bi_2 + v_n \\ i_1 &= Cv_2 + Di_2 + i_n \end{aligned} \tag{7.43}$$

or as shown in Fig. 7.6 as a noiseless circuit and the noise sources referred to the input side. If the input termination, Y_G, produces a noise current, i_G, then the circuit is completed. The polarity markings on the symbols for the noise sources merely point out the distinction between voltage and current sources. Being noise sources, the polarities are actually random. The Thévenin equivalent circuit in Fig. 7.6*b* shows that the short-circuit current at the $1' - 1'$ port is

$$\langle i_{sc}^2 \rangle = \langle i_G^2 \rangle + \langle |i_n + Y_G v_n|^2 \rangle \tag{7.44}$$

$$= \langle i_G^2 \rangle + \langle i_n^2 \rangle + |Y_G|^2 \langle v_n^2 \rangle + Y_G^* \langle v_n^* i_n \rangle + Y_G \langle i_n^* v_n \rangle \tag{7.45}$$

The total output noise power is proportional to $\langle i_{sc}^2 \rangle$, and the noise caused by the input termination source alone is $\langle i_G^2 \rangle$. The noise-free part between $1' - 1'$ and $2 - 2$ is noise free; that is, it adds no additional noise to the output. All the noise sources are referred to the input side so that the noise factor is

$$F = \frac{\langle i_{sc}^2 \rangle}{\langle i_G^2 \rangle} \tag{7.46}$$

Part of the noise current source, i_n, is correlated and part is uncorrelated with the noise voltage v_n. The uncorrelated current is i_u. The rest of the current is correlated with v_n and is given by $i_n - i_u$. This correlated noise current must be proportional to v_n. The proportionality constant is the correlation admittance given by $Y_c = G_c + jB_c$ and is defined so that

$$i_n = i_u + Y_c v_n \tag{7.47}$$

While this defines Y_c, its explicit value in the end will not be needed. The mean value of the product of the correlated and uncorrelated current is, of course, 0. By definition, the average of the product of the noise voltage, v_n, and the uncorrelated noise current, i_u, must also be 0. Using the complex conjugate of the current (which is a fixed phase shift) will not change this fact:

$$\langle v_n i_u^* \rangle = 0 \tag{7.48}$$

Rearranging Eq. (7.47) gives

$$\frac{i_n - i_u}{Y_c} = v_n \tag{7.49}$$

The product of the noise voltage and the uncorrelated current in Eq. (7.48) can be expressed by substitution of Eq. (7.49) into Eq. (7.48):

$$\left\langle (i_n - i_u)\, i_u^* \right\rangle = 0 \tag{7.50}$$

Because $\left\langle v_n i_u^* \right\rangle = 0$ from Eq. (7.48), the product of the noise voltage and the correlated current can be found using Eq. (7.47):

$$\left\langle v_n i_n^* \right\rangle = \langle v_n (i_u + Y_c v_n)^* \rangle = Y_c^* \langle v_n^2 \rangle \tag{7.51}$$

The noise source values are determined by their corresponding resistances:

$$\langle v_n^2 \rangle = 4kT_0 R_n \,\Delta f \tag{7.52}$$

$$\langle i_u^2 \rangle = 4kT_0 G_u \,\Delta f \tag{7.53}$$

$$\langle i_G^2 \rangle = 4kT_0 G_G \,\Delta f \tag{7.54}$$

The resistance, R_n, is the equivalent noise resistance for $\langle v_n^2 \rangle$, and G_u is the equivalent noise conductance for the uncorrelated part of the noise current, $\langle i_u^2 \rangle$. The total noise current is the sum of the uncorrelated current and the remaining correlated current:

$$\begin{aligned} \langle i_n^2 \rangle &= \langle i_u^2 \rangle + \langle |i_n - i_u|^2 \rangle \\ &= \langle i_u^2 \rangle + |Y_c|^2 \langle v_n^2 \rangle \qquad (7.55) \\ &= 4kT_0 \,\Delta f \left(G_u + R_n |Y_c|^2 \right) \qquad (7.56) \end{aligned}$$

The expression for the short-circuit current in Eq. (7.45) can be modified by Eq. (7.51):

$$\langle i_{sc}^2 \rangle = \langle i_G^2 \rangle + \langle i_n^2 \rangle + |Y_G|^2 \langle v_n^2 \rangle + Y_G^* Y_c \langle v_n^2 \rangle + Y_G Y_c^* \langle v_n^2 \rangle \tag{7.57}$$

Furthermore, $\langle i_n^2 \rangle$ can be replaced by Eq. (7.55).

$$\langle i_{sc}^2 \rangle = \langle i_G^2 \rangle + \langle i_u^2 \rangle + |Y_c|^2 \langle v_n^2 \rangle + |Y_G|^2 \langle v_n^2 \rangle + Y_G^* Y_c \langle v_n^2 \rangle + Y_G Y_c^* \langle v_n^2 \rangle \tag{7.58}$$

The noise factor, given by Eq. (7.46), can now be put in more convenient form:

$$F = 1 + \frac{\langle i_u^2 \rangle + \langle v_n^2 \rangle \left(|Y_c|^2 + |Y_G|^2 + Y_G^* Y_c + Y_G Y_G^* \right)}{\langle i_G^2 \rangle} \tag{7.59}$$

$$= 1 + \frac{4kT_0 G_u \, \Delta f + 4kT_0 R_n \, \Delta f \left(|Y_G| + |Y_c|^2 \right)}{4kT_0 G_G \, \Delta f} \tag{7.60}$$

$$= 1 + \frac{G_u}{G_G} + \frac{R_n}{G_G} \left[(G_G + G_c)^2 + (B_G + B_c)^2 \right] \tag{7.61}$$

The value of F is a function of the input termination admittance, Y_G, and reaches a minimum when the source admittance is optimum. In particular, the optimum susceptance is $B_G = B_{opt} = -B_c$. The value for F_{min} is found by setting the derivative of F with respect to G_G to zero and setting $B_G = -B_c$. This will determine the value for $G_G = G_{opt}$ in terms of G_u, R_u, and G_c:

$$\frac{dF}{dG_G} = 0 = -\frac{G_u}{G_G^2} - \frac{R_n}{G_G^2}(G_G + G_c)^2 + \frac{2R_n}{G_G}(G_G + G_c) \tag{7.62}$$

Solution for G_G yields

$$G_G = G_{opt} = \sqrt{\frac{G_u + R_n G_c^2}{R_n}} \tag{7.63}$$

or

$$G_c^2 = G_{opt}^2 - \frac{G_u}{R_n} \tag{7.64}$$

Substituting this into Eq. (7.61) (with susceptance $B_G + B_c = 0$) provides the minimum noise factor, F_{min}:

$$F_{min} = 1 + \frac{1}{G_{opt}} \left[G_u + R_n \left(G_{opt}^2 + 2G_{opt} \sqrt{G_{opt}^2 - \frac{G_u}{R_n}} + G_{opt}^2 - \frac{G_u}{R_n} \right) \right] \tag{7.65}$$

$$= 1 + 2R_n \left[G_{opt} + \sqrt{G_{opt}^2 - \frac{G_u}{R_n}} \right] \tag{7.66}$$

The correlation conductance, G_c of Eq. (7.64) is substituted into the total noise factor expression of Eq. (7.61) to give

$$F = 1 + \frac{G_u}{G_G} + \frac{R_n}{G_G} \times \left[\left(G_G^2 + 2G_G\sqrt{G_{opt}^2 - \frac{G_u}{R_n}} + G_{opt}^2 - \frac{G_u}{R_n} \right) + (B_G - B_{opt})^2 \right]$$

$$= 1 + \frac{R_n}{G_G} \times \left[\left(G_G^2 + 2G_G\sqrt{G_{opt}^2 - \frac{G_u}{R_n}} + G_{opt}^2 - 2G_G G_{opt} + 2G_G G_{opt} \right) + (B_G - B_{opt})^2 \right]$$

$$= 1 + \frac{R_n}{G_G}\left(2G_G G_{opt} + 2G_G\sqrt{G_{opt}^2 - \frac{G_u}{R_n}} \right) + \frac{R_n}{G_G}\left[(G_G - G_{opt})^2 + (B_G - B_{opt})^2 \right] \tag{7.67}$$

The first two terms are the same as F_{min} in Eq. (7.66), so

$$F = F_{min} + \frac{R_n}{G_G}\left[(G_G - G_{opt})^2 + (B_G - B_{opt})^2 \right] \tag{7.68}$$

7.9 FUKUI NOISE MODEL FOR TRANSISTORS

Fukui found an empirically based model that accurately describes the frequency dependence of the noise for high-frequency field-effect transistors [10]. This model reduces to predicting the four noise parameters, F_{min}, R_n, R_{opt}, and X_{opt} where the latter two parameters are formed from the reciprocal of Y_{opt}. For the circuit shown in Fig. 7.7, the Fukui relationships are as follows:

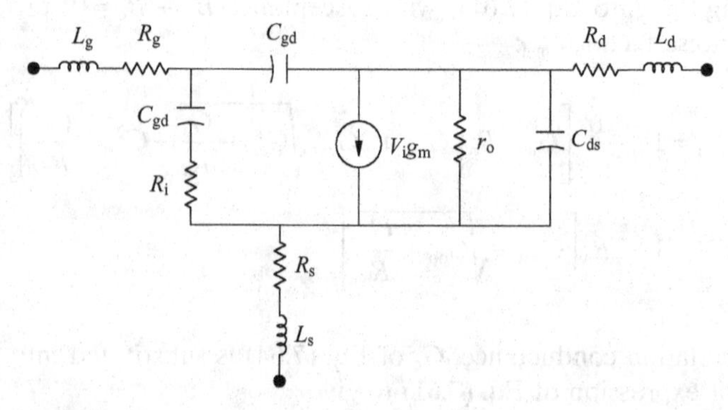

FIGURE 7.7 Equivalent circuit for noise calculation for FET.

$$F_{\min} = 1 + k_1 f C_{gs} \left(\frac{R_g + R_s}{g_m} \right)^{1/2} \quad (7.69)$$

$$R_n = \frac{k_2}{g_m} \quad (7.70)$$

$$R_{opt} = \frac{k_3}{f} \left(\frac{1}{4g_m} + R_s + R_g \right) \quad (7.71)$$

$$X_{opt} = \frac{k_4}{f C_{gs}} \quad (7.72)$$

In these expressions, f is the operating frequency in gigahertz, the capacitance is in picofarads, and the transconductance in siemens. The constants k_1, k_2, k_3, and k_4 are empirically based fitting factors. The expression for R_{opt} in Eq. (7.71) differs from that originally given by Fukui, as modified by Golio [11]. The circuit elements of the equivalent FET model in Fig. 7.7 can be extracted at a particular bias level. The resistance, R_i, is often difficult to obtain, but for purposes of the noise estimation, it may be incorporated with the R_g. The empirically derived fitting factors should be independent of frequency. They are not quite constant, but over a range of 2 to 18 GHz average values for these are shown below [11]:

$$k_1 = 0.0259$$

$$k_2 = 2.966$$

$$k_3 = 14.51$$

$$k_4 = 162.6$$

These values can be used for approximate estimates of noise factor for both metal semiconductor field-effect transistors (MESFETs) as well as high electron mobility transistors (HEMTs).

The transistor itself can be modified to provide either improved noise characteristics or improved power handling capability by adjusting the gate width W. The drain current, I_{ds}, increases with the base width W. Consequently, those equivalent circuit parameters determined by derivatives of I_{ds} will also be proportional to W. Also the capacitance between the gate electrode and the source electrode or between the gate electrode and the drain electrode will be also proportional to W. This is readily seen from the layout of a FET shown in Fig. 7.8. The gate resistance, R_g, scales differently since the gate current flows in the direction of the width. Also, the number of gate fingers, N, will reduce the effective gate resistance. The gate resistance is then proportional to W/N [12]. These relationships may be summarized as follows:

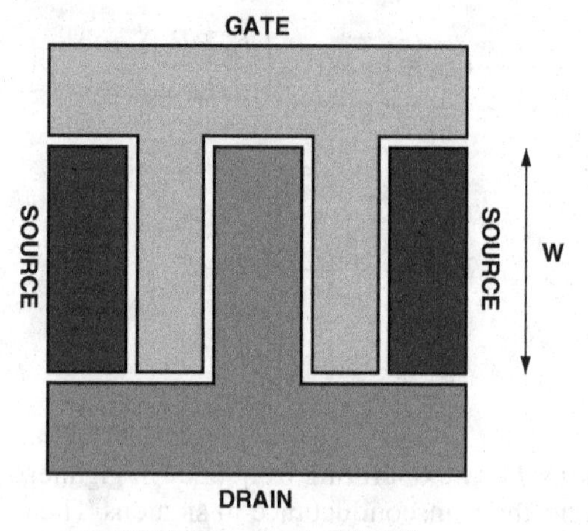

FIGURE 7.8 Typical FET layout.

$$g_m \propto W$$

$$R_{ds} \propto \frac{1}{W}$$

$$C_{gs} \propto W$$

$$C_{gd} \propto W$$

$$R_g \propto \frac{W}{N}$$

These circuit elements can clearly be adjusted by scaling the transistor geometry. This scaling will in turn change the noise characteristics. If a transistor with a given geometry has a known set of noise parameters, then the noise characteristics of a new modified transistor can be predicted. The scaling factors between the new and the old transistor are

$$s_1 = \frac{W'}{W} \tag{7.73}$$

$$s_2 = \frac{W'/N'}{W/N} \tag{7.74}$$

As a result, the new equivalent circuit parameters can be predicted [11]:

$$g'_m = g_m s_1 \tag{7.75}$$

$$R'_s = \frac{R_s}{s_1} \tag{7.76}$$

$$R'_{\rm d} = \frac{R_{\rm d}}{s_1} \tag{7.77}$$

$$C'_{\rm gs} = C_{\rm gs} s_1 \tag{7.78}$$

$$R'_{\rm g} = R_{\rm g} s_2 \tag{7.79}$$

The Fukui equations, (7.69) to (7.72), for the newly scaled equivalent circuit parameters are

$$\begin{aligned} F'_{\min} &= 1 + k_1 f C'_{\rm gs} \left(\frac{R'_{\rm g} + R'_{\rm s}}{g'_{\rm m}} \right)^{1/2} \\ &= 1 + (F_{\min} - 1) \left(\frac{s_1 s_2 R_{\rm g} + R_{\rm s}}{R_{\rm g} + R_{\rm s}} \right)^{1/2} \end{aligned} \tag{7.80}$$

$$R'_{\rm n} = \frac{R_{\rm n}}{s_1} \tag{7.81}$$

$$R'_{\rm opt} = \frac{R_{\rm opt}}{s_1} \left[\frac{1 + 4 g_{\rm m} (R_{\rm s} + R_{\rm g} s_1 s_2)}{1 + 4 g_{\rm m} (R_{\rm s} + R_{\rm g})} \right] \tag{7.82}$$

$$X'_{\rm opt} = \frac{X_{\rm opt}}{s_1} \tag{7.83}$$

Reference should be made to [11] for a much fuller treatment of modeling MESFETs and HEMTs.

A further refinement in the calculation of drain noise current resulting from an N finger gate resistance, each of value $R_{\rm g}$/N, was given in [13]. The total drain noise current is found to be

$$\langle i_{\rm D}^2 \rangle = g_{\rm m}^2 (4kT\,\Delta f) R_{\rm g} \frac{(N+1)(2N-1)}{6N^2} \tag{7.84}$$

and if $N \to \infty$

$$\langle v_{\rm D}^2 \rangle = \frac{\langle i_{\rm D}^2 \rangle}{g_{\rm m}^2} = 4kT\,\Delta f \frac{R_{\rm g}}{3} \tag{7.85}$$

The bipolar transistor has a much different variation of noise with frequency than does the FET type of device. An approximate value for $F_{\min}$ for the bipolar transistor at high frequencies is [14]

$$F_{\min} \approx 1 + h \left(1 + \sqrt{1 + \frac{2}{h}} \right) \tag{7.86}$$

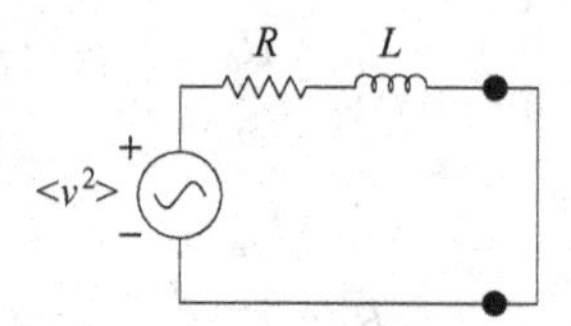

FIGURE 7.9 Noise current developed by series *RL* circuit.

where

$$h \triangleq \frac{qI_c r_b}{kT}\left(\frac{\omega}{\omega_T}\right)^2 \tag{7.87}$$

In this equation, I_c is the dc collector current, r_b is the base resistance, and ω_T is the frequency where the short-circuit current gain is 1. Values for Y_{opt} and R_n are also given in [14], but are rather lengthy. A somewhat more accurate expression is given in [15].

Comparison of Eq. (7.86) with the corresponding expression for FETs, Eq. (7.69), indicates that the bipolar transistor minimum noise factor increases with f^2, while that for the FET it increases only as f. Consequently, designs of low-noise amplifiers at RF and microwave frequencies would tend to favor use of FETs.

PROBLEMS

7.1. Determine the noise power at $T = 290\,K$, $f = 10\,GHz$, and $\Delta f = 1\,Hz$. Determine the noise power at liquid helium temperature (4K). What is the value of the error if the standard Nyquist formula is used?

7.2. What is the noise current from a noise voltage source in a series *RL* circuit shown in Fig. 7.9?

7.3. Derive Eqs. (7.80) to (7.83).

7.4. A MESFET has a base width $W = 350\,\mu m$ and at 3GHz with a given bias is found to have $g_m = 70\,mS$, $R_g = 5\,\Omega$, $R_d = 7\,\Omega$, $R_s = 5\,\Omega$, and $C_{gs} = 0.3\,pF$. What are the four noise parameters F_{min}, R_n, R_{opt}, and X_{opt}? If the base width is changed to $W' = 200\,\mu m$ and the number of base fingers remains unchanged, what are the four noise parameters?

REFERENCES

1. J. L. Plumb and E. R. Chenette, "Flicker Noise in Transistors," *IEEE Trans. Electron Devices*, **ED-10**, pp. 304–308, Sept. 1963.

2. A. Messiah, *Quantum Mechanics*, Amsterdam: North-Holland, Chapter 12, 1964.
3. W. A. Davis, *Microwave Semiconductor Circuit Design*, New York: Van Nostrand, Chapters 8 and 9, 1984.
4. A. E. Siegman, "Zero-Point Energy as the Source of Amplifier Noise," *Proc. IRE*, **49**, pp. 633–634, March 1961.
5. H. Nyquist, "Thermal Agitation of Electronic Charge in Conductors," *Phys. Rev.*, **32**, p. 110, July 1928.
6. H. A. Haus, "IRE Standards on Methods of Measuring Noise in Linear Twoports," *Proc. IRE*, **47**, pp. 66–68, Jan. 1959.
7. K. Kurokawa, "Actual Noise Measure of Linear Amplifiers," *Proc. IRE*, **49**, pp. 1391–1397, March 1961.
8. H. Rohte and W. Danlke, "Theory of Noisy Fourpoles," *Proc. IRE*, **44**, pp. 811–818, June 1956.
9. H. A. Haus, "Representation of Noise in Linear Twoports," *Proc. IRE*, **48**, pp. 69–74, Jan. 1960.
10. H. Fukui, "Design of Microwave GaAs MESFETs for Broad-Band Low-Noise Amplifiers," *IEEE Trans. Electron Devices*, **ED-26**, pp. 1032–1037, July 1979.
11. J. M. Golio, *Microwave MESFETs and HEMTs*, Norwood MA: Artech, Chapter 2, 1991.
12. A. El-Sabban, H. Haddara, and H. F. Ragai, "Validation of RF MOSFET Transistor Layout-Aware Macromodel," *IEEE Int. Conf. Electrical, Electronic Comput. Eng.*, 2004 ICEEC'04, pp. 524–527, Sept. 2004.
13. B. Razavi, R. Yan, and K. F. Lee, "Impact of Distributed Gate Resistance on the Performance of MOS Devices," *IEEE Trans. Circuits Syst.*, **41**, pp. 750–754, Nov. 1994.
14. H. Fukui, "The Noise Performance of Microwave Transistors," *IEEE Trans. Electron Devices*, **ED-13**, pp. 329–341, March 1966.
15. R. J. Hawkins, "Limitations of Nielsen's and Related Noise Equations Applied to Microwave Bipolar Transistors, and a New Expression for the Frequency and Current Dependent Noise Figure," *Solid State Electr.*, **20**, pp. 191–196, March 1977.

CHAPTER EIGHT

Class A Amplifiers

8.1 INTRODUCTION

The class A amplifier is typically used as the first amplification stage of a receiver or a transmitter where minimum noise is desired. This is achieved with a cost of relatively low efficiency. In a receiver the first stage in an amplifier chain handles low power levels, so the low efficiency of the first amplifiers actually wastes little power. Power amplifiers with different class designations are used in later stages. The variety of amplifier classes are described in [1] and will be covered more extensively in Chapter 9. The primary properties of importance to class A amplifier design are gain, bandwidth control, stability, return loss, and noise figure. Noise figure was considered in Chapter 7, but the other topics are described in the present one.

8.2 DEFINITION OF GAIN [2]

In low-frequency circuits, gain is often thought of in terms of voltage or current gain, for example, the ratio of the output voltage across the load to the input applied voltage. At radio frequencies it is difficult to directly measure a voltage, so typically some form of gain is used. But once the notion of power is introduced, there are several definitions of power gain that might be used.

1. *Power Gain* This is the ratio of power dissipated in the load, Z_L, to the power delivered to the input of the amplifier. This definition is independent of the generator impedance, Z_G. Certain amplifiers, especially negative resistance amplifiers, are strongly dependent on Z_G.

Radio Frequency Circuit Design, Second Edition, by W. Alan Davis

2. *Available Gain* This is the ratio of the amplifier output power to the available power from the generator source. This definition depends on Z_G but is independent of Z_L.
3. *Exchangeable Gain* This is the ratio of the output exchangeable power to the input exchangeable power. The exchangeable power of the source is defined as

$$P = \frac{|V|^2}{4\Re\{Z_G\}} \qquad \Re\{Z_G\} \neq 0 \tag{8.1}$$

For negative resistance amplifiers P < 0! Furthermore, this definition is independent of Z_L.
4. *Insertion Gain* This is the ratio of output power to the power that would be dissipated in the load if the amplifier were not present. There is a problem in applying this definition to mixers or parametric upconverters where the input and output frequencies differ.
5. *Transducer Power Gain* This is the ratio of the power delivered to the load to the available power from the source. This definition depends on both Z_G and Z_L. It gives positive gain for negative resistance amplifiers as well. Since the characteristics of real amplifiers change when either the load or generator impedance is changed, it is desirable that the gain definition reflect this characteristic. Thus, the transducer power gain definition is found to be the most useful.

8.3 TRANSDUCER POWER GAIN OF A TWO-PORT NETWORK

The linear two-port circuit in Fig. 8.1 is characterized by its impedance parameters:

$$V_1 = z_{11}I_1 + z_{12}I_2 \tag{8.2}$$

$$V_2 = z_{21}I_1 + z_{22}I_2 \tag{8.3}$$

But the relationship between the port 2 voltage and current is determined by the load impedance as illustrated in Fig. 8.2:

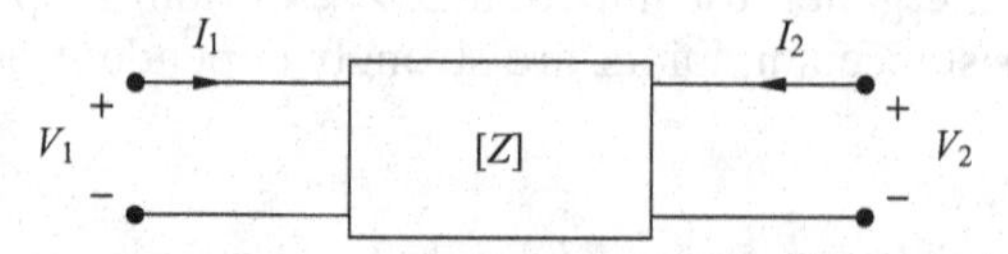

FIGURE 8.1 Two-port circuit expressed in impedance parameters.

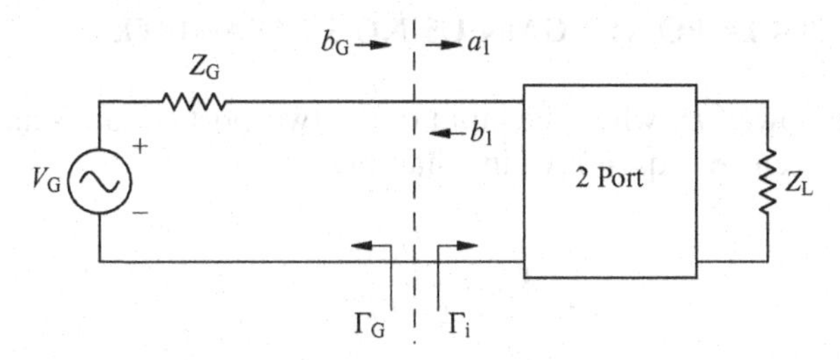

FIGURE 8.2 Equivalent circuit to determine the input available power.

$$V_2 = -I_2 Z_L \tag{8.4}$$

Substitution of this for V_2 in Eq. (8.3) gives the input impedance. This is dependent on both the contents of the two-port circuit itself and also the load:

$$Z_{in} = \frac{V_1}{I_1} = z_{11} - \frac{z_{12} z_{21}}{z_{22} + z_L} \tag{8.5}$$

This will be used to determine the transducer power gain. The power delivered to the load is P_2:

$$P_2 = \tfrac{1}{2} |I_2|^2 \Re\{Z_L\} \tag{8.6}$$

Since the power available from the source at port 1 is

$$P_{1a} = \frac{|V_G|^2}{8\Re\{Z_G\}} \tag{8.7}$$

the transducer power gain can be shown to be

$$G_T = \frac{P_2}{P_{1a}} \tag{8.8}$$

$$= \frac{4\Re\{Z_L\}\Re\{Z_G\}|z_{21}|^2}{|(Z_G + z_{11})(Z_L + z_{22}) - z_{21} z_{12}|^2} \tag{8.9}$$

Similar expressions can be obtained for y, h, or g parameters by simply replacing the corresponding z_{ij} with the desired matrix elements and by replacing the Z_G and Z_L with the appropriate termination. However, for RF and microwave circuits, scattering parameters are the most readily measured quantities. The transducer power gain will be found in terms of the scattering parameters in the following section.

8.4 TRANSDUCER POWER GAIN USING S PARAMETERS

The available power, P_a, when the input of the two-port circuit is matched with $\Gamma_i = \Gamma_G^*$, was given by Eq. (4.170) in Chapter 4:

$$P_{1a} = \frac{\frac{1}{2}|b_G|^2}{1-|\Gamma_G|^2} \tag{8.10}$$

At the output side of the circuit, the power delivered to the load is given by the following:

$$P_L = \tfrac{1}{2}|b_2|^2\left(1-|\Gamma_L|^2\right) \tag{8.11}$$

The transducer gain is simply the ratio of Eq. (8.11) to Eq. (8.10):

$$G_T = \frac{|b_2|^2}{|b_G|^2}\left(1-|\Gamma_L|^2\right)\left(1-|\Gamma_G|^2\right) \tag{8.12}$$

As this stands, b_2 and b_G are not very meaningful. However, this ratio can be expressed entirely in terms of the known S parameters of the two-port circuit. From the description of the S parameters as a matrix corresponding to forward- and backward-traveling waves, the two-port circuit can be represented in terms of a flow graph. Each branch of the flow graph is unidirectional and the combination describes the S matrix completely. The presumption is that the circuit is linear. The problem of finding b_2/b_G can be done using either algebra or some flow graph reduction technique. The classical method developed for linear systems represented as flow graphs uses Mason's nontouching loop rules. The method shown below is easier to remember, but it is more complicated to administer to complex circuits that require a computer analysis. For the relatively simple graph shown in Fig. 8.3 the simpler method works well. This method of flow graph reduction is based on four rules:

1. The cascade of two branches in series can be reduced to one branch with the value equal to the product of the two original branches (Fig. 8.4*a*).

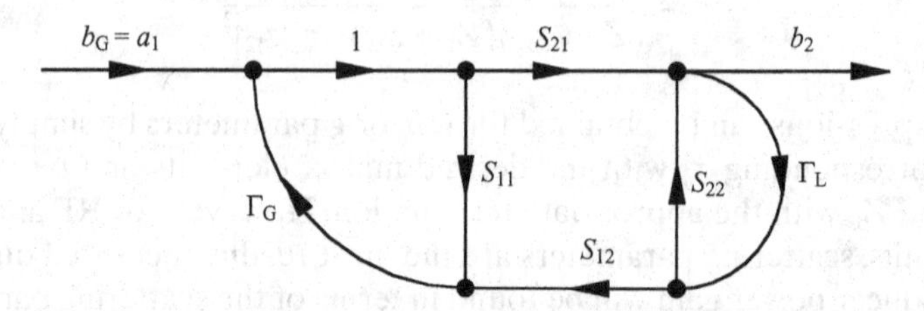

FIGURE 8.3 Flow graph equivalent of two-port circuit in Fig. 8.2.

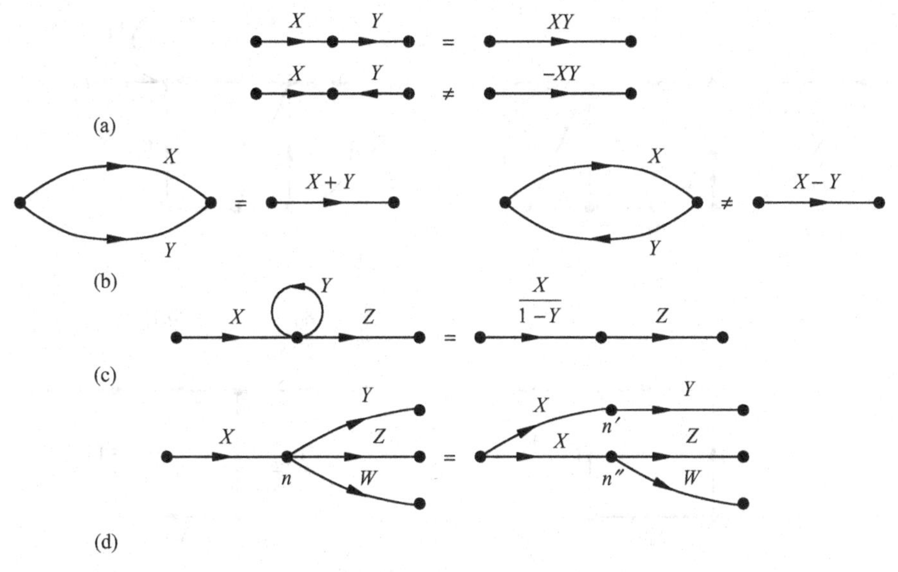

FIGURE 8.4 Flow graph reduction rules for (*a*) two series branches, (*b*) two shunt branches, (*c*) a self-loop, and (*d*) splitting a node.

2. Two parallel branches can be reduced to one branch whose value is the sum of the two original branches (Fig. 8.4*b*).
3. As illustrated in Fig. 8.4*c*, a self-loop with value Y with an incoming branch X can be reduced to a single line of value

$$\frac{X}{1-T} \tag{8.13}$$

4. The transfer function remains unchanged if a node with one input branch and N output branches can be split into two nodes. The input branch goes to each of the new nodes. Similarly, the transfer function remains unchanged if a node with one output branch and N input branches can be split into two nodes. The output branch goes to each of the new nodes (Fig. 8.4*d*).

These rules can be used to finish the calculation of the transducer power gain of expression (8.12) by finding b_2/b_G. The first step in this reduction is the splitting of two nodes shown in Fig 8.5*a* by use of rule 4. This forms a self-loop in the right-hand side of the circuit. The lower left-hand node is also split into two nodes (Fig. 8.5*b*). The incoming branches to the self-loop on the right-hand side are modified by means of rule 3 (Fig. 8.5*c*). In the same figure, another self-loop is made evident on the left-hand side. In this case there are two incoming branches modified by the self-loop. Use of rule 3 produces Fig. 8.5*d*. Splitting the node by means of rule 4 results in Fig. 8.5*e*. The resulting self-loop modifies

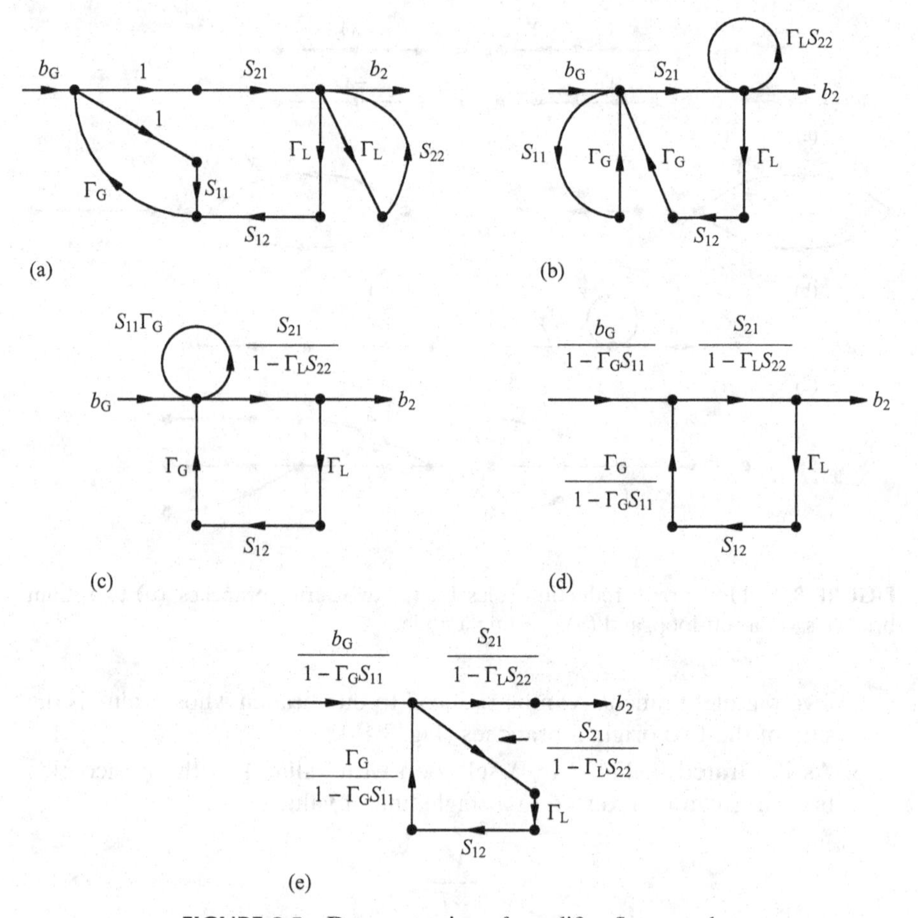

FIGURE 8.5 Demonstration of amplifier flow graph.

the incoming branch on the left-hand side (rule 3). The result is three branches in series (rule 1) so the transfer function can now be written by inspection:

$$b_2 = \frac{\dfrac{b_G}{1-\Gamma_G S_{11}}}{1-\dfrac{S_{21}S_{12}\Gamma_L\Gamma_G}{(1-\Gamma_L S_{22})(1-\Gamma_G S_{11})}}\left(\frac{S_{21}}{1-\Gamma_L S_{22}}\right)$$

$$\frac{b_2}{b_G} = \frac{S_{21}}{(1-\Gamma_L S_{22})(1-\Gamma_G S_{11})-S_{12}S_{21}\Gamma_G\Gamma_L} \tag{8.14}$$

This ratio can be substituted into the transducer power gain expression Eq. (8.12). Thus, the transducer power gain is known in terms of the scattering parameters of the two-port circuit and the terminating reflection coefficients:

$$G_T = \frac{|S_{21}|^2(1-|\Gamma_G|^2)(1-|\Gamma_L|^2)}{|(1-\Gamma_L S_{22})(1-\Gamma_G S_{11}) - S_{12}S_{21}\Gamma_G\Gamma_L|^2} \tag{8.15}$$

This is the full equation for the transducer power gain. Other expressions making use of approximations are strictly speaking a fiction, though this fiction is sometimes used to characterize certain transistors. For example, unilateral power gain is found by setting $S_{12} = 0$. In real transistors S_{12} should be small, but it is never actually 0. The maximum unilateral power gain is found by setting $S_{12} = 0$, $\Gamma_G = S_{11}^*$, and $\Gamma_L = S_{22}^*$:

$$G_{u,max} = \frac{|S_{21}|^2}{(1-|S_{11}|^2)(1-|S_{22}|^2)} \tag{8.16}$$

8.5 SIMULTANEOUS MATCH FOR MAXIMUM POWER GAIN

Maximum gain is obtained when both the input and output ports are simultaneously matched. One way to achieve this is to guess at a Γ_L and calculate Γ_i (Fig. 8.6). The generator impedance then is made to match the complex conjugate of Γi. With this new value of Γ_G, a new value of Γ_o is found. Matching this to Γ_L means the that Γ_L changes. This iterative process continues until both sides of the circuit are simultaneously matched.

A better way is to recognize this as basically a problem with two equations and two unknowns. Simultaneous match forces the following two requirements:

$$\Gamma_i = \Gamma_G^* = S_{11} + \frac{S_{21}S_{12}\Gamma_L}{1-\Gamma_L S_{22}} \tag{8.17}$$

$$\Gamma_o = \Gamma_L^* = S_{22} + \frac{S_{21}S_{12}\Gamma_G}{1-\Gamma_G S_{11}} \tag{8.18}$$

Since both of these equations have to be satisfied simultaneously, finding Γ_G and Γ_L requires solution of two equations with two unknowns. These can be written in terms of the determinate of the S matrix, Δ, as follows:.

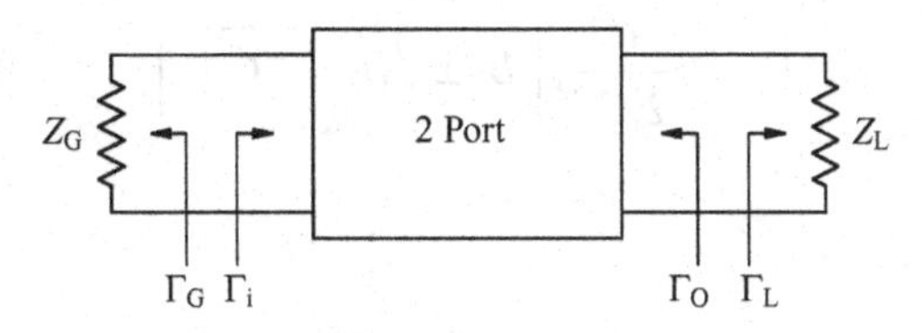

FIGURE 8.6 Definition of reflection coefficients for two-port circuit.

$$\Gamma_G^* = \frac{S_{11} - \Gamma_L S_{11} S_{22} + S_{12} S_{21} \Gamma_L}{1 - \Gamma_L S_{22}} = \frac{S_{11} - \Gamma_L \Delta}{1 - \Gamma_L S_{22}} \tag{8.19}$$

$$\Gamma_L^* = \frac{S_{22} - \Gamma_G \Delta}{1 - \Gamma_G S_{11}} \tag{8.20}$$

Substitution of Eq. (8.20) into Eq. (8.19) eliminates Γ_L:

$$\Gamma_G^* = \frac{S_{11}\left(1 - \Gamma_G^* S_{11}^*\right) - \Delta\left(S_{22}^* - \Gamma_G^* \Delta^*\right)}{1 - \Gamma_G^* S_{11}^* - |S_{22}|^2 + S_{22}\Delta^* \Gamma_G^*} \tag{8.21}$$

This expression can be rearranged in the usual quadratic form. After taking the complex conjugate, this yields the following:

$$\Gamma_G^2\left(S_{22}^*\Delta - S_{11}\right) + \Gamma_G\left(1 - |S_{22}|^2 + |S_{11}|^2 - |\Delta|^2\right) - S_{11}^* + \Delta^* S_{22} = 0 \tag{8.22}$$

This equation can be rewritten in the form

$$0 = -\Gamma_G^2 C_1 + \Gamma_G B_1 - C_1^* \tag{8.23}$$

where

$$C_1 = S_{11} - \Delta S_{22}^* \tag{8.24}$$

$$B_1 = 1 + |S_{11}|^2 - |S_{22}|^2 - |\Delta|^2 \tag{8.25}$$

In the solution of Eq. (8.23) the required generator reflection coefficient for maximum gain is

$$\Gamma_{Gm} = \frac{C_1^*}{2|C_1|^2}\left[B_1 \pm \sqrt{B_1^2 - 4|C_1|^2}\right] \tag{8.26}$$

In a similar fashion the load reflection coefficient for maximum gain is

$$\Gamma_{Lm} = \frac{C_2^*}{2|C_2|^2}\left[B_2 \pm \sqrt{B_2^2 - 4|C_2|^2}\right] \tag{8.27}$$

where

$$C_2 = S_{22} - \Delta S_{11}^* \tag{8.28}$$

$$B_2 = 1 + |S_{22}|^2 - |S_{11}|^2 - |\Delta|^2 \tag{8.29}$$

The parameters B_i and C_i are determined solely from the scattering parameters of the two-port circuit. The – sign is used when $B_i > 0$, and the + sign is used when $B_i < 0$. Once the terminating reflection coefficients are known, the corresponding impedances may be determined:

$$Z_G = Z_0 \frac{1+\Gamma_G}{1-\Gamma_G} \tag{8.30}$$

$$Z_L = Z_0 \frac{1+\Gamma_L}{1-\Gamma_L} \tag{8.31}$$

8.6 STABILITY

A stable amplifier is an amplifier where there are no unwanted oscillations anywhere. Instability outside the operating band of the amplifier can still cause unwanted noise and even device burn out. Oscillations can only occur when there is some feedback path from the output back to the input. This feedback can be a result of an external circuit, external feedback parasitic circuit elements, or by an internal feedback path such as through C_μ in a common emitter bipolar transistor. Of these three sources, the last is usually the most troublesome. The following sections describe a method for determining transistor stability and some procedures to stabilize an otherwise unstable transistor.

8.6.1 Stability Circles

The criteria for unconditional stability require that $|\Gamma_i| \le 1$ and $|\Gamma_o| \le 1$ for any passive terminating loads. A useful amplifier may still be made if the terminating loads are carefully chosen to stay out of the unstable regions. It is helpful to find the borderline between the stable and the unstable regions. For the input side, this is done by finding the locus of points of Γ_L that will give $|\Gamma_i| = 1$. The borderline between stability and instability is found from Eq. (8.19) when $\Gamma_i = \Gamma_G^*$ and $|\Gamma_i| = 1$:

$$1 = \left| \frac{S_{11} - \Delta\Gamma_L}{1 - \Gamma_L S_{22}} \right| \tag{8.32}$$

This can be squared and then split up into its complex conjugate pairs:

$$(1-\Gamma_L S_{22})\left(1-\Gamma_L^* S_{22}^*\right) = (S_{11} - \Delta\Gamma_L)\left(S_{11}^* - \Delta^*\Gamma_L^*\right) \tag{8.33}$$

The coefficients of the different forms of Γ_L are collected together:

$$|\Gamma_L|^2\left(|S_{22}|^2 - |\Delta|^2\right) + \Gamma_L\left(\Delta S_{11}^* - S_{22}\right) + \Gamma_L^*\left(S_{11}\Delta^* - S_{22}^*\right) = |S_{11}|^2 - 1 \tag{8.34}$$

$$|\Gamma_L|^2 + \Gamma_L\left(\frac{\Delta S_{11}^* - S_{22}}{|S_{22}|^2 - |\Delta|^2}\right) + \Gamma_L^*\left(\frac{S_{11}\Delta^* - S_{22}^*}{|S_{22}|^2 - |\Delta|^2}\right) = \frac{|S_{11}|^2 - 1}{|S_{22}|^2 - |\Delta|^2} \tag{8.35}$$

Equation (8.35) can be put in a form that can be factored by completing the square. The value $|m|^2$, defined below, is added to both sides of this equation:

$$(\Gamma_L + m^*)(\Gamma_L^* + m) = |m|^2 + \frac{|S_{11}|^2 - 1}{|S_{22}|^2 - \Delta^2} \tag{8.36}$$

where

$$m \triangleq \frac{\Delta S_{11}^* - S_{22}}{|S_{22}|^2 - |\Delta|^2} \tag{8.37}$$

Substitution of Eq. (8.37) into Eq. (8.36) and upon simplification yields the following factored form:

$$\left(\Gamma_L + \frac{\Delta^* S_{11} - S_{22}^*}{|S_{22}|^2 - |\Delta|^2}\right)\left(\Gamma_L^* + \frac{\Delta S_{11}^* - S_{22}}{|S_{22}|^2 - |\Delta|^2}\right) = \frac{|S_{12}S_{21}|^2}{\left(|S_{22}|^2 - |\Delta|^2\right)^2} \tag{8.38}$$

This is the equation of a circle whose center is

$$C_L = \frac{S_{11}\Delta^* - S_{22}^*}{|\Delta|^2 - |S_{22}|^2} \tag{8.39}$$

The radius of the load stability circle is

$$r_L = \left|\frac{S_{21}S_{12}}{|\Delta|^2 - |S_{22}|^2}\right| \tag{8.40}$$

The center and radius for the generator stability circle can be found by symmetry:

$$C_G = \frac{S_{22}\Delta^* - S_{11}^*}{|\Delta|^2 - |S_{11}|^2} \tag{8.41}$$

$$r_G = \left|\frac{S_{21}S_{12}}{|\Delta|^2 - |S_{11}|^2}\right| \tag{8.42}$$

These two circles, one for the load and one for the generator, represent the borderline between stability and instability. These two circles can be overlaid on a Smith chart. The center of the circle is located at the vectorial position relative to the center of the Smith chart. The "dimensions" for the center and radius are normalized to the Smith chart radius (whose value is unity).

The remaining issue is which side of these circles is the stable region. Consider first the load stability circle shown in Fig. 8.7. If a matched transmission line with $Z_0 = 50\,\Omega$ were connected directly to the output port of the

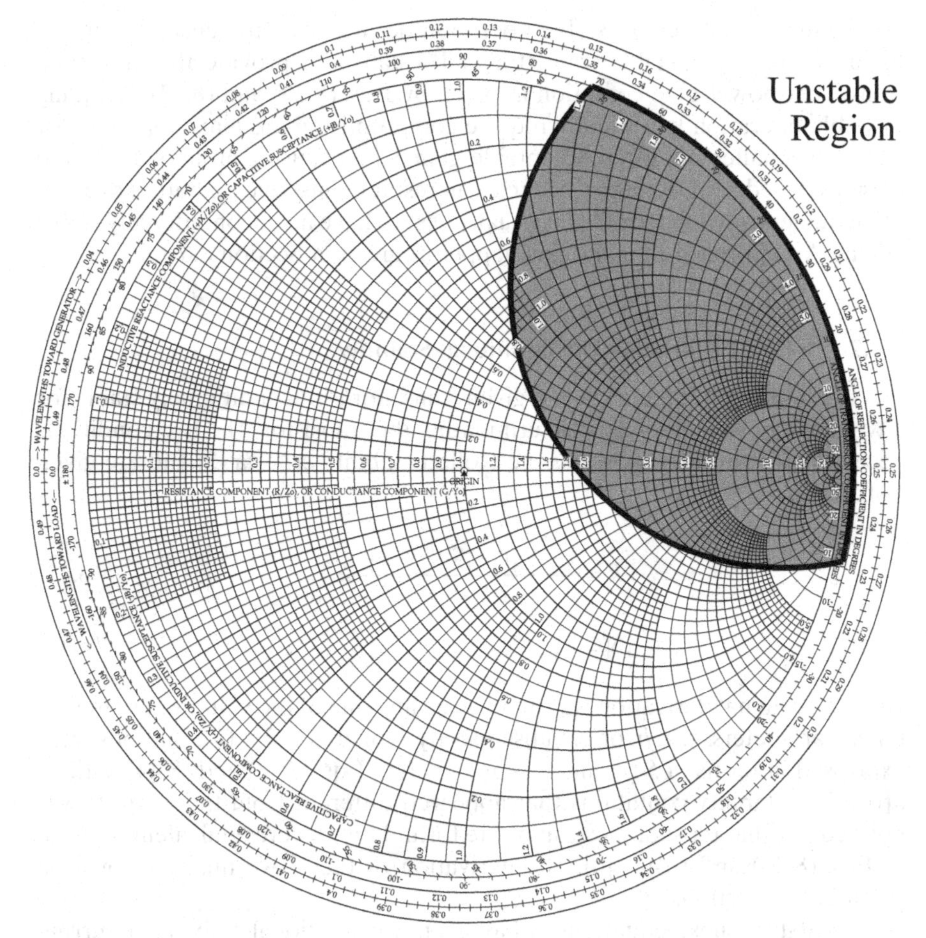

FIGURE 8.7 Illustration of stability circles where shaded region is unstable.

two-port circuit, then $\Gamma_L = 0$. This load would be located in the center of the Smith chart. Under this condition, Eq. (8.17) indicates that $\Gamma_i = S_{11}$. If the known value of $|S_{11}| < 1$, then $|\Gamma_i| < 1$ when the load is at the center of the Smith chart. If one point on one side of the stability circle is known to be stable, then all points on that side of the stability circle are also stable. The same rule would apply to the load side when Z_G is replaced by a matched load $=Z_0$. Then from Eq. (8.18) $\Gamma_o = S_{22}$.

Unconditional stability requires that both $|\Gamma_i| < 1$ and $|\Gamma_o| < 1$ for any passive load and generator impedances attached to the ports. In this case if $|S_{11}| < 1$ and $|S_{22}| < 1$, the stability circles would lie completely outside the Smith chart. Conditional stability occurs when at least one of the stability circles intersects the Smith chart. As long as the load and source impedances are on the stable side of the stability circle, stable operation occurs. Avoiding unstable regions for a potentially unstable transistor will usually not provide the maximum transducer power gain condition as given by Eqs. (8.26) and (8.27). Avoiding unstable operation will usually require compromising the maximum gain for a slightly smaller but often acceptable gain. Clearly, using an impedance too close to the edge of the stability circle can result in unstable operation because of manufacturing tolerances. It is usually best to have an unconditionally stable circuit. An approach to make this happen is given in Section 8.6.3.

8.6.2 Rollett Criteria for Unconditional Stability

It is often useful to determine if a given transistor is unconditionally stable for any pair of passive impedances terminating the transistor. The two conditions necessary for this are known as the Rollett stability criteria [3] and are given as follows:

$$k = \frac{1 - |S_{11}|^2 - |S_{22}|^2 + |\Delta|^2}{2|S_{12}S_{21}|} \geq 1 \tag{8.43}$$

$$|\Delta| \leq 1 \tag{8.44}$$

Rollett's original derivation was done using any one of the volt-ampere immittance parameters, z, y, h, or g. Subsequently, Rollett's stability equations were expressed in terms of S parameters as shown in Eqs. (8.43) and (8.44). Others arrived at stability conditions that appeared different from these, but it was pointed out that most of these alternate formulations were equivalent to those in Eqs. (8.43) and (8.44) [4]. The derivation of these two quantities will be given in this section.

The first of these equations is based on unconditional stability occurring when the load stability circle lies completely outside the Smith chart and when $|S_{11}| < 1$, that is,

$$|C_L| - r_L \geq 1 \tag{8.45}$$

or

$$r_L - |C_L| \geq 1 \tag{8.46}$$

where Eq. (8.46) describes the case where the stability circle encompasses the entire Smith chart within it. Substitution of Eqs. (8.39) and (8.40) into Eq. (8.45) gives

$$\frac{|S_{22} - S_{11}^*\Delta| - |S_{12}S_{21}|}{\left||\Delta|^2 - |S_{22}|^2\right|} \geq 1 \tag{8.47}$$

Squaring Eq. (8.47) gives the following:

$$\left||\Delta|^2-|S_{22}|^2\right|^2 \le \left[|S_{22}-S_{11}^*\Delta|-|S_{12}S_{21}|\right]^2 \tag{8.48}$$

$$\left||\Delta|^2-|S_{22}|^2\right|^2 \le |S_{22}-S_{11}^*\Delta|^2 - 2|S_{12}S_{21}||S_{22}-S_{11}^*\Delta|+|S_{12}S_{21}|^2 \tag{8.49}$$

$$2|S_{12}S_{21}||S_{22}-S_{11}^*\Delta| \le -\left||\Delta|^2-|S_{22}|^2\right|^2 + |S_{12}S_{21}|^2+|S_{22}-S_{11}^*\Delta|^2 \tag{8.50}$$

The last term on the right-hand side of Eq. (8.50) can be expanded:

$$\begin{aligned}|S_{22}-S_{11}^*\Delta|^2 &= (S_{22}-S_{11}^*\Delta)(S_{22}^*-S_{11}\Delta^*)\\ &= |S_{22}|^2 - S_{11}S_{22}\Delta^* - S_{11}^*S_{22}^*\Delta + |S_{11}|^2|\Delta|^2 \quad (8.51)\\ &= |S_{22}|^2+|\Delta|^2|S_{11}|^2-|S_{11}S_{22}|^2\\ &\quad + (S_{11}S_{22}S_{12}^*S_{21}^* - |S_{11}S_{22}|^2 + S_{11}^*S_{22}^*S_{12}S_{21}) \quad (8.52)\end{aligned}$$

Expansion of $|\Delta|^2$ gives

$$\begin{aligned}|\Delta|^2 &= (S_{11}S_{22}-S_{12}S_{21})(S_{11}^*S_{22}^*-S_{12}^*S_{21}^*)\\ &= |S_{11}S_{22}|^2+|S_{12}S_{21}|^2 - S_{11}S_{22}S_{12}^*S_{21}^* - S_{11}^*S_{22}^*S_{12}S_{21} \quad (8.53)\end{aligned}$$

By subtracting $|S_{12}S_{21}|^2$ inside the parenthesis in Eq. (8.52) and adding the same value outside the parenthesis, the quantity inside the parenthesis is equivalent to $|\Delta|^2$ given in Eq. (8.53). Thus Eq. (8.52) can be factored as shown below:

$$\begin{aligned}|S_{22}-S_{11}^*\Delta|^2 &= |S_{22}|^2+|\Delta|^2|S_{11}|^2-|S_{11}S_{22}|^2+|S_{12}S_{21}|^2-|\Delta|^2\\ &= |S_{12}S_{21}|^2+(1-|S_{11}|^2)(|S_{22}|^2-|\Delta|^2) \quad (8.54)\\ &= \delta+\alpha\beta \quad (8.55)\end{aligned}$$

where Eq. (8.55) is based on the temporary definitions:

$$\alpha \triangleq (1-|S_{11}|^2) \tag{8.56}$$

$$\beta \triangleq (|S_{22}|^2-|\Delta|^2) \tag{8.57}$$

$$\delta \triangleq |S_{12}S_{21}|^2 \tag{8.58}$$

The original inequality, Eq. (8.50), written in terms of these new variables is

$$2\sqrt{\delta}\sqrt{\alpha\beta+\delta} \le (\alpha\beta+\delta)+\delta-\beta^2 \tag{8.59}$$

By first squaring both sides and then canceling terms, Eq. (8.59) can be greatly simplified:

$$4\delta(\alpha\beta+\delta)\le\left[(\alpha\beta+2\delta)-\beta^2\right]^2$$

$$4\delta(\alpha\beta+\delta)\le(\alpha\beta+2\delta)^2-2\beta^2(\alpha\beta+2\delta)+\beta^4$$

$$4\delta(\alpha\beta+\delta)\le(\alpha\beta)^2+4\delta(\alpha\beta)+4\delta^2-2\beta^2(\alpha\beta+2\delta)+\beta^4$$

$$0\le(\alpha\beta)^2-2\beta^2(\alpha\beta+2\delta)+\beta^4$$

$$0\le\left[(\alpha-\beta)^2-4\delta\right]\beta^2$$

$$1\le\frac{(\alpha-\beta)^2}{4\delta} \tag{8.60}$$

Taking the square root of Eq. (8.60) yields

$$1\le\frac{\alpha-\beta}{2\sqrt{\delta}}=\frac{1-|S_{11}|^2-|S_{22}|^2+|\Delta|^2}{2|S_{12}S_{21}|}=k \tag{8.61}$$

This is the same as Eq. (8.43), which has now been demonstrated. Since the value of k is symmetrical on interchange of ports 1 and 2, the same result would occur with either the generator or load port stability circle.

The second condition for unconditional stability, Eq. (8.44), can also be demonstrated based on the requirement that $|\Gamma_i| < 1$. The second term of the right-hand side of Eq. (8.17) can be modified by multiplying it by 1 ($= S_{22}/S_{22}$) and adding 0 ($= S_{12}S_{21} - S_{12}S_{21}$) to the numerator. This results in the following:

$$|\Gamma_i|=\left|S_{11}+\frac{\Gamma_L S_{12}S_{21}S_{22}+(S_{12}S_{21}-S_{12}S_{21})}{(1-\Gamma_L S_{22})S_{22}}\right|$$

$$=\frac{1}{|S_{22}|}\left|\frac{S_{11}S_{22}(1-\Gamma_L S_{22})-S_{12}S_{21}(1-\Gamma_L S_{22})+S_{12}S_{21}}{1-\Gamma_L S_{22}}\right|$$

$$=\frac{1}{|S_{22}|}\left|\Delta+\frac{S_{12}S_{21}}{1-\Gamma_L S_{22}}\right|<1 \tag{8.62}$$

The complex quantity $1 - \Gamma_L S_{22}$ can be written in polar form as $1 - |\Gamma_L S_{22}|e^{j\theta}$. Any passive load must lie within the unit circle $|\Gamma_L| \le 1$, so $|\Gamma_L|$ is set to 1. As described in [5], the quantity

$$\frac{1}{1-|S_{22}|e^{j\theta}}$$

which appears in Eq. (8.62), is a circle as pictured in Fig. 8.8 centered at

$$\frac{1}{2}\left(\frac{1}{1-|S_{22}|}+\frac{1}{1+|S_{22}|}\right)=\frac{1}{1-|S_{22}|^2}$$

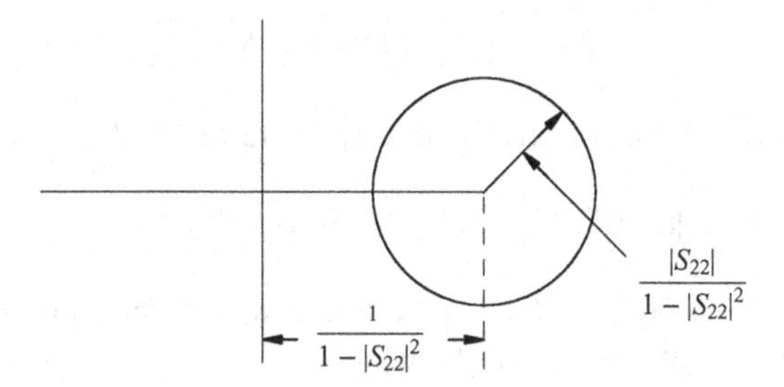

FIGURE 8.8 Representation of circle with $|\Gamma_L| = 1$.

and with radius

$$\frac{1}{2}\left(\frac{1}{1-|S_{22}|}-\frac{1}{1+|S_{22}|}\right)=\frac{|S_{22}|}{1-|S_{22}|^2}$$

Equation (8.62) is expressed in terms of this circle:

$$\frac{1}{|S_{22}|}\left|\Delta+\frac{S_{12}S_{21}}{1-|S_{22}|^2}+\frac{S_{12}S_{21}|S_{22}e^{j\theta}}{1-|S_{22}|^2}\right|<1 \tag{8.63}$$

The phase of the load is chosen so that it maximizes the left-hand side of Eq. (8.63). However, it must still obey the stated inequality. This means that Eq. (8.63) can be written as the sum of the two magnitudes without violating the inequality condition:

$$\frac{1}{|S_{22}|}\left|\Delta+\frac{S_{12}S_{21}}{1-|S_{22}|^2}\right|+\frac{|S_{12}S_{21}|}{1-|S_{22}|^2}<1$$

$$0<\frac{1}{|S_{22}|}\left|\Delta+\frac{S_{12}S_{21}}{1-|S_{22}|^2}\right|<1-\frac{|S_{12}S_{21}|}{1-|S_{22}|^2}$$

Comparison of the far right-hand side of this expression with 0 results in the following inequality:

$$1-|S_{22}|^2>|S_{12}S_{21}| \tag{8.64}$$

If the process had begun with the condition that $|\Gamma_o| < 1$, then the result would be the same as Eq. (8.64) with the 1's and 2's interchanged:

$$1-|S_{11}|^2>|S_{12}S_{21}| \tag{8.65}$$

When Eqs. (8.64) and (8.65) are added together,

$$2-|S_{11}|^2-|S_{22}|^2>2|S_{12}S_{21}| \tag{8.66}$$

However, from the definition of the determinate of the S parameter matrix,

$$|\Delta|=|S_{11}S_{22}-S_{12}S_{21}|<|S_{11}S_{22}|+|S_{12}S_{21}| \tag{8.67}$$

When the term $|S_{12}S_{21}|$ in Eq. (8.67) is replaced with something larger, as given in Eq. (8.66), the inequality is still true:

$$|\Delta|<|S_{11}S_{22}|+1-\tfrac{1}{2}\left(|S_{11}|^2+|S_{22}|^2\right)$$

$$|\Delta|<1-\tfrac{1}{2}(|S_{11}|-|S_{22}|)^2<1 \tag{8.68}$$

An alternate, but equivalent, set of requirements for stability are [4]

$$k>1 \tag{8.69}$$

and either

$$B_1>0 \tag{8.70a}$$

or

$$B_2>0 \tag{8.70b}$$

The requirement of Eq. (8.70a) or (8.70b) is equivalent to $|\Delta|<1$.

8.6.3 Stabilizing a Transistor Amplifier

There are a variety of approaches to stabilizing an amplifier. In Section 8.6.1, it was suggested that stability could be achieved from a potentially unstable transistor by making sure the chosen amplifier terminating impedances remain inside the stable regions at all frequencies as determined by the stability circles.

Another method would be to load the amplifier with an additional shunt or series resistor on either the generator or load side. The resistor is incorporated as part of the the two-port parameters of the transistor. If the condition for unconditional stability is achieved for this expanded transistor model, then optimization can be performed for the other circuit elements in order to achieve the desired gain and bandwidth. It is usually better to try loading the output side rather than the input side in order to minimize increasing the amplifier noise figure.

Another approach that is sometimes useful is to introduce an external feedback path that can neutralize the internal feedback of the transistor. The most widely used scheme is the shunt–shunt feedback circuit shown in Fig. 8.9.

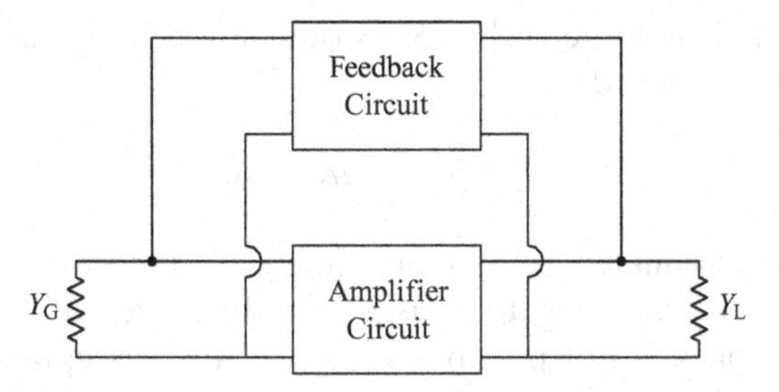

FIGURE 8.9 Shunt–shunt feedback for stabilizing a transistor.

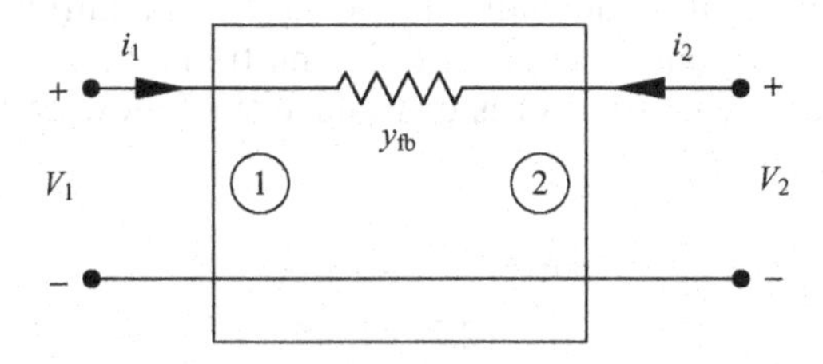

FIGURE 8.10 Two-port representation of feedback circuit.

The y parameters for the composite circuit are simply the sum of the y parameters of the amplifier and feedback two-port circuits:

$$[Y_c]=[Y_a]+[Y_f] \tag{8.71}$$

To use this method, the transistor scattering parameters must be converted to admittance parameters (Appendix D). The y parameters for a simple series admittance, y_{fb} can be found from circuit theory (Fig. 8.10):

$$y_{11f} = y_{22f} = \left.\frac{i_1}{v_1}\right|_{v_2=0} = y_{fb} \tag{8.72}$$

$$y_{12f} = y_{21f} = \left.\frac{i_2}{v_1}\right|_{v_2=0} = -y_{fb} \tag{8.73}$$

Consequently, the composite y parameters are

$$y_{11c} = y_{11a} + y_{11f} = y_{11a} + y_{fb} \tag{8.74}$$

$$y_{12c} = y_{12a} + y_{12f} = y_{12a} - y_{fb} \tag{8.75}$$

$$y_{21c} = y_{21a} + y_{21f} = y_{21a} - y_{fb} \tag{8.76}$$

$$y_{22c} = y_{22a} + y_{22f} = y_{22a} + y_{fb} \tag{8.77}$$

If y_{12c} could be made to be zero, then S_{12c} would also be zero and unconditional stability could be achieved:

$$g_{12a} + jb_{12a} = g_{fb} + jb_{fb} \tag{8.78}$$

Since the circuit parameter $g_{12a} < 0$ the value $g_{fb} < 0$ must be true also. Since it is not possible to have a negative passive conductance, complete removal of the internal feedback is not possible. However, the susceptance, b_{12a} can be canceled by a passive external feedback susceptance. Although total removal of y_{12a} cannot be achieved, yet progress toward stabilizing the amplifier can often be achieved. There is no guarantee that neutralization will provide a composite y matrix that is unconditionally stable. In addition, neutralization of the feedback susceptance occurs at only one frequency.

As an example, consider a transistor to have the following S parameters at a given frequency:

$$\begin{aligned} S_{11a} &= 0.73\angle{-102^\circ} \\ S_{21a} &= 2.21\angle 104^\circ \\ S_{12a} &= 0.10\angle 48^\circ \\ S_{22a} &= 0.47\angle{-48^\circ} \end{aligned} \tag{8.79}$$

For this transistor, $k = 0.752$ and $|\Delta| = 0.294$ as found from Eqs. (8.43) and (8.44). Conversion of the S parameters given by Eq. (8.79) to y parameters gives

$$\begin{aligned} y_{11a} &= 5.5307\times10^{-3} + j1.9049\times10^{-2} \quad \text{S} \\ y_{12a} &= 3.9086\times10^{-4} - j2.3092\times10^{-3} \quad \text{S} \\ y_{21a} &= 4.7114\times10^{-2} - j2.1376\times10^{-2} \quad \text{S} \\ y_{22a} &= 5.4445\times10^{-3} + j5.1841\times10^{-3} \quad \text{S} \end{aligned} \tag{8.80}$$

Nothing can be done about g_{12a}, but b_{12a} can be removed by setting $b_{fb} = b_{12a} = -2.3092 \times 10^{-3}$. The composite admittance matrix becomes

$$\begin{aligned} y_{11c} &= 5.5307\times10^{-3} + j1.6739\times10^{-2} \quad \text{S} \\ y_{12c} &= 3.9086\times10^{-4} - j0 \quad \text{S} \\ y_{21c} &= 4.7114\times10^{-2} - j1.9067\times10^{-2} \quad \text{S} \\ y_{22c} &= 5.4445\times10^{-3} + j2.8750\times10^{-3} \quad \text{S} \end{aligned} \tag{8.81}$$

The composite scattering parameters can now be found and the stability factor calculated yielding $k = 2.067$ and $|\Delta| = 0.4037$. The transistor with the feedback

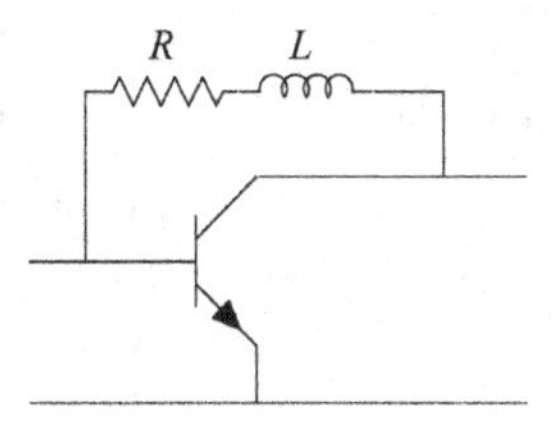

FIGURE 8.11 Broadband feedback stabilization.

circuit is unconditionally stable at the given frequency. This stability has been achieved by adding inductive susceptance in shunt with the transistor input and output ports.

Broadband stability can be achieved by replacing the feedback inductor with an inductor and resistor as shown in Fig. 8.11. A starting value for the inductor can be found as described for the single-frequency analysis. The resistor is typically in the 200- to 800-Ω range, but optimum values for *R* and *L* are best found by computer optimization.

8.7 CLASS A POWER AMPLIFIERS

Class A amplifiers, whether for small-signal or large-signal operation, are intended to amplify the incoming signal in a linear fashion. This type of amplifier will not introduce significant distortion in the amplitude and phase of the signal. A linear class A power amplifier will introduce low-amplitude harmonic frequency components and low intermodulation distortion (IMD). An example of IMD can be described in terms of a double sideband suppressed carrier wave that is represented as

$$\frac{V}{2}\cos(\omega_c + \omega_m)t + \frac{V}{2}\cos(\omega_c - \omega_m)t \tag{8.82}$$

where ω_c is the high-frequency carrier frequency and ω_m is the low-frequency modulation frequency. Intermodulation distortion would result in frequencies at $\omega_c \pm n\omega_m$ and harmonic distortion would cause frequency generation at $k\omega_c \pm n\omega_m$. The later harmonic distortion can usually be filtered out, but the third-order IMD is more difficult to handle because the distortion frequencies are near if not actually inside the system pass band. Clearly, this distortion in a class A amplifier is a greater problem for power amplifiers than for small-signal amplifiers. Reduction of IMD depends on efficient power combining methods and careful design of the transistors themselves.

A transistor acting in the class A mode remains in its active state throughout the complete cycle of the signal. Two examples of common emitter class A

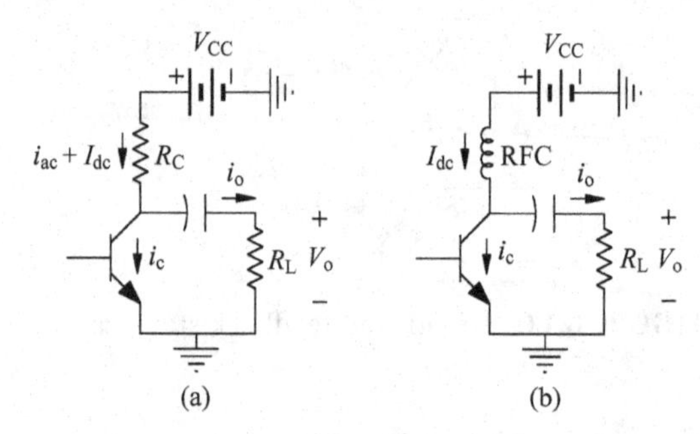

FIGURE 8.12 Class A amplifiers with (*a*) collector resistor and (*b*) collector inductor.

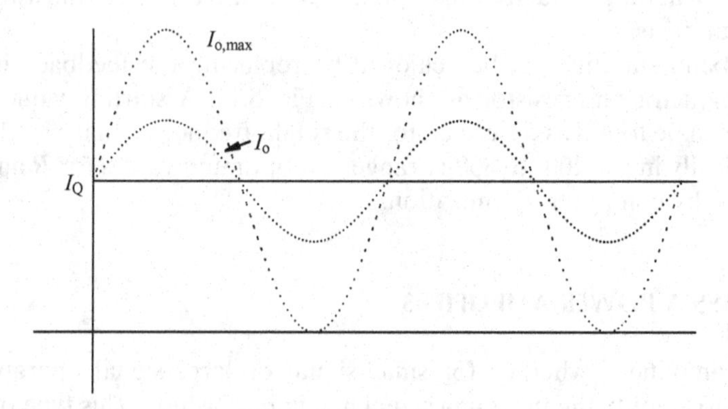

FIGURE 8.13 Magnitude of output current and quiescent current of class A amplifier.

amplifiers are shown in Fig. 8.12. The maximum efficiency of the class A amplifier in Fig. 8.12*a* has been shown to be 25% (e.g., see [6]). However, if an RF coil can be used in the collector (Fig. 8.12*b*), the efficiency can be increased to almost 50%. This can be shown by recognizing first that there is no ac flow in the bias source and no dc flow in the load, R_L. The total current flowing in the transistor collector is

$$i_c = I_Q - I_o \sin \omega t \tag{8.83}$$

and the total collector voltage is

$$V_{CE} = V_{CC} + V_o \sin \omega t \tag{8.84}$$

Both the quiescent current, I_Q, and the output current, I_o, are defined in Fig. 8.13. The quiescent current, I_Q, is the direct current flowing through the collector, which sets the ac operating point. When the load is drawing the maximum instantaneous power,

$$I_{o,max} = I_Q = I_{dc} \tag{8.85}$$

At this point, the maximum output voltage is

$$V_{o,max} = I_{o,max} R_L \tag{8.86}$$

where $|V_{o,max}| \approx V_{CC}$.

The dc power source supplies

$$P_{dc} = I_{dc} V_{CC} = \frac{V_{CC}^2}{R_L} \tag{8.87}$$

The maximum average power delivered to the load can now be written in terms of the supply voltage:

$$P_o = \frac{|V_{o,max}|^2}{2R_L} \approx \frac{V_{CC}^2}{2R_L} \tag{8.88}$$

The collector efficiency is

$$n_c = \frac{P_o}{P_{dc}} \tag{8.89}$$

This definition is meaningful for high-gain amplifiers where $P_i \ll P_o$. For the class A amplifier the maximum collector efficiency is $\eta_c \approx 50\%$. The $V_{o,max}$ will always be slightly less than the supply voltage because of V_{be} or $V_{ce,sat}$. However, it should be noted that many times high-power amplifiers do not have high gain, so the power-added efficiency offers a more useful quality factor for a transistor than if P_i were neglected. Power-added efficiency will be used in Section 8.9.2 for analysis of multistage amplifiers.

8.8 POWER COMBINING OF POWER AMPLIFIERS

Design of power FET amplifiers requires use of large-gate periphery devices. However, eventually, the large gate periphery causes other problems, such as impedance matching especially at RF and microwave frequencies. Bandwidth improvement can be obtained by combining several transistors, often on a single chip. An example of combining two transistors is shown in Fig. 8.14 [7, 8]. The separation of the transistors may induce odd-order oscillations in the circuit even if the stability factor of the individual transistors (even-order stability) indicate they are stable. This odd-order instability can be controlled by adding R_{odd} between the two drains to damp out such oscillations. This resistor is typically less than 400Ω. Symmetry indicates no power dissipation

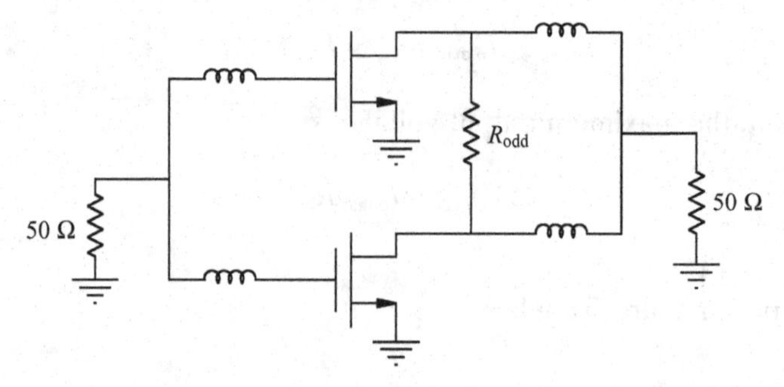

FIGURE 8.14 Power combining two transistors [7, 8].

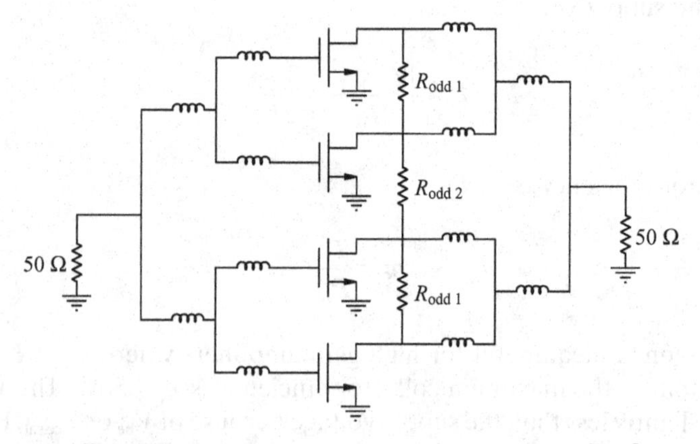

FIGURE 8.15 Power combining four transistors [7, 8].

when the outputs of the two transistors are equal and in phase. An example of a four-transistor combining circuit is shown in Fig. 8.15, which now includes resistors R_{odd1} and R_{odd2} to help suppress odd-order oscillations.

8.9 PROPERTIES OF CASCADED AMPLIFIERS

An ideal amplifier is completely unilateral so that there is no feedback signal returning from the output to the input side. Under this condition, analysis of cascaded amplifiers results in some interesting properties related to noise figure and efficiency. The results obtained will be approximately valid for almost unilateral amplifiers, even if some of the "amplifiers" are mixers or attenuators. The following two sections deal with the total noise figure and total efficiency, respectively, of a cascade of unilateral amplifiers.

8.9.1 Friis Noise Formula

The most critical part for achieving low noise in a receiver is the noise figure and gain of the first stage. This is intuitively clear since the magnitude of the noise in the first stage will be a much larger percentage of the incoming signal than it will be in subsequent stages where the signal amplitude is much larger. Although both signal and noise get amplified the same amount, the difference between the signal and noise increases. For a receiver with n unilateral stages, the total noise factor for all n stages is [9]

$$F_{\mathrm{T},n} = \frac{N_{\mathrm{T},n}}{kT_0\ \Delta f\ G_1 G_2 \cdots G_n} \tag{8.90}$$

where $N_{\mathrm{T},n}$ is the total noise power delivered to the load. This can be expressed in terms of the sum of the noise added by the last stage, N_n, and that of all the previous stages multiplied by the gain of the last stage:

$$N_{\mathrm{T},n} = N_n + G_n N_{\mathrm{T}(n-1)} \tag{8.91}$$

If the nth stage were removed, and its noise factor measured alone, then its noise factor would be

$$F_n = \frac{kT_0 G_n\ \Delta f + N_n}{kT_0 G_n\ \Delta f} \tag{8.92}$$

$$F_n - 1 = \frac{N_n}{kT_0 G_n\ \Delta f} \tag{8.93}$$

By substitution Eq. (8.91) into Eq. (8.90) an expression for the noise factor is obtained that separates the contributions of the noise coming from the last stage only from the previous $n - 1$ stages:

$$F_{\mathrm{T},n} = \frac{N_n}{(kT_0\ \Delta f G_n) G_1 G_2 \cdots G_{n-1}} + \frac{G_n N_{\mathrm{T}(n-1)}}{kT_0\ \Delta f\ G_1 G_2 \cdots G_{n-1} G_n} \tag{8.94}$$

Canceling the G_n in the second term and substituting (8.93) yields

$$F_{\mathrm{T},n} = \frac{F_n - 1}{G_1 G_2 \cdots G_{n-1}} + \frac{N_{\mathrm{T}(n-1)}}{kT_0\ \Delta f\ G_1 G_2 \cdots G_{n-1}} \tag{8.95}$$

The second term in Eq. (8.95) is the same as Eq. (8.90) except that n has been reduced to $n - 1$. This process is repeated n times, giving what is known as the Friis formula for the noise factor for a cascade of unilateral gain stages:

$$F_{\mathrm{T},n} = F_1 + \frac{F_2 - 1}{G_1} + \frac{F_3 - 1}{G_1 G_2} + \cdots + \frac{F_n - 1}{G_1 G_2 \cdots G_{n-1}} \tag{8.96}$$

Clearly, the noise factor of the first stage is the most important contributor to the overall noise factor of the system. If the first stage has reasonable gain, the subsequent stages can have much higher noise factor without affecting the overall noise factor of the receiver.

8.9.2 Multistage Amplifier Efficiency

For a multistage amplifier, the overall power efficiency can be found that will correspond in some way with the overall noise figure expression. Unlike the noise figure, however, the efficiency of the last stage will be found to be most important. Again, this would appear logical since the last amplifying stage handles the greatest amount of power so that poor efficiency here would waste the most amount of power. For the kth stage of an n-stage amplifier chain, the power-added efficiency is

$$\eta_k = \frac{P_{o,k} - P_{i,k}}{P_{d,k}} \tag{8.97}$$

where $P_{o,k}$ = output power of the kth stage
$P_{i,k}$ = input power to the kth stage
$P_{d,k}$ = source of power, which is typically the dc bias for the kth stage

If the input power to the first stage is $P_{i,1}$, then

$$P_{i,k} = P_{i,1} G_1 G_2 G_3 \cdots G_{k-1} \tag{8.98}$$

and for the kth stage alone

$$P_{o,k} = P_{i,k} G_k \tag{8.99}$$

When Eqs. (8.98) and (8.99) are substituted into the efficiency equation, Eq. (8.97), the power from the power source can be found:

$$P_{d,k} = \frac{G_1 G_2 \cdots G_{k-1}(G_k - 1)}{\eta_k} P_{i,1} \tag{8.100}$$

The total added power for a chain of n amplifiers is

$$P_{o,n} - P_{i,1} = P_{i,1}(G_1 G_2 G_3 \cdots G_n - 1) \tag{8.101}$$

The efficiency for the whole amplifier chain is clearly given by the following:

$$\eta_T = \frac{P_{o,n} - P_{i,1}}{\sum_{i=1}^{n} P_{d,k}} \tag{8.102}$$

Replacing the power levels in Eq. (8.102) with their explicit expression gives the value for the overall efficiency of a chain of unilateral amplifiers:

$$\eta_T = \frac{G_1 G_2 G_3 \cdots G_n - 1}{\dfrac{G_1 - 1}{\eta_1} + \dfrac{G_1(G_2 - 1)}{\eta_2} + \cdots + \dfrac{G_1 G_2 \cdots G_{n-1}(G_n - 1)}{\eta_n}} \tag{8.103}$$

When each amplifier stage has a gain sufficiently greater than 1, the overall efficiency becomes

$$\eta_T \approx \frac{1}{\dfrac{1}{G_2 G_3 \cdots G_n \eta_1} + \cdots + \dfrac{1}{G_n \eta_{n-1}} + \dfrac{1}{\eta_n}} \tag{8.104}$$

This final equation highlights the assertion that it is most important to make the final stage the most efficient one.

8.10 AMPLIFIER DESIGN FOR OPTIMUM GAIN AND NOISE

The gain for a nonunilateral amplifier was previously given as Eq. (8.18) is repeated here:

$$G_T = \frac{|S_{21}|^2 \left(1 - |\Gamma_G|^2\right)\left(1 - |\Gamma_L|^2\right)}{|(1 - \Gamma_G S_{11})(1 - \Gamma_L S_{22}) - S_{12} S_{21} \Gamma_G \Gamma_L|^2} \tag{8.105}$$

If S_{12} is set to zero, thereby invoking the unilateral approximation for the amplifier gain, then

$$G_T \approx \frac{1 - |\Gamma_G|^2}{|1 - \Gamma_G S_{11}|^2} |S_{21}|^2 \frac{1 - |\Gamma_L|^2}{|1 - \Gamma_L S_{22}|^2} \tag{8.106}$$

This approximation, of course, removes the possibility of analytically determining the transistor stability conditions. Using this expression, a set of constant gain circles can be drawn on a Smith chart for a given transistor. That is, for a given set of device scattering parameters and for a fixed load impedance, a set of constant gain circles can be drawn for a range of generator impedances expressed here in terms of Γ_G.

The noise figure was previously found in Eq. (7.40). As was done for the constant gain circles, constant noise figure circles can be drawn for a range of values for the generator admittance, $Y_G = G_G + jB_G$. The optimum gain occurs when $\Gamma_G = S_{11}^*$ and the minimum noise figure occurs when $Y_G = Y_{opt}$. These two source impedances are rarely the same, but a procedure is available to at least

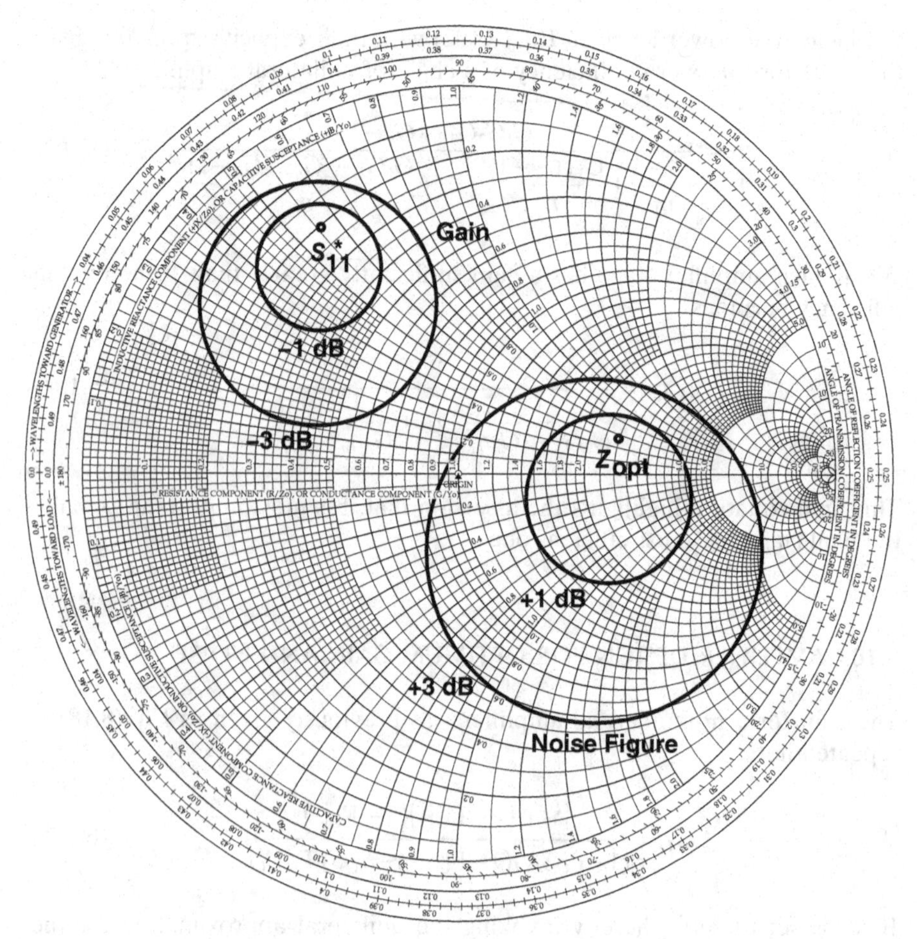

FIGURE 8.16 Constant gain and noise figure circles.

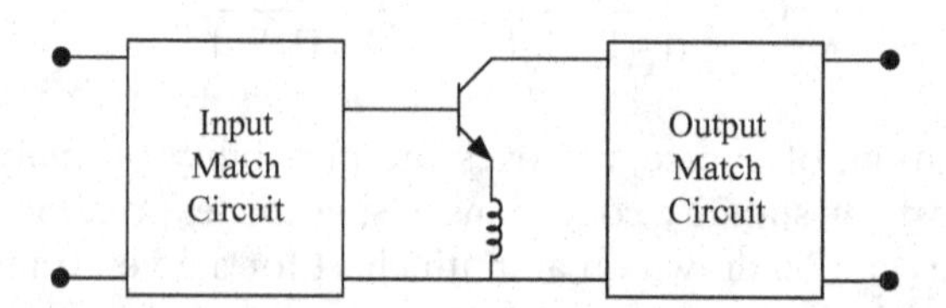

FIGURE 8.17 Series inductive feedback can be used to lower noise figure.

optimize for both of these parameters simultaneously [10]. As seen on the Smith chart in Fig. 8.16, it appears possible that the least damage to either the gain or the noise figure is obtained if the actual chosen generator impedance lies on a line between S_{11}^* and Y_{opt}. It has been found that addition of series emitter inductance, such as shown in Fig. 8.17, will lower the minimum noise

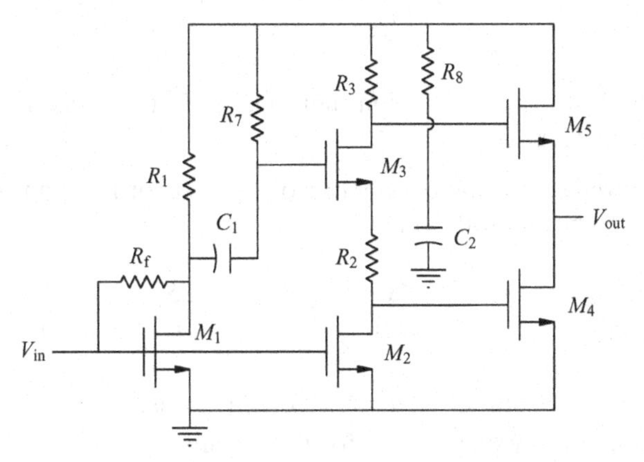

FIGURE 8.18 Feed-forward noise cancellation low-noise amplifier described in [12].

figure of the circuit because it lowers the effective F_{min} and r_n. This series inductance also increases the real part of the input impedance. The output impedance does not affect the noise figure but can be manipulated to adjust the gain.

Low-noise, wide-band amplifiers have been reported using CMOS technology. Wide bandwidth can be achieved using negative feedback by using a resistor between the gate and source (or base and emitter). Since the gain is reduce by the factor of $1 + T$ where T is the loop gain, then the bandwidth increases by $1 + T$. However, even a larger bandwidth and a better noise figure can be achieved by using inductive source (or emitter) degeneracy along with an added input inductor and shunt capacitor [11]. Inductors are problematic in integrated circuit (IC) designs because of their size and relatively poor Q. A wide-bandwidth CMOS circuit with a noise figure below 4.8 dB has been reported that uses no inductors. Low noise is achieved in Fig. 8.18 by using a feed-forward technique that effectively cancels noise [12]. This circuit provides not only noise cancellation but enhanced gain (11 dB in this case) that former feed-forward designs lacked. Noise coming out of the input matching transistor, M_1, is canceled by M_2 and M_3, and delivered to M_4 and M_5 through two feed-forward paths. The capacitor C_2 is large enough to provide an effective RF ground.

8.11 CONCLUSION

This chapter began with a definition and graphical derivation of the most useful form of power gain, the transducer power gain. A linear analysis of an amplifier focuses on gain, bandwidth, stability, and noise. An analytical treatment for bandwidth is usually intractable, so this aspect is normally done by computer simulation. An introduction to class A power amplifiers is given along with a schematic of at least one integrated circuit low-noise amplifier that can be implemented with CMOS technology.

PROBLEMS

8.1. Using the flow graph reduction method, verify the reflection coefficient found in Eq. (8.17).

8.2. The measured scattering parameters of a transistor in an amplifier circuit are found to be the following:

$\lvert S_{11}\rvert$	$\angle S_{11}$	$\lvert S_{21}\rvert$	$\angle S_{21}$	$\lvert S_{12}\rvert$	$\angle S_{12}$	$\lvert S_{22}\rvert$	$\angle S_{22}$
0.65	–40	3.6	–140	0.05	74	0.90	–20

a. Determine the stability factor, k, for this transistor.
b. Determine the y parameters for this circuit.
c. Determine the circuit at 1.5 GHz that would neutralize (almost unilateralize) the circuit. While this procedure does not guarantee stability in all cases, it usually helps lead toward greater stability.
d. Determine the new scattering parameters for the neutralized circuit.
e. Determine the generator and load impedances that would give maximum transducer power gain (not unilateral power gain).
f. What is the value for the maximum transducer power gain?

8.3. Determine the transfer function for the flow graph in Fig. 8.19.

8.4. A certain transistor has the following S parameters:

$S_{11} = 1.2 \quad S_{21} = 4.0 \quad S_{12} = 0 \quad S_{22} = 0.9$

Determine whether this transistor is unconditionally stable.

8.5. Verify Eq. (8.38).

8.6. A three-stage amplifier consists of three individual unilateral amplifiers. The first one (the input stage) has a gain $G_1 = 10\,\text{dB}$, a noise factor $F_1 = 1.5$, and an efficiency $\eta_1 = 1\%$. For stage 2, $G_2 = 20\,\text{dB}$, $F_2 = 10$, and $\eta_2 = 5\%$. For stage 3, $G_3 = 5\,\text{dB}$, $F_3 = 15$, and $\eta_3 = 50\%$. What is the overall total noise factor for the cascaded amplifier? What is the overall total efficiency for the cascaded amplifier?

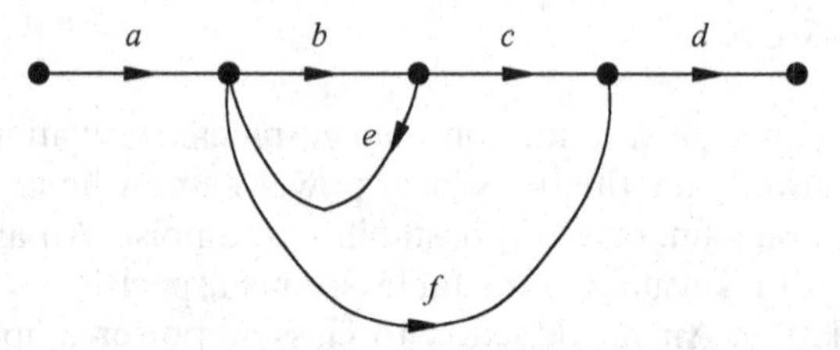

FIGURE 8.19 Flow graph for Problem 8.3.

REFERENCES

1. H. L. Krauss, C. W. Bostian, and F. H. Raab, *Solid State Radio Engineering*, New York: Wiley, pp. 371–382, 1980.
2. P. J. Khan, private communication, 1971.
3. J. M. Rollett, "Stability and Power-Gain Invariants of Linear Twoports," *IRE Trans. Circuit Theory*, **CT-9**, pp. 29–32, March 1962.
4. D. Woods, "Reappraisal of the Unconditional Stability Criteria for Active 2-Port Networks in Terms of S Parameters" *IEEE Trans. Circuits Syst.*, **CAS-23**, pp. 73–81, Feb. 1976.
5. T. T. Ha, *Solid State Microwave Amplifier Design*, New York: Wiley, 1981.
6. S. G. Burns and P. R. Bond, *Principles of Electronic Circuits*, Boston: PWS Publishing, 1997.
7. R. G. Reitag, S. H. Lee, and D. M. Krafcsik, "Stability and Improved Circuit Modeling Considerations for High Power MMIC Amplifiers," *1988 IEEE Microwave Theory Tech. Symp. Digest*, pp. 175–178, 1988.
8. R. G. Freitag "A Unified Analysis of MMIC Power Amplifier Stability," *1992 IEEE Microwave Theory Tech. Symp. Digest*, pp. 297–300, 1992.
9. H. T. Friis, "Noise Figures of Radio Receivers," *Proc. IRE*, **32**, pp. 419–422, July 1944.
10. R. E. Lehmann and D. D. Heston, "X-Band Monolithic Series Feedback LNA," *IEEE Trans. Microwave Theory Tech.*, **MTT-33**, pp. 1560–1566, Dec. 1985.
11. A. Bevilacqua and A. M. Niknejad, "An Ultrawideband CMOS Low-Noise Amplifier for 3.1–10.6-GHz Wireless Receivers," *IEEE J. Solid State Circuits*, **39**, pp. 2259–2268, Dec. 2004.
12. Q. Li and Y. P. Zhang, "A 1.5-V 2–9.6-GHz Inductorless Low-Noise Amplifier in 0.13-μm CMOS," *IEEE Trans. Microwave Theory Tech.*, **55**, Oct. 2007.

CHAPTER NINE

RF Power Amplifiers

9.1 TRANSISTOR CONFIGURATIONS

In Chapter 8, class A amplifiers were treated. Some discussion was given to its application as a power amplifier. While class A amplifiers are used in power applications where linearity is of primary concern, they do so at the cost of efficiency. This chapter describes power amplifiers that provide higher efficiency than the class A amplifier. Before describing these in detail, it should be recalled that a single-transistor amplifier can be installed in one of four different ways: common emitter, common base, common collector (or emitter follower), and common emitter with emitter degeneracy. Although there are always exceptions, the common emitter circuit is used in amplifiers where high-voltage gain is required. The common base amplifier is used when low input impedance and high output impedance is desired. This is accompanied with a current gain ≈ 1. The emitter follower is used when high input impedance and low output impedance is desired. This is accompanied with a voltage gain ≈ 1. The common emitter with emitter degeneracy configuration is used when improved repeatability is needed with respect to differences in the transistor short-circuit current gain (β). The emitter resistance provides negative feedback and thus causes some loss in voltage gain. These configurations (minus bias circuits) are illustrated in Fig. 9.1. These properties are described in detail in most electronics texts.

The transistor itself can be in one of four different states: saturation, forward active, cutoff, and reverse active. It is in the forward active region, when for the bipolar transistor, the base–emitter junction is forward biased and the

Radio Frequency Circuit Design, Second Edition, by W. Alan Davis

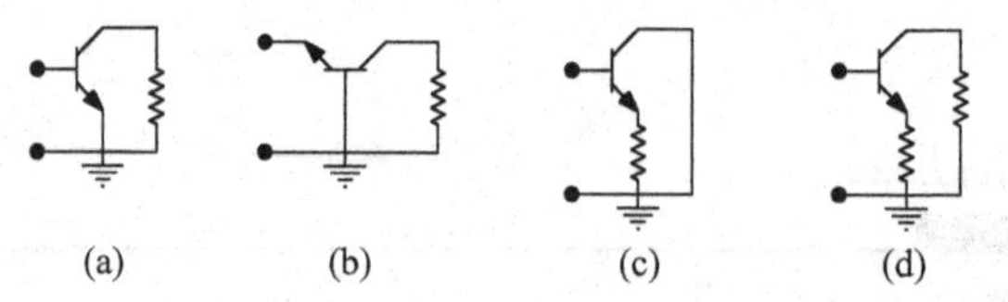

FIGURE 9.1 Connections for (*a*) common emitter, (*b*) common base, (*c*) common collector or emitter follower, and (*d*) common emitter with emitter degeneracy.

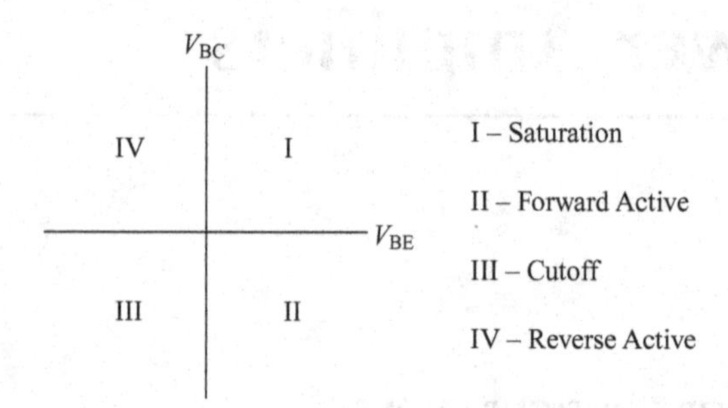

FIGURE 9.2 Four bias regions for *npn* bipolar transistor.

base–collector junction is reverse biased. These states are illustrated in Fig. 9.2 for an *npn* transistor. An actual bipolar transistor requires a base–emitter voltage greater than 0.6 to 0.8 V for it to go into the active state. For MOSFETs the gate–source voltage must exceed the threshold voltage, v_t.

The voltage swing of a class A amplifier will remain in the forward active region throughout its entire cycle. If the output signal current is

$$i_o(\omega t) = \hat{I}_C \sin \omega t \tag{9.1}$$

and the dc bias current is I_{dc}, then the total instantaneous current is

$$I_{dc} + \hat{I}_C \sin \omega t \tag{9.2}$$

The quiescent current, I_Q, is the current around which the alternating current flows. For the class A amplifier, $\hat{I}_C < I_Q$, so that the entire waveform of the ac signal is amplified without distortion. The conduction angle is 360°. For the power amplifiers under consideration in this chapter, the transistor(s) will be operating during part of their cycle in either cutoff or saturation or both.

9.2 CLASS B AMPLIFIER

The class B amplifier is biased so that the transistor is on only during half of the cycle of the incoming signal. The other half of the cycle is amplified by

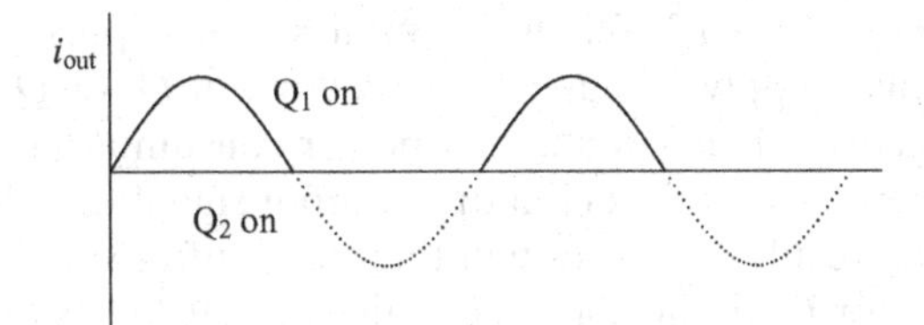

FIGURE 9.3 Reconstituted waveform of class B amplifier.

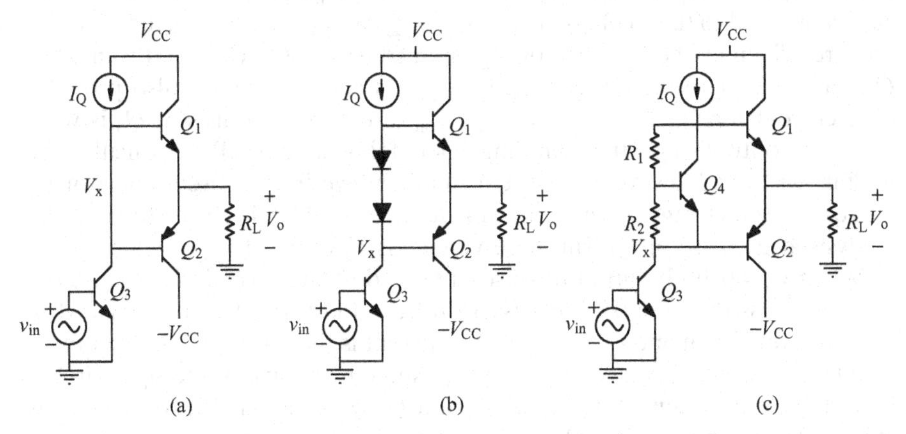

FIGURE 9.4 Complementary class B amplifier: (*a*) the basic amplifier, (*b*) compensation to reduce crossover distortion, and (*c*) amplifier with V_{BE} multiplier to reduce crossover distortion.

another transistor so that at the output the full wave is reconstituted. This is illustrated in Fig. 9.3. While each transistor is clearly operating in a nonlinear mode, the total input wave is directly replicated at the output. The class B amplifier is therefore classed as a linear amplifier. In this case, the quiescent current, $I_Q = 0$. Since only one of the transistors is cut off when the total voltage is less than 0, only the positive half of the wave is amplified. The conduction angle is 180°. The term class AB amplifier is sometimes used to describe the case when the direct quiescent current is much smaller than the signal amplitude, $\hat{I}_C$, but still greater than 0. In this case,

$$180° < \text{conduction angle} \ll 360°$$

9.2.1 Complementary (*npn/pnp*) Class B Amplifier

The illustration in Fig. 9.4*a* shows a complementary type of class B amplifier. In this case transistor Q_1 is biased so that it is in the active mode when the input voltage is greater than the transistor turn-on voltage, $v_{in} > 0.7$, and cut off when the input signal $v_{in} < 0.7$. The other half of the signal is amplified by transistor Q_2 when $v_{in} < -0.7$. When no input signal is present, no power is drawn from the bias supply through the collectors of Q_1 or Q_2 so the class B operation is attractive when low standby power consumption is an important consideration. There is a small region of the input signal for which neither Q_1 nor Q_2 are on. The resulting output will therefore suffer some distortion.

The *npn* transistor Q_1 in the class B circuit in Fig. 9.4*a* has its collector connected to the positive power supply, V_{CC}, and its emitter connected to the load, R_L. The collector of the *pnp* output transistor, Q_2, has its collector connected to the negative supply voltage V_{EE}, which is often equal to $-V_{CC}$, and its emitter also connected to the load, R_L. The bases of Q_1 and Q_2 are connected together and are driven by the collector of the input transistor Q_3. The input transistor, Q_3, has a bias current source, I_{bias}, feeding its collector, which also provides base current for Q_1. The input voltage, v_{in} to the input transistor Q_3 is what drives the output stage. It is tempting when doing hand or SPICE calculations to start with v_{in}. However, because a small change in the base voltage of Q_3 makes a large change in the collector voltage of Q_3, it is easier to start the analysis at the base of Q_2. This base voltage is V_x in Fig. 9.4.

When $V_x = 0$, both output transistors Q_1 and Q_2 are turned off because the voltage is less than the 0.7V necessary to turn the transistors on. If $V_x > 0.7$, then Q_1 (*npn*) is on and Q_2 (*pnp*) is off. Current is then drawn from the power supply, V_{CC}, through Q_1 to produce the positive half-wave of the signal in the load. If $V_x < 0.7$, then Q_1 (*npn*) is off and Q_2 (*pnp*) is on. The voltage V_x is made negative by turning Q_3 on, thus bringing the collector voltage of Q_3 closer to V_{EE}, which is less than zero. An extreme positive or negative input voltage puts the turned-on output transistor (either Q_1 or Q_2) into saturation. The maximum positive output voltage is

$$V_o^+ = V_{CC} - V_{CE1(sat)} \tag{9.3}$$

and the maximum negative output voltage is

$$V_o^- = -V_{EE} + V_{EC2(sat)} \tag{9.4}$$

Typically, the value for $V_{CE(sat)} \approx 0.2V$ for a bipolar transistor. More design details are available from a variety of sources, such as [1].

9.2.2 Elimination of the Dead Band

The 1.2- to 1.4-V range in the base voltages of Q_1 and Q_2 can be substantially compensated by addition of two diodes in series between the bases of Q_1 and Q_2 (Fig. 9.4*b*). These diodes are named, respectively, D_4 and D_5. For purposes of calculation, let V_x stand for the voltage at the collector of the driver transistor Q_3, which is the same as the base voltage for the *pnp* output transistor Q_2. To get to the base of Q_1 from V_x now requires going through the two series-connected diodes "backward" from cathode to anode. If $V_x > 0$ but not so high as to turn off the diodes D_4 and D_5, then Q_1 is on as inferred from Fig. 9.4*a*. The voltage across the load is

$$V_o^+ = V_x + V_{D4} + V_{D3} - V_{BE1} \tag{9.5}$$

to make $V_x < 0$. The input voltage to the driver Q_3 must be a positive voltage. The *npn* output transistor Q_1 is turned off, and the excess bias current from I_{bias} flows through the diodes D_4 and D_5 and then through the now turned-on Q_3. In this case the output voltage is not now affected directly by the diode:

$$V_o^- = V_{EB2} + V_x \tag{9.6}$$

Under this condition, the value of V_x is actually a negative number. In the middle where $V_x = 0$, the output voltage across R_L is

$$V_o^+ = V_{D4} + V_{D5} - V_{BE1} \approx V_{BE} \tag{9.7}$$

and

$$V_o^- = V_{EB2} = V_{BE} \tag{9.8}$$

In either case, the output voltage is the voltage drop across one *pn* junction. If the forward diode voltage drops are equal to the base–emitter drops of the transistors, there is no discontinuity in V_o in going from negative to positive input voltages.

In actual production circuits, tight specifications are needed on diodes D_4 and D_5 since they are in the base circuit of the output transistors and consequently carry much less current than the output power devices. The discrepancy between the high-power and low-power devices can be alleviated by using the V_{BE} multiplier shown in Fig. 9.4*c*. In this circuit the base–emitter voltage of Q_4 sets the current through R_2:

$$I_{R2} = \frac{V_{BE4}}{R_2} \tag{9.9}$$

Assuming the base current of Q_4 is negligible, the voltage drop between the bases of the output transistors Q_1 and Q_2 is

$$V_{CE4} = I_{R2}(R_1 + R_2) = V_{BE4}\left(\frac{R_1}{R_2} + 1\right) = V_{BE1} + V_{EB2} \tag{9.10}$$

When the voltage at the base of Q_2 is positive, the load voltage is

$$V_o^+ = V_x + V_{BE4}\left(\frac{R_1}{R_2} + 1\right) - V_{BE1} \tag{9.11}$$

and when $V_x < 0$,

$$V_o^- = V_{EB2} + V_x \tag{9.12}$$

In the middle when $V_x = 0$, the V_o^+ and V_o^- can be forced to be equal by adjustment of the resistors R_1 and R_2:

$$V_o = V_{BE4}\left(\frac{R_1}{R_2}+1\right) - V_{BE1} = V_{BE2} \tag{9.13}$$

In addition to reducing or eliminating the dead-band zone, the compensation circuits in Figs. 9.4*b* and 9.4*c* also provide for temperature stability, since a change in the temperature changes the transistor V_{BE} value. The compensation circuit and the power transistors vary in the same way with temperature since they are physically close together.

Another aspect that deserves attention is the actual value of the current source, I_{bias}. Since this supplies the base current for the *npn* output transistor Q_1, I_{bias} must be large enough to not "starve" Q_1 when it is drawing the maximum current through its collector. This means that $I_Q \geq I_{C1}/\beta_1$.

9.2.3 Composite *pnp* Transistor

One of the primary problems in using this type of class B amplifier is the requirement for obtaining two equivalent complementary transistors. Fundamentally, the problem arises because of the greater mobility of electrons by over a 3:1 factor over that of holes in silicon. The symmetry of the gain in this circuit depends on the two output transistors having the same small-signal short-circuit base-to-collector current gain, $\beta = i_c/i_b$. When it is not possible to obtain a high-β *pnp* transistor, it is sometimes possible to use a composite transistor connection. A high-power *npn* transistor, Q_1, is connected to a low-power low-β *pnp* transistor, Q_2, as shown in Fig. 9.5. Normally, the base–emitter junctions of the composite and single *pnp* transistor are forward biased so that

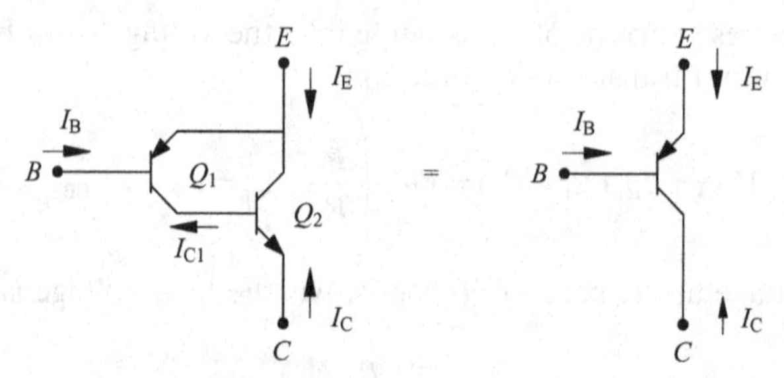

FIGURE 9.5 Composite connection for *pnp* transistor.

the Shockley diode equation may be used to describe the bias currents. For Q_2 in the composite circuit,

$$I_{C2} = -I_S e^{qV_{EB}/kT} \tag{9.14}$$

The collector current for Q_1 in the composite circuit is the same as the collector current for the single *pnp* transistor:

$$I_C = (\beta_1 + 1) I_{C2} = -(\beta_1 + 1) I_S e^{V_{BE}/kT} \tag{9.15}$$

The composite circuit has the polarity of a *pnp* transistor with potentially the gain of an *npn* transistor.

9.2.4 Small-Signal Analysis

The three fundamental parameters that characterize an amplifier are its voltage gain, A_v, input resistance, R_i, and output resistance, R_o. In the circuit shown in Fig. 9.4*a*, neither Q_1 nor Q_2 are on simultaneously. If Q_1 is on, Q_2 is an open circuit and need not be considered as part of the ac analysis. A small-signal hybrid π model (Fig. 9.6) for a bipolar transistor consists of a base resistance, r_b, base–emitter resistance, r_π, collector–emitter resistance, r_o, transconductance, g_m, and short-circuit current gain $\beta = g_m r_\pi$. There are in addition high-frequency effects caused by reactive parasitic elements within the device. Since the voltage gain of an emitter follower is ≈1, the voltage gain of the Q_3 and Q_1 combination is

$$A_v = -g_{m3} R_{L(eff)} \tag{9.16}$$

The effective load resistance $R_{L(eff)}$ seen by the first transistor, Q_3, is the same as the input resistance of the emitter follower circuit Q_1. Circuit analysis of the low-frequency transistor hybrid model shown in Appendix F gives

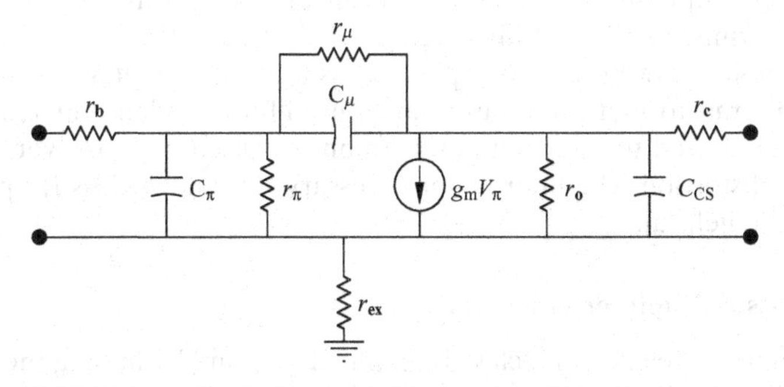

FIGURE 9.6 Small signal hybrid π model of bipolar transistor.

$$R_{L(eff)} = r_{\pi 1} + r_{b1} + (\beta_1 + 1)(r_{o1} \| R_L) \quad (9.17)$$

$$\approx \beta_1 R_L \quad (9.18)$$

The voltage gain is then found by substitution:

$$A_v^+ = -g_{m3}[r_{\pi 1} + r_{b1} + (\beta_1 + 1)(r_{o1} \| R_L)] \quad (9.19)$$

$$\approx -g_{m3}\beta_1 R_L \quad (9.20)$$

The low-frequency input resistance to the actual class B amplifier is given by $R_{L(eff)}$ in Eq. (9.17), and the output resistance is

$$R_o = \frac{r_\pi + r_{bb} + r_b}{1+\beta} \quad (9.21)$$

Thus, the input resistance is high and the output resistance is low for a class B amplifier, which enables it to drive a low-impedance load with high efficiency.

9.2.5 All-*npn* Class B Amplifier

The complementary class B amplifier shown in Fig. 9.4 needs to have symmetrical *npn* and *pnp* devices. In addition this circuit also requires complementary power supplies. These two problems can be alleviated by using the totem pole or all *npn* transistor class B amplifier. This circuit requires only one power supply and has identical *npn* transistors that amplify both the positive and negative halves of the signal. However, it requires that the two transistors operate with an input phase differential of 180°. This circuit is illustrated in Fig. 9.7. Clearly, the cost of the all-*npn* transistor amplifier is the added requirement of two center-tapped transformers. These are necessary to obtain 180° phase difference between Q_1 and Q_2. The center-tapped transformer also provides dc isolation for the load. When the input voltage is positive, Q_1 is on and Q_2 is off. When the input voltage is negative, the input transformer induces a positive voltage at the "un-dotted" secondary winding, which turns Q_2 on. The output of Q_2 will induce on the output transformer a positive voltage on the un-dotted terminal and a negative voltage on the "dotted" terminal. The negative input voltage swing is thus replicated as a negative voltage swing at the output. The transformer turns ratio can be used for impedance matching. The output filter is used to filter out any harmonics caused by crossover or other sources of distortion. The filter is not necessary to achieve class B operation, but it can be helpful.

9.2.6 Class B Amplifier Efficiency

The maximum efficiency of a class B amplifier is found by finding the ratio of the output power delivered to the load to the required dc power from the bias

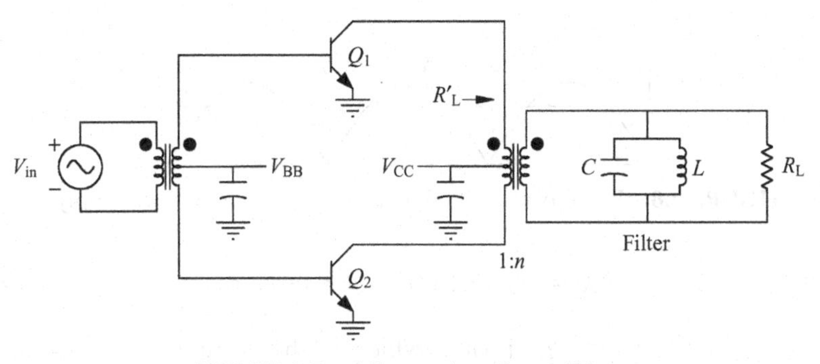

FIGURE 9.7 All *npn* class B amplifier.

voltage supply. In determining efficiency in this way, power losses caused by nonzero base currents and crossover distortion compensation circuits used in Figs. 9.4*b* and 9.4*c* are neglected. Furthermore, the power efficiency (or collector efficiency) rather than the power-added efficiency is calculated so as to form a basis for comparison for alternative circuits. It is sufficient to do the calculation during the part of the cycle when Q_1 is on and Q_2 is off. The load resistance in Fig. 9.7 is transformed through to the primary side of the output transformer, loading the transistors with a value of R'_L. Since the transformer is assumed lossless, referring the load resistance to the primary side, R'_L will not change the efficiency.

The peak magnitude of the collector current that flows into R'_L is $\hat{I}_C$. The alternating current is

$$i_o(\omega t) = \hat{I}_C \sin(\omega t) \tag{9.22}$$

and the peak voltage is

$$v_o(\omega t) = \hat{I}_C R'_L \sin(\omega t) \tag{9.23}$$

Since the collector–base voltage must remain positive to avoid the danger of burning out the transistor, $\hat{V}_C = \hat{I}_C R'_L < V_{CC}$. The maximum allowable output power delivered to the load is

$$P_o = \frac{\hat{V}_o^2}{2R'_L} \tag{9.24}$$

A determination of the direct current supplied by the bias supply is needed. The magnitude of the current delivered by the bias supply to the load by Q_1 is

$$i_{BB1} = \hat{I}_C \sin(\omega t) \qquad 0 < \omega t < \pi \tag{9.25}$$

and for Q_2

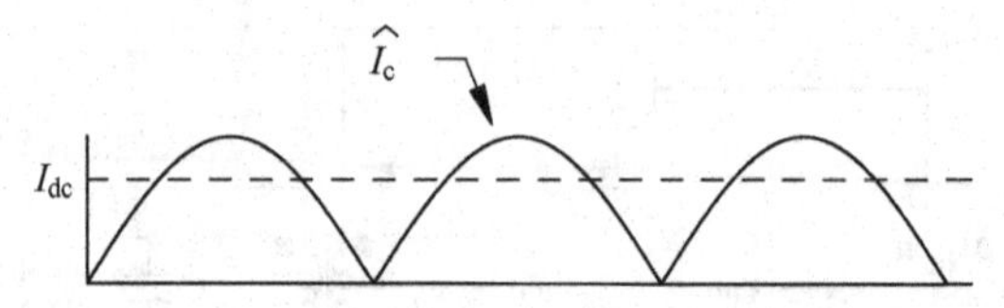

FIGURE 9.8 Waveform for finding average dc from power supply.

$$i_{BB2} = -\hat{I}_C \sin(\omega t) \qquad \pi < \omega t < 2\pi \tag{9.26}$$

The total current is then $\hat{I}_C \sin(\omega t)$, which is shown in Fig. 9.8. The direct current from the bias sources is found by finding the average current:

$$\begin{aligned} I_{dc} &= \frac{2}{T}\int_0^{T/2} \hat{I}_{C1} \sin \omega t \, dt \\ &= -\frac{2\hat{I}_{C1}}{\omega t} \cos \omega t \Big|_0^{T/2} \\ &= -\frac{2\hat{I}_{C1}}{(2\pi/T)T}\left[\cos\left(\frac{2\pi}{T}\frac{T}{2}\right) - 1\right] \\ &= -\frac{\hat{I}_{C1}}{\pi}[-1-1] \end{aligned} \tag{9.27}$$

$$I_{dc} = \frac{2\hat{I}_{C1}}{\pi} = \frac{2}{\pi}\frac{\hat{V}_o}{R'_L} \tag{9.28}$$

The power drawn from both of the power supplies by both of the output transistors is

$$P_{dc} = V_{CC} I_{dc} = \frac{2}{\pi}\frac{\hat{V}_o}{R'_L} V_{CC} \tag{9.29}$$

Thus, the output power is proportional to $\hat{V}_o$ and is the average power drawn from the power supply. The power delivered to the load is

$$P_o = \frac{\left|\hat{V}_o\right|^2}{2R'_L} \tag{9.30}$$

The efficiency is the ratio of these latter two values:

$$\eta = \frac{P_o}{P_{dc}} = \frac{\left|\hat{V}_o\right|^2}{2R'_L}\frac{\pi}{2}\frac{R'_L}{V_{CC}V_o} \tag{9.31}$$

$$= \frac{\pi}{4}\frac{\hat{V}_o}{V_{CC}} \tag{9.32}$$

The maximum output power occurs when the output voltage is $V_{CC} - V_{CE(sat)}$:

$$P_{o(max)} = \frac{1}{2}\frac{(V_{CC} - V_{CE(sat)})^2}{R_L'} \tag{9.33}$$

$$\eta_{max} = \frac{\pi}{4}\frac{V_{CC} - V_{CE(sat)}}{V_{CC}} \approx 78.5\% \tag{9.34}$$

This efficiency for the class B amplifier should be compared with the maximum efficiency of a class A amplifier where $\eta_{max} = 25\%$ when the bias to the collector is supplied through a resistor and $\eta_{max} = 50\%$ when the bias to the collector is supplied through an RF choke.

9.3 CLASS C AMPLIFIER

The class C amplifier is useful for providing a high-power continuous wave (CW) output. When it is used in amplitude modulation schemes, the output variation is done by varying the bias supply [2]. There are several characteristics that distinguish the class C amplifier from the class A or B amplifier. First of all, it is biased so that the transistor conduction angle is <180°. Consequently, the class C amplifier is clearly nonlinear in that it does not directly replicate a broadband input signal like the class A and B amplifiers do (at least in principle). The class A amplifier requires one transistor, the class B amplifier requires two transistors, and the class C amplifier uses one transistor. Topologically, it looks similar to the class A except for the dc bias level at the transistor input. It was noted in the class B amplifier, an output filter was used optionally to help clean up the output signal. In the class C amplifier, such a tuned output is necessary in order to recover the sine wave. Finally, class C operation is capable of higher efficiency than either of the previous two cases, so when dealing with the appropriate signal types, they become very attractive as power amplifiers.

The class C amplifier in Fig. 9.9 shows the output circuit for an *n*-channel enhancement mode MOSFET. A power bipolar transistor might also be used

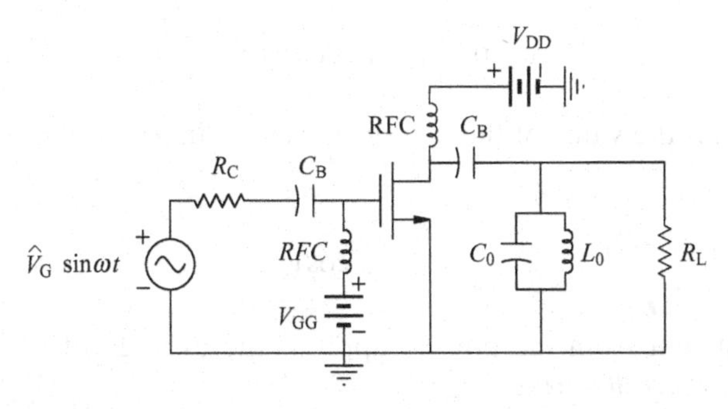

FIGURE 9.9 Simple class C amplifier where V_{GG} determines conduction angle.

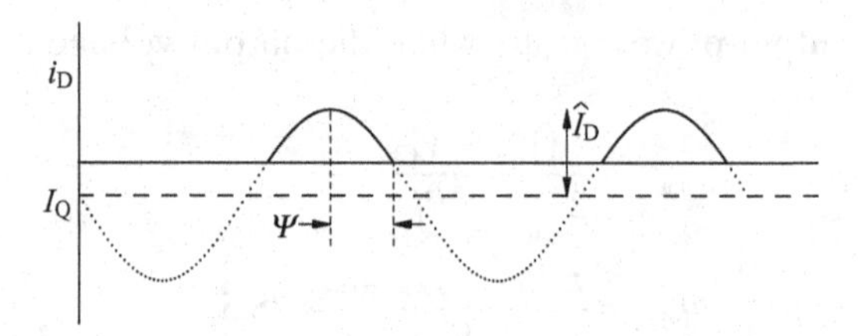

FIGURE 9.10 Drain current waveform for class C operation.

here. The choice between the two lies in the specific design goals for the amplifier. Actually, the analysis that follows for the class C amplifier is based on triode vacuum tubes [3], which were later incorporated by [2]. The Q of the tuned circuit will determine the bandwidth of the amplifier. The large inductance RF coil in the drain voltage supply ensures that only dc current flows there. During that part of the input cycle when the transistor is on, the bias supply current flows through the transistor and the output voltage is approximately 90% of V_{DD}. When the transistor is off, the supply current flows into the blocking capacitor. The current waveform at the drain can be modeled as the waveform shown in Fig. 9.10:

$$i_D(\omega t) = \begin{cases} I_Q - \hat{I}_D \sin(\omega t) & 3\pi/2 - \psi \le \omega t \le 3\pi/2 + \psi \\ 0 & \text{otherwise} \end{cases} \tag{9.35}$$

Other approximations have been made for this current waveform based on different device characteristics, but Eq. (9.35) will provide a reasonable analytical solution.

For class C operation, the magnitude of the quiescent current is $|I_Q| < \hat{I}_C$. The point where the total current = 0 is

$$i_D\left(\frac{3\pi}{2} \pm \psi\right) = 0 = I_Q - \hat{I}_D \sin\left(\frac{3\pi}{2} \pm \psi\right)$$

$$0 = I_Q + \hat{I}_D \cos\psi \tag{9.36}$$

This determines the value of the quiescent current in terms of the conduction angle 2ψ:

$$I_Q = -\hat{I}_D \cos\psi \tag{9.37}$$

The direct current from the power supply is the average of the total drain current $i_D(\theta)$ where $\theta = \omega t$:

$$
\begin{aligned}
I_{\mathrm{dc}} &= \frac{1}{2\pi}\int_0^{2\pi} i_{\mathrm{D}}(\theta)\,d\theta \\
&= \frac{1}{2\pi}\int_{3\pi/2-\psi}^{3\pi/2+\psi}\left(I_{\mathrm{Q}} - \hat{I}_{\mathrm{D}}\sin\theta\right)d\theta \\
&= \frac{1}{\pi}\left(\psi I_{\mathrm{Q}} + \hat{I}_{\mathrm{D}}\sin\psi\right) \qquad (9.38)
\end{aligned}
$$

Evaluation of the integral makes use of the trigonometric identity, $\cos(\alpha \pm \beta) = \cos\alpha\cos\beta \pm \sin\alpha\sin\beta$. From Eq. (9.36) the direct current is

$$I_{\mathrm{dc}} = \frac{\hat{I}_{\mathrm{D}}}{\pi}(\sin\psi - \psi\cos\psi) \qquad (9.39)$$

This gives the average of the actual conducting current of the clipped sine wave in Fig. 9.10. The quiescent current, I_{Q}, is the negative bias current around which the alternating current flows if it could. This gives the direct current from the power supply in terms of $\hat{I}_D$ and the conduction angle ψ, so that power supplied by the source is

$$P_{\mathrm{in}} = V_{\mathrm{DD}} I_{\mathrm{dc}} \qquad (9.40)$$

The ac component of the current flows through the blocking capacitor and into the load. Harmonic current components are shorted to ground by the tuned circuit. The magnitude of the output voltage at the fundamental frequency is found using the Fourier method. The minus sign is used because the drain current convention draws current up from the load:

$$\hat{V}_{\mathrm{o}} = -\frac{1}{\pi}\int_0^{2\pi} i_{\mathrm{D}}(\theta) R_{\mathrm{L}}\sin\theta\,d\theta \qquad (9.41)$$

$$= -\frac{R_{\mathrm{L}}}{\pi}\int_{(3\pi/2)-\psi}^{(3\pi/2)+\psi}\left(I_{\mathrm{Q}} - \hat{I}_{\mathrm{D}}\sin\theta\right)\sin\theta\,d\theta$$

$$= \frac{R_{\mathrm{L}}}{\pi}\left[I_Q(-\cos\theta)\Big|_{(3\pi/2)-\psi}^{(3\pi/2)+\psi} + \frac{\hat{I}_{\mathrm{D}}}{2}\left(\theta - \frac{\sin 2\theta}{2}\right)\Bigg|_{(3\pi/2)-\psi}^{(3\pi/2)+\psi}\right] \qquad (9.42)$$

$$= \frac{R_{\mathrm{L}}}{\pi}\left(2I_{\mathrm{Q}}\sin\psi + \frac{\hat{I}_{\mathrm{D}}}{2}\left\{(2\psi) - \frac{1}{2}[\sin(3\pi + 2\psi) - \sin(3\pi - 2\psi)]\right\}\right) \qquad (9.43)$$

The quiescent current term, I_{Q}, is replaced using Eq. (9.37) again, and the trigonometric identity for $\sin\alpha\cos\beta$ is used:

$$\hat{V}_o = \frac{R_L \hat{I}_D}{\pi}\left[-\sin 2\psi + \psi - \frac{1}{4}(-\sin 2\psi - \sin 2\psi)\right] \tag{9.44}$$

$$= \frac{R_L \hat{I}_D}{2\pi}(2\psi - \sin 2\psi) \tag{9.45}$$

The ac output power delivered to the load is

$$P_o = \frac{\hat{V}_o^2}{2R_L} \tag{9.46}$$

The efficiency (neglecting the input power) is simply the ratio of the output ac power to the input dc power. The maximum output power occurs when $\hat{V}_o = V_{DD}$. The maximum efficiency is then

$$\eta_{max} = \frac{P_{o(max)}}{P_{dc}} = \left(\frac{V_{DD}^2}{2R_L}\right)\left(\frac{1}{V_{DD} I_{dc}}\right) \tag{9.47}$$

$$= \frac{2\psi - \sin 2\psi}{4(\sin\psi - \psi\cos\psi)} \tag{9.48}$$

A plot of this expression (Fig. 9.11) clearly illustrates the efficiency in terms of the conduction angle for class A, B, and C amplifiers. The increased efficiency of the class C amplifier is a result of the drain current flowing for less than a

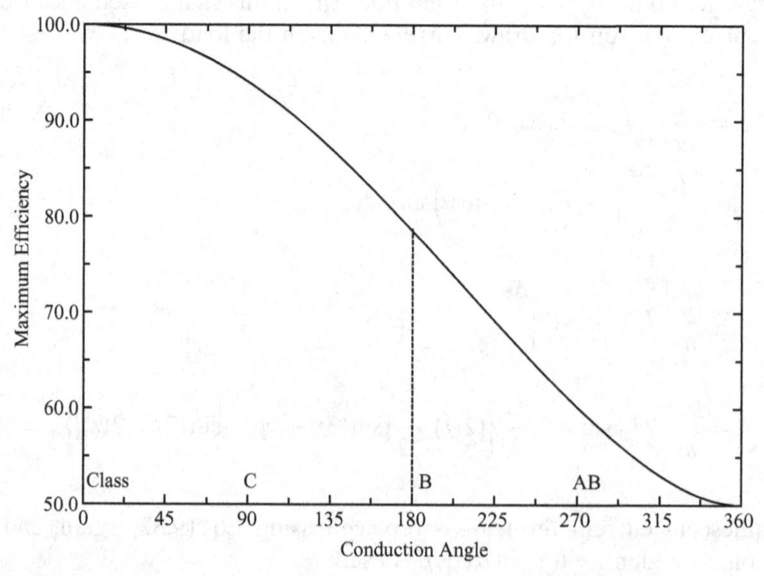

FIGURE 9.11 Power efficiency for class A, B, and C amplifiers.

half cycle. When the drain current is maximum, the drain voltage is minimum, so the power dissipation is inherently lower than class B or class A operation.

Another important parameter for the power amplifier is the ratio of the maximum average output power where $\hat{V}_o = V_{DD}$, to the peak instantaneous output power:

$$r = \frac{P_{o(max)}}{V_{D(max)} i_{D(max)}} \tag{9.49}$$

The maximum average output power occurs when $\hat{V}_o = V_{DD}$ and is given by

$$P_{o(max)} = \frac{V_{DD}^2}{2R_L} \tag{9.50}$$

$$= \frac{\hat{I}_D^2 R_L^2}{4\pi^2} \frac{(2\psi - \sin 2\psi)^2}{2R_L} \tag{9.51}$$

The maximum voltage at the drain is the peak output voltage plus the dc bias voltage:

$$V_{D(max)} = 2V_{DD} = 2\frac{\hat{I}_D R_L}{2\pi}(2\psi - \sin 2\psi) \tag{9.52}$$

The maximum current based on Eq. (9.37) is

$$i_{D(max)} = I_Q + \hat{I}_D = -\hat{I}_D \cos\psi + \hat{I}_D \tag{9.53}$$

The ratio of the maximum average power to the peak power from Eq. (9.49) is

$$r = \frac{2\psi - \sin 2\psi}{8\pi(1 - \cos\psi)} \tag{9.54}$$

A plot of this ratio as a function of conduction angle in Fig. 9.12 shows that maximum efficiency of the class C amplifier occurs when there is no output power. Figures 9.11 and 9.12 show the trade-offs in choosing the appropriate conduction angle for class C operation.

9.4 CLASS C INPUT BIAS VOLTAGE

Device SPICE models for RF power transistors are relatively rare. Manufacturers often do supply optimum generator and load impedances that have been found to provide the rated output power for the designated frequency. The circuit shown in Fig. 9.9 is a generic example of a class C amplifier in which the manufacturer has determined empirically the optimum generator and load impedances. The Q of the output tuned circuit is

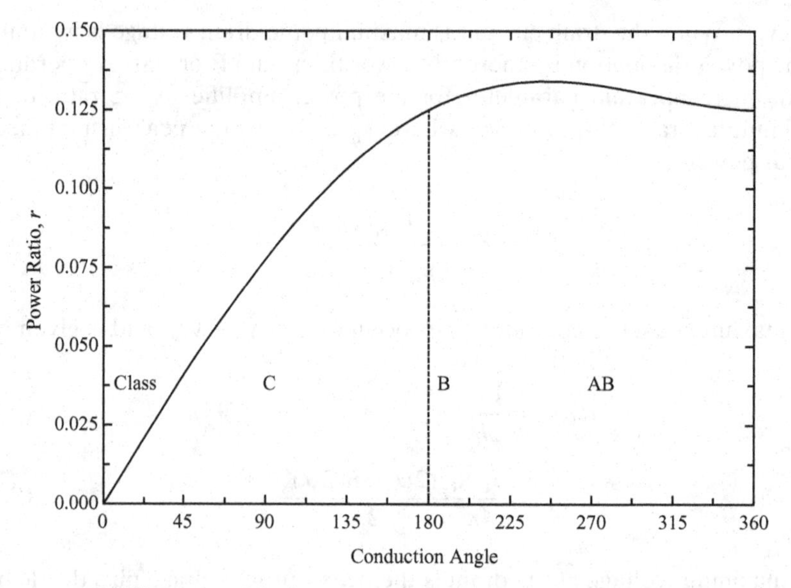

FIGURE 9.12 Maximum output power to peak power ratio.

$$Q = \frac{f_0}{\Delta f} \tag{9.55}$$

The Q then determines the inductance and capacitance of the output shunt resonant circuit:

$$C = \frac{Q}{\omega_0 R_L} \tag{9.56}$$

$$L = \frac{R_L}{\omega_0 Q} \tag{9.57}$$

Furthermore, if the desired output power is $P_{Q(max)}$, the drain voltage source, V_{DD}, and the maximum drain current is $i_{D(max)}$, then the average to peak power ratio, r, is found from Eqs. (9.49) and (9.54). Iterative solution of Eq. (9.54) gives the value for the conduction angle, ψ. This will then allow for estimation of the maximum efficiency from Eq. (9.48). Alternatively, for a given desired efficiency, the conduction angle, ψ, can be obtained by iterative solution of Eq. (9.48). Numerically, it is useful to take the natural logarithm of Eq. (9.48) before searching iteratively for a solution:

$$\ln \eta_{max} = \ln(2\psi - \sin 2\psi) - \ln[4(\sin\psi - \psi\cos\psi)] \tag{9.58}$$

The efficiency expression can be modified to account for the nonzero drain–source voltage:

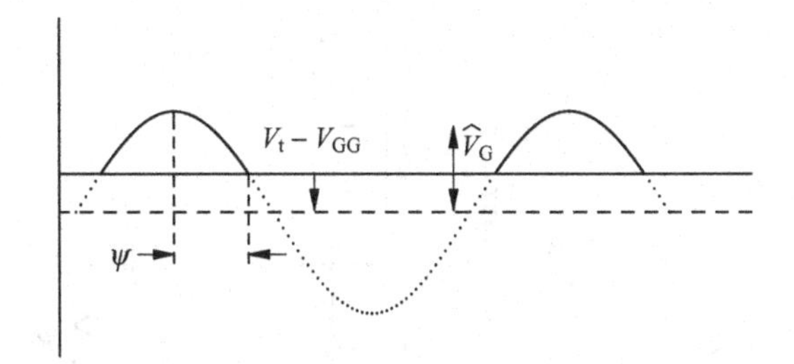

FIGURE 9.13 Conduction angle dependence on V_{BB}.

$$\eta_{max} = \frac{2\psi - \sin 2\psi}{4(\sin\psi - \psi\cos\psi)}\left(\frac{V_{DD} - V_{DS(on)}}{V_{DD}}\right) \tag{9.59}$$

To achieve the desired conduction angle, the gate–source voltage must be biased so that the transistor will be in conduction for the desired portion of the input signal. Drain current flows when the transistor is above cutoff, $V_{GS} > V_t$. First, it is necessary to determine the required generator voltage amplitude, $\hat{V}_G$, that will produce the desired maximum output current. This is illustrated in Fig. 9.13 where the input ac signal is superimposed on the gate bias voltage. The input voltage commences to rise above the turn on voltage of the transistor at

$$-V_{GG} = \hat{V}_G \cos\psi - V_t \tag{9.60}$$

In this way the gate bias voltage is determined.

9.5 CLASS D POWER AMPLIFIER

Inspection of the efficiency and output power of a class C amplifier reveals that 100% efficiency only occurs when the input power is zero. A modification of class B operation shown in [4] indicates that a judicious choice of bias voltages and circuit impedances provide a clipped voltage waveform at each output of the transistors while retaining the half sine wave output current. In the limit the clipped voltage waveform becomes a square wave. This is no longer linear and, thus, is distinguished from the class B amplifier.

The class D amplifier shown in Fig. 9.14 superficially looks like a class B amplifier except for the input side bias. In class D operation, the transistors act as near ideal switches that are on half of the time and off half of the time. The transistors may be pulse width modulated to produce an output that does not have a 50% duty cycle. However, in this discussion the input is excited by a square wave. If the transistor switching time is near zero, then drain current flows while the drain–source voltage $V_{DS} = 0$ and V_{DS} is nonzero when $I_D = 0$.

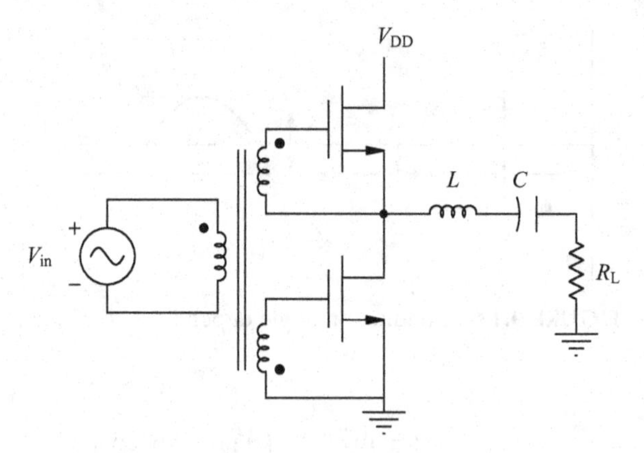

FIGURE 9.14 Class D power amplifier.

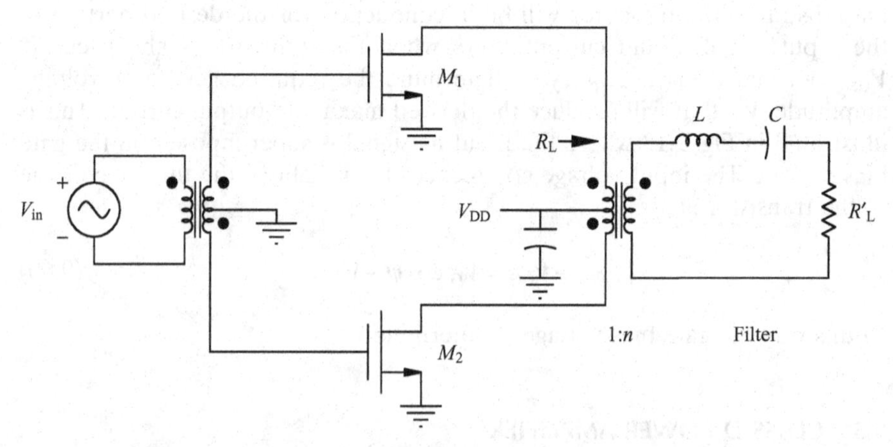

FIGURE 9.15 Practical voltage mode class D amplifier.

As a result 100% efficiency is theoretically possible. In practice, the switching speed of a transistor is not sufficiently fast at high frequencies to produce square waves using this design.

For the circuit shown in Fig. 9.14, the voltage is either at V_{DD} or 0, depending on the phase of the input signal. However, only current at the resonant frequency of the *LC* resonator can pass on to the load. Since one transistor is on while the other is off, the transistors must either be complementary devices or make use of a center-tapped transformer as shown. This ensures the required 180° phase difference between the inputs of the two transistors. A more practical class D circuit is shown in Fig. 9.15. Both of these circuits are described as *voltage mode* circuits since the voltage at the input of the filter is

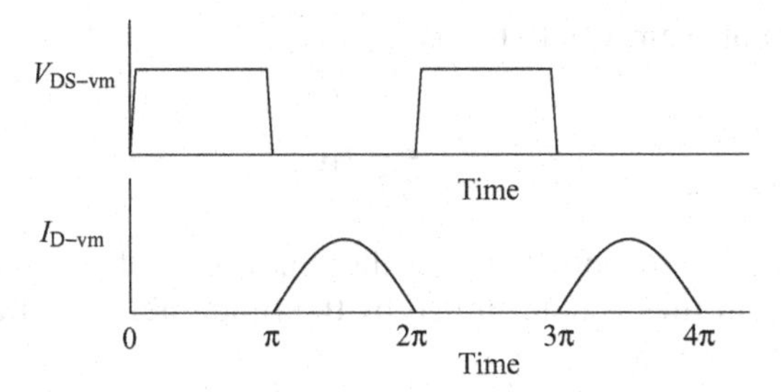

FIGURE 9.16 Voltage and current waveforms for voltage mode (vm) amplifier.

approximately a square wave. The waveform of the voltage mode circuit is shown in Fig. 9.16.

9.5.1 Class D Amplifier Efficiency

The analysis begins by finding the ac power delivered to the load. The Fourier series expansion of the square wave voltage at the input to the LC filter is

$$V_{DS} = V_{DD}\left\{\frac{1}{2}+\sum_{n=1}\frac{1}{n\pi}\left[1-(-1)^n\right]\sin n\theta\right\} \tag{9.61}$$

where $\theta = \omega t$. The current going through the resonant LC circuit is

$$i_L(\theta)=\frac{2V_{DD}}{\pi R_L}\sin\theta \tag{9.62}$$

Consequently, the RF power delivered to the load is

$$P_o = \frac{1}{2}\left[\frac{2V_{DD}}{\pi R_L}\right]^2 R_L \tag{9.63}$$

$$= V_{DD}^2\frac{2}{\pi^2 R_L} \tag{9.64}$$

The rectified voltage mode current wave shown in Fig. 9.16 is given as

$$I_D = \begin{cases} 0 & 0 \le \theta < \pi \\ \hat{I}_D \sin\theta & \pi \le \theta < 2\pi \end{cases} \tag{9.65}$$

A Fourier expansing of Eq. (9.65) results in

$$I_D = \hat{I}_D\left\{\frac{1}{\pi}+\frac{1}{2}\sin\theta+\sum_{n=1}\frac{1}{\pi(n^2-1)}\left[1-(-1)^{n+1}\right]\cos n\theta\right\} \tag{9.66}$$

The dc component from Eq. (9.66) is

$$I_{dc} = \frac{1}{\pi}\hat{I}_D \tag{9.67}$$

The amplitude, $\hat{I}_D$, is proportional to the amplitude of the rectified current sine wave, which in turn is determined by the fundamental voltage amplitude, $2V_{DD}/\pi$. From Eq. (9.67)

$$I_{dc} = \frac{1}{\pi}\left(\frac{2V_{DD}}{\pi R_L}\right) \tag{9.68}$$

The dc input power is

$$\begin{aligned} P_{dc} &= I_{dc}V_{DD} \\ &= \frac{2V_{DD}^2}{\pi^2 R_L} \end{aligned} \tag{9.69}$$

The ratio of the RF power in the load from Eq. (9.64) and the dc power from Eq. (9.69) gives an efficiency of 100%.

The efficiency of the class D amplifier falls short of the promised 100% because of the presence of parasitic lead inductance and especially the shunt capacitance, C_{DS}. The switching transistors can be modeled as an ideal switch plus these parasitic reactances (Fig. 9.17) [5]. The energy stored in the inductor, $Li^2/2$ and in the capacitor, $Cv^2/2$, is dissipated during each cycle. The current is not completely zero when the voltage starts to turn on, and the voltage is not completely zero when the current starts to turns on. This overlap gets worse as the operating frequency increases.

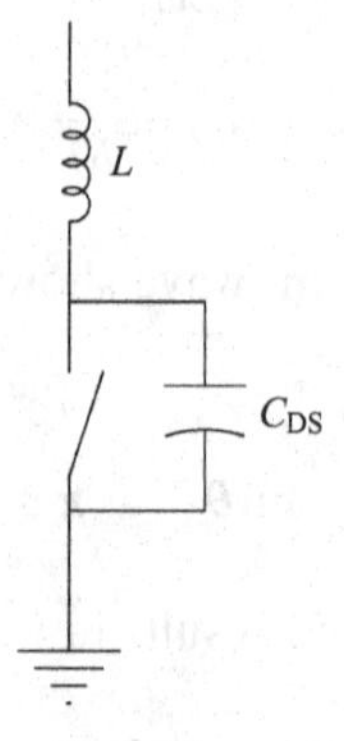

FIGURE 9.17 Switching transistor model.

9.5.2 Current Mode Class D Amplifier

Voltage mode operation occurs when the voltage is a square wave and the current a rectified sine wave. Reversing this as suggested by Fig. 9.18 so that the current is a square wave and the voltage a sine wave can increase the operating frequency range significantly. This circuit design was described in [5] and later in a thesis [6]. A circuit schematic for the current mode class D amplifier is shown in Fig. 9.19. The basic idea is to use the parasitic inductance or capacitance (Fig. 9.17) as part of the resonant circuit. The voltage mode circuit can absorb the inductance while the current mode circuit can absorb the capacitance. Since the capacitance is a far larger problem than the inductance, the current mode design is in practice capable of a much higher frequency range.

Current mode class D waveforms are shown in Fig. 9.18 and the circuit is shown in Fig. 9.19. The two transistors must be either a complementary pair or they must be excited out of phase with one another. The drain bias for the transistors goes through a choke that permits only a dc current to flow into the transformer center tap. This steady current is switched between the top transistor and the bottom transistor, thereby producing a current square wave. The *LC* filter will only allow the fundamental ac voltage to appear across the load. All higher voltage harmonic components are shorted to ground by the capacitor, *C*. Furthermore, *C* and the C_{DS} from the transistors appear in shunt with one another so that C_{DS} can be made to be part of the resonant circuit.

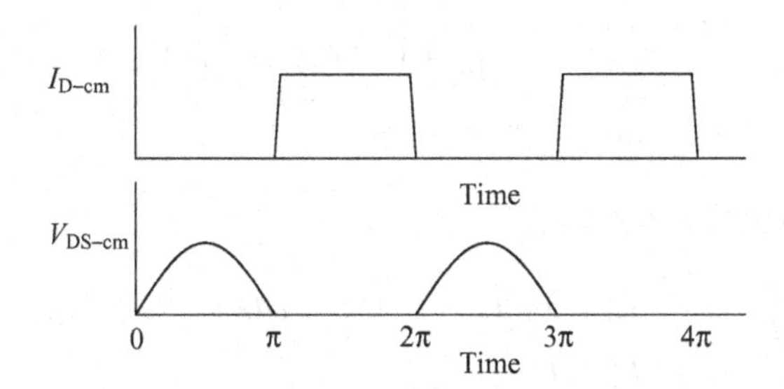

FIGURE 9.18 Current mode waveforms.

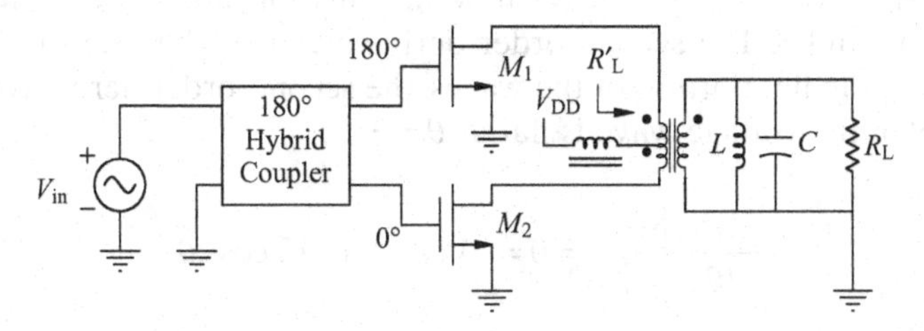

FIGURE 9.19 Current mode class D amplifier.

The square current waveform and the rectified sine voltage waveform can be expanded in a Fourier series much like the voltage mode waveforms were. The choice of origin will change the sign of the harmonic components from that obtained for the voltage mode:

$$V_{\mathrm{DS}} = V_{\mathrm{DD}}\left\{\frac{1}{\pi}+\frac{1}{2}\sin\theta+\sum_{n=1}\frac{1}{\pi(n^2-1)}\left[(-1)^{n+1}-1\right]\cos n\theta\right\} \tag{9.70}$$

$$I_{\mathrm{D}} = \hat{I}_{\mathrm{D}}\left\{\frac{1}{2}+\sum_{n=1}\frac{1}{n\pi}\left[(-1)^{n}-1\right]\sin n\theta\right\} \tag{9.71}$$

These two equations are complementary to Eqs. (9.66) and (9.61), respectively. The efficiency can be calculated when the current waveform is less than a perfect square wave. If only the fundamental and one harmonic is considered, the first three terms of these two expressions give

$$V_{\mathrm{DS}} = V_{\mathrm{DD}} + V_1\sin\theta - V_2\cos 2\theta \tag{9.72}$$

$$I_{\mathrm{D}} = I_{\mathrm{dc}} - I_1\sin\theta - I_3\sin 3\theta \tag{9.73}$$

Rather than consider the last two equations as a truncated series, they are considered to be the total voltage and current with the restriction that neither one is ever less than zero. This is the technique used in [5]. From the voltage waveform pictured in Fig. 9.18, at $\theta = 3\pi/2$:

$$V_{\mathrm{DS}} = 0 = V_{\mathrm{DD}} + V_1\cdot(-1) - V_2\cdot(-1) \tag{9.74}$$

$$V_{\mathrm{DD}} = V_1 - V_2 \tag{9.75}$$

and for the current at $\theta = \pi/2$

$$I_{\mathrm{D}} = 0 = I_{\mathrm{dc}} - I_1\cdot(1) - I_3\cdot(-1) \tag{9.76}$$

$$I_{\mathrm{dc}} = I_1 - I_3 \tag{9.77}$$

The first-order derivatives of the waveforms are zero at $V_{\mathrm{DS}}(\pi/2)$, $I_{\mathrm{D}}(\pi/2)$, and $I_{\mathrm{D}}(3\pi/2)$. However, this yields no new information about the voltage and current amplitudes. The second-order derivative gives the rate of change of the slope. In the flat portion of the waves, the second-order derivative is zero. In particular, the voltage wave is flat at $\theta = 3\pi/2$:

$$\left.\frac{d^2V_{\mathrm{DS}}}{d\theta^2}\right|_{\theta=3\pi/2} = 0 = -V_1\sin\theta + 4V_2\cos 2\theta \tag{9.78}$$

$$V_1 = 4V_2 \tag{9.79}$$

This along with Eq. (9.75) yields

$$V_1 = \frac{4}{3} V_{DD} \tag{9.80}$$

$$V_2 = \frac{1}{3} V_{DD} \tag{9.81}$$

For the current wave, there are two flat spots. Evaluating the second-order derivative at either of these places gives the same result:

$$\left. \frac{d^2 I_D}{d\theta^2} \right|_{\pi/2, 3\pi/2} = 0$$

$$= I_1 \sin\theta + 9 I_3 \sin 3\theta \tag{9.82}$$

or

$$I_1 = 9 I_3 \tag{9.83}$$

This with Eq. (9.77) yields

$$I_1 = \frac{9}{8} I_{dc} \tag{9.84}$$

$$I_3 = \frac{1}{8} I_{dc} \tag{9.85}$$

The above values for the voltage and current components using one harmonic can be used to determine the efficiency. When the V_{DS} from transistor M_1 in Fig. 9.19 is high, the fundamental voltage across R_L is high. When the V_{DS} from M_2 is high the fundamental voltage across R_L is low. The ac voltage swing around the input to the center-tapped transformer is $2V_1$ centered at V_{DD}. For the purposes of calculating efficiency, the transformer is considered lossless. The load resistance transformed to the primary side is R_L'. The total dc current from the power supply to the two transistors is

$$I_{dc-t} = 2 I_{dc} = \frac{16}{9} I_1 \tag{9.86}$$

and the dc power is

$$P_{dc} = I_{dc-t} V_{DD}$$

$$= \frac{16}{9} I_1 V_{DD} \tag{9.87}$$

Since

$$I_1 = \frac{2V_1}{R_L'} = \frac{8}{3} \frac{V_{DD}}{R_L'} \tag{9.88}$$

then

$$P_{dc} = \frac{128}{27} \frac{V_{DD}^2}{R_L'} \tag{9.89}$$

The fundamental ac power delivered to the load, R_L' is

$$P_o = \frac{1}{2} I_1^2 R_L' = \frac{1}{2}\left(\frac{8V_{DD}}{3R_L'}\right)^2 R_L' \tag{9.90}$$

$$= \frac{32}{9}\frac{V_{DD}^2}{R_L'} \tag{9.91}$$

Thus, the drain efficiency is

$$\eta = \frac{P_o}{P_{dc}} = \frac{3}{4} \tag{9.92}$$

This calculation was done using only one harmonic above the fundamental to approximate the current square wave. As the number of harmonics are increased, the more square the current waveform becomes and the closer to 100% efficiency is achieved. Indeed, an analytical expression using two harmonics above the fundamental has been provided in [5]. The major contribution of this study is to show how current mode class D amplifiers can incorporate C_{DS} so that high-efficiency, high-power (in the tens of watts) operation can be practically achieved in the low-gigahertz frequency range [6].

9.6 CLASS E POWER AMPLIFIER

The class E amplifier, like the class D, assumes the transistor acts as an ideal switch rather than a current source. Both the class D and the class E nonlinear amplifiers are capable of achieving 100% efficiency. The major weakness of the class D amplifier is that the frequency of operation is limited by the rise and fall times of the transistor output pulse. The nonzero switching time is a result of parasitic capacitances in the circuit and transistor. Even the current mode class D amplifier will eventually have this difficulty at high frequencies. The class E design turns this liability into a feature, so that high-efficiency operation can be achieved at higher frequencies than that available in the class D circuit.

The class E amplifier concept was first introduced by Ewing [7] in 1964 and later independently rediscovered by Sokal and Sokal [8] in 1975. In 1977 and 1978, Raab derived design equations for the class E amplifier [9, 10]. This work was later expanded to include a relationship between maximum output power and maximum efficiency. The derivation in [9] was done for an arbitrary duty cycle, but it is simpler to assume the duty cycle is 50% as done in [11] and [12] for the simple reason that high-frequency switching is difficult and counterproductive at a duty cycle other than 50%.

The material in this section on class E amplifiers is largely the result of work done by Cantrell as described in [11] and [12]. The basic feature of the class

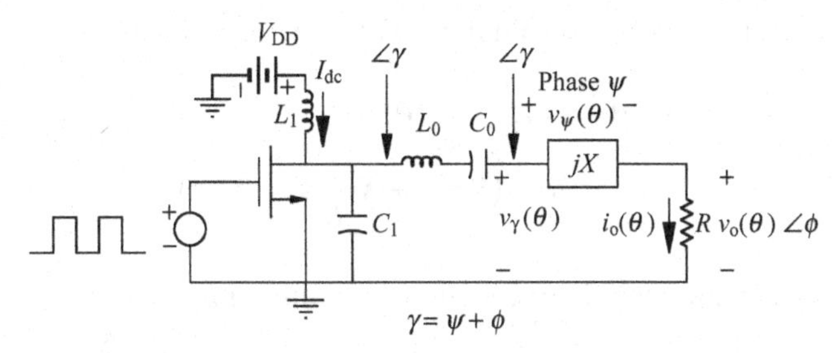

FIGURE 9.20 Class E amplifier with switching transistor.

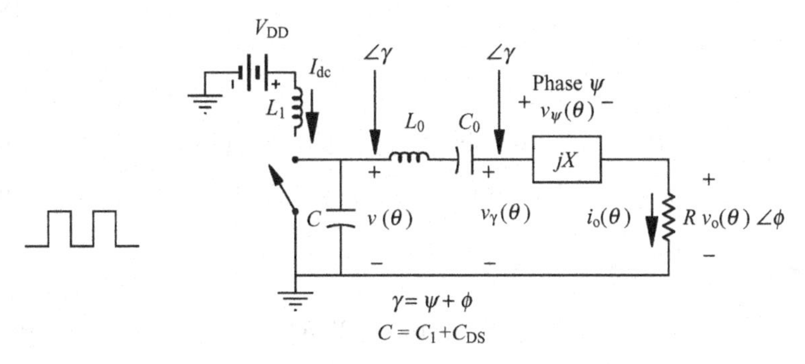

FIGURE 9.21 Ideal class E amplifier with ideal switch used for analysis.

E amplifier (Fig. 9.20) is to produce a voltage wave across the capacitance shunting the transistor that has zero amplitude and zero slope when the transistor is turned on. This relieves the switching speed requirement of the class D amplifier and yet maintains the possibility of 100% efficiency. The circuit in Fig. 9.20 consists of a transistor that is switched between on and off by the input gate voltage, a choke inductor, L_1, to maintain dc current flow, a high Q series resonant circuit, an additional excess reactance, jX, and an external drain to source capacitance, C_1. An idealized equivalent circuit is shown in Fig. 9.21. For analysis purposes, the transistor has been replaced by an ideal switch.

The analysis starts with a pure sine wave at the output and proceeds back toward the input. The output voltage with amplitude a and unspecified phase ϕ is

$$v_o(\theta) = a\sin(\omega t + \phi) = a\sin(\theta + \phi) \tag{9.93}$$

and the output current is

$$i_o(\theta) = \frac{a}{R}\sin(\theta + \phi) \tag{9.94}$$

The voltage, $v_\gamma(\theta)$, is also sinusoidal but with a phase offset γ:

$$\begin{aligned} v_\gamma(\theta) &= v_o(\theta) + v_\psi(\theta) \\ &= a\sin(\theta+\phi) + X\frac{a}{R}\cos(\theta+\phi) \end{aligned} \tag{9.95}$$

These two terms may be added together by defining b and ψ as

$$\frac{Xa}{R} = b\sin\psi \tag{9.96}$$

$$a = b\cos\psi \tag{9.97}$$

then

$$v_o(\theta) = b\cos\psi\sin(\theta+\phi) + b\sin\psi\cos(\theta+\phi) \tag{9.98}$$

$$= b\sin(\theta+\gamma) \tag{9.99}$$

where

$$\gamma \triangleq \phi + \psi = \phi + \arctan(X/R) \tag{9.100}$$

and

$$b = a\sqrt{1+(X/R)^2} \triangleq a\rho \tag{9.101}$$

When the switch is turned off, the voltage across the switch results from charging the capacitance C with $i_c(\theta)$. This voltage is not sinusoidal:

$$v(\theta) = \frac{1}{\omega C}\int_0^\theta i_c(\xi)\,d\xi \tag{9.102}$$

$$= \frac{1}{B}\int_0^\theta \left[I_{dc} - \frac{a}{R}\sin(\xi+\phi) \right] d\xi \tag{9.103}$$

$$= \frac{I_{dc}}{B}\theta + \frac{a}{BR}\cos(\theta+\phi) - \frac{a}{BR}\cos\phi \tag{9.104}$$

In this expression, the susceptance is $B = \omega C$.

At the fundamental frequency the tuned circuit is resonant, so at this frequency $v(\theta) = v_\gamma(\theta)$, and the fundamental amplitude of the switch voltage is

$$b = \frac{1}{\pi}\int_0^\pi v(\theta)\sin(\theta+\gamma)\,d\theta \tag{9.105}$$

During the second half of the cycle, the switch is closed so that no voltage component appears during this time. This expression can be expanded into four terms by substitution from Eq. (9.104):

$$b = \int_0^{\pi} \left[\frac{I_{dc}}{\pi B}(\theta+\gamma) - \frac{I_{dc}}{\pi B}\gamma + \frac{a}{\pi BR}\cos(\theta+\phi) - \frac{a}{\pi BR}\cos\phi \right] \sin(\theta+\gamma)\,d\theta \quad (9.106)$$

$$\triangleq b_1 + b_2 + b_3 + b_4 \quad (9.107)$$

Each component of (9.107) is simplified using the identities $\sin(\alpha \pm \beta) = \sin\alpha\cos\beta \pm \sin\beta\cos\alpha$ and $\cos(\alpha \pm \beta) = \cos\alpha\cos\beta \pm \sin\alpha\sin\beta$. The value for b_1 is found by using the change of variables, $\xi = \theta + \gamma$ and integration by parts:

$$b_1 = \frac{I_{dc}}{\pi B}\int_{\gamma}^{\gamma+\pi} \xi \sin\xi\, d\xi \quad (9.108)$$

$$= \frac{I_{dc}}{\pi B}\left(-\xi\cos\xi \Big|_{\gamma}^{\gamma+\pi} + \int_{\gamma}^{\gamma+\pi} \cos\xi\, d\xi \right)$$

$$= \frac{I_{dc}}{\pi B}(\pi\cos\gamma + 2\gamma\cos\gamma - 2\sin\gamma) \quad (9.109)$$

$$b_2 = -\frac{I_{dc}}{\pi B}\gamma[-\cos(\theta+\gamma)]\Big|_0^{\pi}$$

$$= -\frac{I_{dc}}{\pi B}2\gamma\cos\gamma \quad (9.110)$$

$$b_3 = \frac{a}{2\pi BR}\int_0^{\pi}[\sin(2\theta+\gamma+\phi) + \sin(\gamma-\phi)]\,d\theta \quad (9.111)$$

$$= \frac{a}{2BR}\sin\psi \quad (9.112)$$

$$b_4 = \frac{a}{\pi BR}\cos\phi\cos(\theta+\gamma)\Big|_0^{\pi} \quad (9.113)$$

$$= -\frac{a}{\pi BR}2\cos\gamma\cos\phi \quad (9.114)$$

The sum of these four terms gives

$$b = \rho a = \frac{I_{dc}}{B}\cos\gamma - \frac{2I_{dc}}{\pi B}\sin\gamma + \frac{\alpha}{2BR}\sin\psi - \frac{2a}{\pi BR}\cos\gamma\cos\phi \quad (9.115)$$

This can be solved for the value of a, which will provide a relationship between the dc power supply current and the amplitude of the output signal:

$$a = I_{\rm dc}R\frac{\pi\cos\gamma - 2\sin\gamma}{\pi BR\rho + 2\cos\phi\cos(\phi+\psi) - (\pi/2)\sin\psi} \tag{9.116}$$

$$\triangleq (I_{\rm dc}R)h(\phi, \psi, \gamma, B, R, \rho) \tag{9.117}$$

At the fundamental frequency, the quadrature component of $v(\theta)$ must be zero. This gives a second relationship for a. Again for the half period when $v(\theta) \neq 0$,

$$0 = \frac{1}{\pi}\int_0^{\pi} v(\theta)\cos(\theta+\gamma)\,d\theta \tag{9.118}$$

$$= \int_0^{\pi}\left[\frac{I_{\rm dc}}{B}(\theta+\gamma) - \frac{I_{\rm dc}}{B}\gamma + \frac{a}{BR}\cos(\theta+\phi) - \frac{a}{BR}\cos\phi\right]\cos(\theta+\gamma)\,d\theta \tag{9.119}$$

$$\triangleq c_1 + c_2 + c_3 + c_4 \tag{9.120}$$

The calculation is done in a way analogous to what was done for Eqs. (9.106) to (9.114). Thus,

$$c_1 = \frac{I_{\rm dc}}{B}[-\pi\sin\gamma - 2\gamma\sin\gamma - 2\cos\gamma] \tag{9.121}$$

$$c_2 = \frac{2I_{\rm dc}\gamma}{B}\sin\gamma \tag{9.122}$$

$$c_3 = \frac{a\pi}{2BR}\cos\psi \tag{9.123}$$

$$c_4 = \frac{2a}{BR}\cos\phi\sin\gamma \tag{9.124}$$

which when summed together gives

$$0 = \frac{I_{\rm dc}}{B}[-\pi\sin\gamma - 2\cos\gamma] + \frac{a}{BR}\left[\left(\frac{\pi}{2}\right)\cos\psi + 2\sin\gamma\cos\phi\right] \tag{9.125}$$

This is the same result as Eq. (9.115) (with the factor of π reinserted) where the phase angle has been advanced by 90° or where $\cos\gamma \to -\sin\gamma$, $\sin\gamma \to \cos\gamma$, and $\sin\psi \to \cos\psi$. Solving this for a gives

$$a = I_{\rm dc}R\frac{2\cos\gamma + \pi\sin\gamma}{(\pi/2)\cos\psi + 2\cos\phi\sin\gamma} \tag{9.126}$$

$$\triangleq (I_{\rm dc}R)g(\theta, \psi, \gamma) \tag{9.127}$$

The average value of the voltage, $v(\theta)$, across the capacitance, C, is the same as the dc power supply, V_{DD}. The power supply dc load resistance is R_{dc}. Thus, the value for R_{dc} is found from taking the average of $V(\theta)$ of Eq. (9.104):

$$V_{DD} = I_{dc}R_{dc} = \frac{1}{2\pi}\int_0^{\pi} v(\theta)d\theta \tag{9.128}$$

$$= \frac{1}{2\pi}\int_0^{\pi}\left[\frac{I_{dc}}{B}\theta + \frac{a}{BR}\cos(\theta+\phi) - \frac{a}{BR}\cos\phi\right]d\theta \tag{9.129}$$

$$= \left(\frac{I_{dc}}{2\pi B}\right)\left(\frac{\pi^2}{2}\right) + \frac{a}{2\pi BR}[\sin(\pi+\phi) - \sin\phi] - \frac{a\pi}{2\pi BR}\cos\phi \tag{9.130}$$

and

$$R_{dc} = \frac{1}{2\pi B}\left(\frac{\pi^2}{2} - 2g\sin\phi - \pi g\cos\phi\right) \tag{9.131}$$

where g is defined by Eq. (9.127). The dc input power is

$$P_{dc} = V_{DD}I_{dc} \tag{9.132}$$

and using $a = I_{dc}Rg$ and Eq. (9.93) the output power is

$$P_o = \frac{\hat{V}_o^2}{2R} = \frac{a^2}{2R} = \frac{I_{dc}^2 Rg^2}{2} \tag{9.133}$$

This gives the (drain) efficiency:

$$\eta = \frac{P_o}{P_{dc}} = \frac{I_{dc}Rg^2}{2V_{DD}} = \frac{Rg^2}{2R_{dc}} \tag{9.134}$$

To achieve 100% efficiency, the drain-to-source voltage, $v(0)$ and $v(\pi)$, must be zero. Because of the transient response of the capacitance, the slope of this voltage must also be zero, that is, $dv(\theta)/d\theta = 0$ at $\theta = \pi$. The class E amplifier does not depend on having a zero rise and fall time for a pulse. Equation (9.104) clearly shows that $v(0) = 0$. Making $v(\pi) = 0$ determines the value for ϕ:

$$0 = v(\pi) = \frac{I_{dc}}{B}\pi + \frac{a}{BR}\cos(\pi+\phi) - \frac{a}{BR}\cos\phi \tag{9.135}$$

$$= \frac{I_{dc}}{B}\pi - \frac{2a}{BR}\cos\phi \tag{9.136}$$

Replacing a from Eq. (9.127) gives

$$\cos\phi = \frac{\pi}{2g} \tag{9.137}$$

To this is now added the constraint that the voltage slope is zero at $\theta = \pi$:

$$0 = \left.\frac{dv(\theta)}{d\theta}\right|_{\theta=\pi} = \frac{I_{\text{dc}}}{B} - \frac{a}{BR}\sin(\pi + \phi) \tag{9.138}$$

$$\sin\phi = -\frac{1}{g} \tag{9.139}$$

There are now two equations and two unknowns, g and ϕ. The ratio of Eqs. (9.139) to (9.137) determines ϕ:

$$\tan\phi = \frac{-2}{\pi} \tag{9.140}$$

Mathematically, there is a sign ambiguity for the value of g. However, since g is related to the voltage magnitude, a, $g > 0$. Figure 9.22 clearly shows how g is found and hence ϕ:

$$\sin\phi = \frac{-1}{g} = \frac{-2}{\sqrt{4+\pi^2}} \tag{9.141}$$

$$\cos\phi = \frac{\pi}{\sqrt{4+\pi^2}} \tag{9.142}$$

$$g = \tfrac{1}{2}\sqrt{4+\pi^2} \tag{9.143}$$

For 100% efficiency

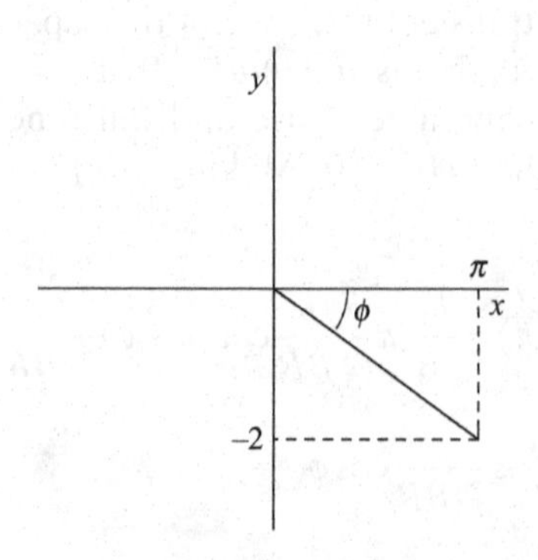

FIGURE 9.22 Determination of value for ϕ.

$$1 = \eta = \frac{g^2 R}{2R_{dc}} = \frac{1}{8}\left(4+\pi^2\right)\frac{R}{R_{dc}} \tag{9.144}$$

or

$$R_{dc} = \frac{V_{DD}}{I_{dc}} = R\frac{1}{8}\left(4+\pi^2\right) = 1.734R \tag{9.145}$$

This gives a relationship between the amplifier load resistance, R, and the dc power source voltage-to-current ratio.

The circuit parameters that remain unknown at this point are the shunt capacitive susceptance and the series reactances. Now that ϕ and g are known, Eqs. (9.131) and (9.145) gives

$$R_{dc} = \frac{1}{\pi B} = R\frac{1}{8}\left(4+\pi^2\right) \tag{9.146}$$

or

$$B = \frac{8}{\pi R\left(4+\pi^2\right)} \tag{9.147}$$

This represents the optimum total shunt susceptance that incorporates C_{DS}, the parasitic capacitance, and the additional circuit capacitance needed to provide 100% efficiency. What remains is the magnitude and phase of the series reactance, $|X| < \psi$. The phase part can be extracted from the expression for g implied by Eqs. (9.126) and (9.127):

$$g(\phi, \psi) = \frac{2\cos(\phi+\psi) + \pi\sin(\phi+\psi)}{(\pi/2)\cos\psi + 2\cos\phi\sin(\phi+\psi)} \tag{9.148}$$

whose numerical value is $\frac{1}{2}\sqrt{4+\pi^2}$ from Eq. (9.143). To solve for ψ, this function is expanded using the trigonometric double-angle formulas:

$$\begin{aligned} &2(\cos\phi\cos\psi - \sin\phi\sin\psi) + \pi(\sin\phi\cos\psi + \sin\psi\cos\phi) \\ &= g\left[\frac{\pi}{2}\cos\psi + 2\cos\phi(\sin\phi\cos\psi + \sin\psi\cos\phi)\right] \end{aligned} \tag{9.149}$$

Next, the coefficients of $\sin\psi$ and that of $\cos\psi$ are each combined together:

$$\begin{aligned} 0 = &\sin\psi\left[-2\sin\phi + \pi\cos\phi - 2g\cos^2\phi\right] \\ &+ \cos\psi\left[2\cos\phi + \pi\sin\phi - g\frac{\pi}{2} - 2g\sin\phi\cos\phi\right] \end{aligned} \tag{9.150}$$

$$\tan\psi = \frac{2\cos\phi + \pi\sin\phi - g\pi/2 - 2g\sin\phi\cos\phi}{2\sin\phi - \pi\cos\phi + 2g\cos^2\phi} \tag{9.151}$$

Substituting for $\cos\phi$, $\sin\phi$, and g gives

$$\tan\psi = \frac{\pi}{4}\left(\frac{\pi^2}{4} - 1\right) \tag{9.152}$$

so that

$$\psi = 49.052° \tag{9.153}$$

Since $\tan\psi = X/R$,

$$X = \frac{\pi}{4}\left(\frac{\pi^2}{4} - 1\right)R = 1.153R \tag{9.154}$$

This optimum series reactance must be inductive. It should be noted that in practice the series resonant circuit absorbs this additional reactance, X, so the total output network does not operate at resonance when operating at maximum efficiency.

The required power supply voltage necessary to provide the desired output voltage amplitude, a, can be found by equating Eqs. (9.132) and (9.133) since efficiency is 100%:

$$P_{dc} = I_{dc}V_{DD} = P_o = \frac{a^2}{2R} \tag{9.155}$$

and replace I_{dc} with Eq. (9.127):

$$V_{DD} = \frac{a}{4}\sqrt{4 + \pi^2} = 0.931a \tag{9.156}$$

In summary, the design process for the maximum efficiency class E amplifier begins with knowing the desired output voltage amplitude, a, and the load resistance, R. The shunt susceptance, B, is found from Eq. (9.147), the series reactance, X, from Eq. (9.154), and the power supply voltage, V_{DD}, from Eq. (9.156). The resonant tank circuit, L_0 and C_0, must be of sufficiently high Q to block harmonic frequencies. The series Q is proportional to L_0, and the capacitance is determined by the resonant frequency. The value for L_1 is chosen to be high enough to provide a dc current with negligible ac content. The size must be tempered with the practical problem of parasitic capacitance in a high value inductor. A summary of the design process is given in Table 9.1.

TABLE 9.1 Class E Design Summary

Formula	Circuit Value
Choose	R
Choose	a
Eq. (9.147)	B
Eq. (9.154)	X
Eq. (9.156)	V_{DD}
Choose	Q
RQ/ω_0	L_0
$1/\omega_0^2 L_0$	C_0
Choose	L_1

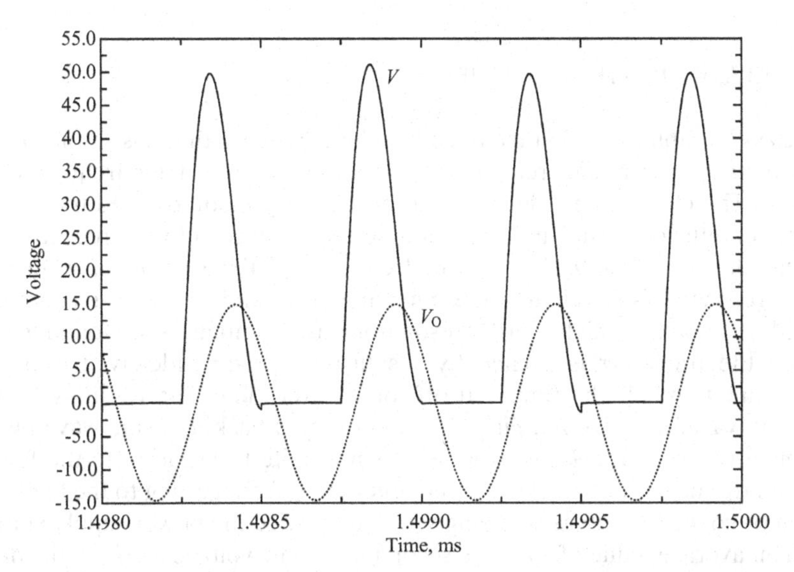

FIGURE 9.23 Voltage, $\upsilon(\theta)$, at switch and the voltage, $\upsilon_o(\theta)$ at the load. Phase difference between these two is ϕ.

A numerical example illustrates the characteristics of this amplifier. Consider the design of the class E amplifier in Fig. 9.21 at $f = 2\,\text{MHz}$, $R = 15\,\Omega$, $P_o = 7.5\,\text{W}$, and tuned circuit Q = 50. Then $a = \sqrt{2pR} = 15\,\text{V}$, $C = 974\,\text{pF}$, $L = 1.37\,\mu\text{H}$, $V_{DD} = 13.97\,\text{V}$, $L_0 = 59.68\,\mu\text{H}$, and $C_0 = 106.1\,\text{pF}$. The RF choke inductance was chosen to be 200 μH. Too small an inductance will allow ac currents to flow into the power supply. A SPICE analysis of this design is shown in Fig. 9.23 where the voltage across the switch, $\upsilon(\theta)$, and the output voltage at the load, $\upsilon_o(\theta)$, are displayed. The phase difference between the switch voltage and the load voltage shown in Fig. 9.23 is ϕ. The SPICE analysis is an approximation to the ideal circuit model, since convergence requirements determine that the switch have a nonzero on resistance and the switching time must be greater than zero.

The assumptions that this analysis was based on are (1) the device capacitance is considered to be independent of voltage amplitude, since it is linear, (2) the gate to drain capacitance, C_{GD} is neglected, and (3) the load at harmonic frequencies is considered to be infinite. The latter could be enhanced by replacing the series-tuned circuit with a multipole filter. Another enhancement for the microwave frequency range is the use of two 45° transmission lines to provide the required shunt susceptance, B, while ensuring an open circuit at the output of the switch at the second-harmonic frequency [13]. This circuit provided a 0.94W of output power with a drain efficiency of 75% at 1GHz. Control of multiple harmonics leads to the class F amplifier described in the following section.

9.7 CLASS F POWER AMPLIFIER

"A class F amplifier is characterized by a load network that has resonances at one or more harmonic frequencies as well as at the carrier frequency" [2, p. 454]. The class F amplifier was one of the early methods used to increase amplifier efficiency and has attracted some renewed interest recently. The circuit shown in Fig. 9.24 is a three-frequency peaking amplifier where the shunt resonator is resonant at the fundamental, f_0, and the series resonator at $3f_0$. More details for higher order resonator class F amplifiers are found in [14]. When the transistor is excited by a sinusoidal source, ideally it is on for approximately half the time and off for half the time. The resulting output current waveform given in Fig. 9.25*a* is converted back to a sine wave by the resonator, L_1, C_1. The L_3, C_3 resonator is not quite transparent to the fundamental frequency, but blocks the frequency at $3f_0$ from getting to the load. The drain or collector voltage will range from 0 to twice the power supply voltage with an average value of V_{CC}. The second harmonic voltage at $3f_0$ on the drain or collector, if it has the appropriate amplitude and phase, will tend to make this device voltage more square in shape. This will make the transistor act more like a switch with the attendant high efficiency.

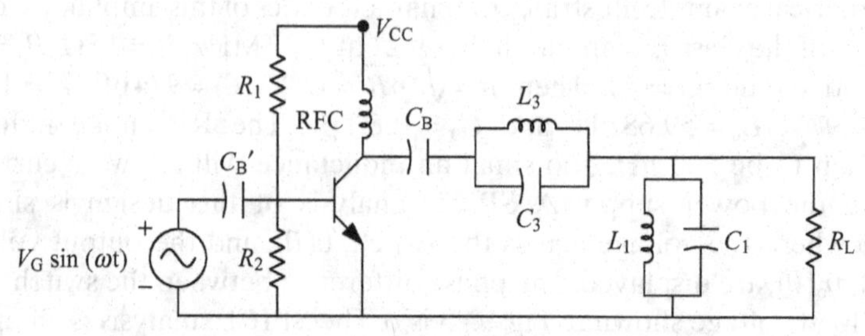

FIGURE 9.24 Class F three-frequency peaking power amplifier.

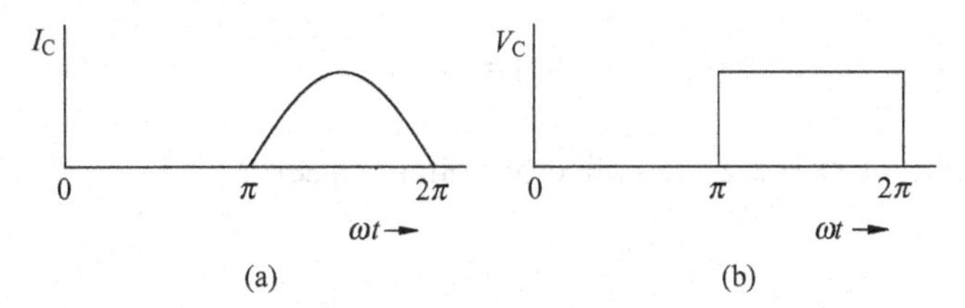

FIGURE 9.25 (*a*) Class F collector current and (*b*) ideal switching square wave.

9.7.1 Three-Frequency Peaking Class F Amplifier

The Fourier expansion of a square wave illustrated in Fig. 9.25*b* with amplitude from +1 to −1 and period 2π is

$$\frac{1}{2}+\frac{2}{\pi}\left(-\sin x-\frac{\sin 3x}{3}-\frac{\sin 5x}{5}\ldots\right) \tag{9.157}$$

Consequently, to produce a square wave voltage waveform at the transistor terminal, the impedance must be a short at even harmonics and large at odd harmonics. Ordinarily, only the fundamental, first-harmonic and second-harmonic impedances are determined. In the typical class F amplifier shown in Fig. 9.24, the L_1C_1 tank circuit is resonant at the output frequency, f_0, and the L_3C_3 tank circuit is resonant at $3f_0$. It has been pointed out [15] that the blocking capacitor, C_B, could be used to provide a short to ground at $2f_0$ rather than simply acting as a dc block.

The design of the class F amplifier proceeds by first determining the output voltage from the desired output power and load resistance requirement. Of course, the load resistance can be transformed to a standard value by use of an impedance transformer. The resulting output voltage is

$$\hat{V}_o = \sqrt{2R_oR_L} \tag{9.158}$$

A square switching waveform at the collector can be approximated with the fundamental and two harmonics. Based on Eq. (9.157), the transistor collector voltage would be of the form

$$V_C(\omega t) = V_{CC} - \hat{V}_C \sin \omega t - \hat{V}_{C3} \sin 3\omega t \tag{9.159}$$

Setting the change of the slope to be zero at $\omega t = 3\pi/2$ is done by

$$\left.\frac{d^2V_C(\omega t)}{d(\omega t)^2}\right|_{3\pi/2} = 0 = -\hat{V}_C + 9\hat{V}_{C3} \tag{9.160}$$

or

$$\hat{V}_{C3} = \frac{1}{9}\hat{V}_C \tag{9.161}$$

Furthermore, at $\omega t = \pi/2$ the collector voltage is zero:

$$V_C\left(\frac{\pi}{2}\right) = 0 = V_{CC} - \hat{V}_C + \hat{V}_{C3}$$

$$0 = V_{CC} - \hat{V}_C + \frac{V_C}{9}$$

$$\hat{V}_C = \frac{9}{8}V_{CC} \tag{9.162}$$

The expected maximum efficiency can be done much like the calculation done for the class B amplifier. The direct current for the class F amplifier is a half-wave-rectified current rather than a full-wave-rectified current. One might expect from the class B analysis that the direct current would be $\hat{V}_o/(R_L\pi)$ where $\hat{V}_o$ is the voltage across the load. Since there is a short between the collector and the load at the fundamental, $\hat{V}_o = \hat{V}_C$. For the class F amplifier the peak value of the load current is $\hat{V}_o/R_L$. However, the current entering the blocking capacitor, C_B, has a peak value of $2\hat{V}_o/R_L$ and a minimum value of zero. Thus, the average of the half wave current entering C_B is

$$I_{dc} = \frac{1}{2\pi}\int_{\pi}^{2\pi}\frac{2\hat{V}_o}{R_L}(-\sin\omega t)\,d\omega t$$

$$= \frac{2\hat{V}_o}{\pi R_L} \tag{9.163}$$

During the time the current is not flowing into the load through C_B, it flows through the transistor to ground. The power from the supply is

$$P_{dc} = I_{dc}V_{CC}$$

$$= \frac{2\hat{V}_o}{\pi R_L}V_{CC}$$

and maximum power occurs for $\hat{V}_o = V_{CC}$:

$$P_{dc} = 2\cdot\left(\frac{9}{8}\frac{V_{CC}^2}{\pi R_L}\right)$$

The maximum output power is

$$P_o = \left(\frac{9}{8}V_{CC}\right)^2\frac{1}{2R_L}$$

Thus, the maximum efficiency is

$$\eta = \frac{9}{8}\frac{\pi}{4} = 83.4\% \tag{9.164}$$

The determination of the reactive circuit begins by finding C_1 from the desired amplifier bandwidth. The circuit Q is assumed to be determined solely by L_1, C_1, and R_L. Thus,

$$Q = \omega_0 C_1 R_L = \frac{\omega_0}{\Delta\omega}$$

or

$$C_1 = \frac{1}{R_L \Delta\omega} \tag{9.165}$$

Once C_1 is determined, the inductance must be that which resonates the tank at f_0:

$$L_1 = \frac{1}{\omega_0^2 C_1} \tag{9.166}$$

At $2f_0$, the L_1C_1 tank circuit has negative reactance and the L_3C_3 tank circuit has positive reactance. The capacitances, C_B and C_3, can be set to provide a short to ground:

$$X(2\omega_0) = -\frac{1}{2\omega_0 C_B} + \frac{2\omega_0 L_3}{1-(2\omega_0)^2 L_3C_3} + \frac{2\omega_0 L_1}{1-(2\omega_0)^2 L_1C_1} = 0 \tag{9.167}$$

Since $L_3C_3 = \frac{1}{9\omega_0^2}$, and $L_1C_1 = 1/\omega_0^2$, Eq. (9.167) reduces to

$$0 = -\frac{1}{C_B} + \frac{4/(9C_3)}{1-4/9} + \frac{4/C_1}{1-4}$$

or

$$\frac{1}{C_B} = \frac{4}{5C_3} - \frac{4}{3C_1} \tag{9.168}$$

which is the requirement for series resonance at $2f_0$. In addition, at the fundamental frequency, C_B and the L_3C_3 tank circuit can be tuned to provide no reactance between the transistor and the load, R_L. This eliminates the approximation that the L_3C_3 has zero reactance at the fundamental:

$$0 = -\frac{1}{\omega_0 C_B} + \frac{\omega_0 L_3}{1 - \omega_0^2 L_3 C_3} \tag{9.169}$$

$$C_B = 8C_3 \tag{9.170}$$

This value for C_B can be substituted back into Eq. (9.168) to give a relationship between C_3 and C_1:

$$C_3 = \frac{81}{160} C_1 \tag{9.171}$$

In summary, C_1 is determined by the bandwidth Eq. (9.165), L_1 by Eq. (9.166), C_3 from Eq. (9.171), L_3 from its requirement to resonate C_3 at $3f_0$, and finally C_B from Eq. (9.170). In addition, interstage networks are presented in [15] that aim at reducing the spread in circuit element values and hence help make circuit design practical.

As a numerical example, an amplifier is to be designed to deliver 10W of power to a 25-Ω load at 900MHz. From (9.158), $\hat{V}_o = 22.36$ V and using Eq. (9.161) the required dc supply voltage is 19.87V, so $V_{CC} = 20$V is chosen. Knowledge of $\hat{V}_o$ gives the direct current from the power supply as $I_{dc} = 569.4$mA from P_{dc}/V_{CC}. Assume the desired bandwidth requires $Q = 100$. Then $C_1 = 707.35$pF, $L_1 = 44.21$pH, $C_3 = 358.10$pF, $L_3 = 9.70$pH, and $C_B = 2.86$nF. A SPICE analysis using the default SPICE bipolar transistor model gives the collector current (Fig. 9.26) and the voltages at the collector and load (Fig. 9.27).

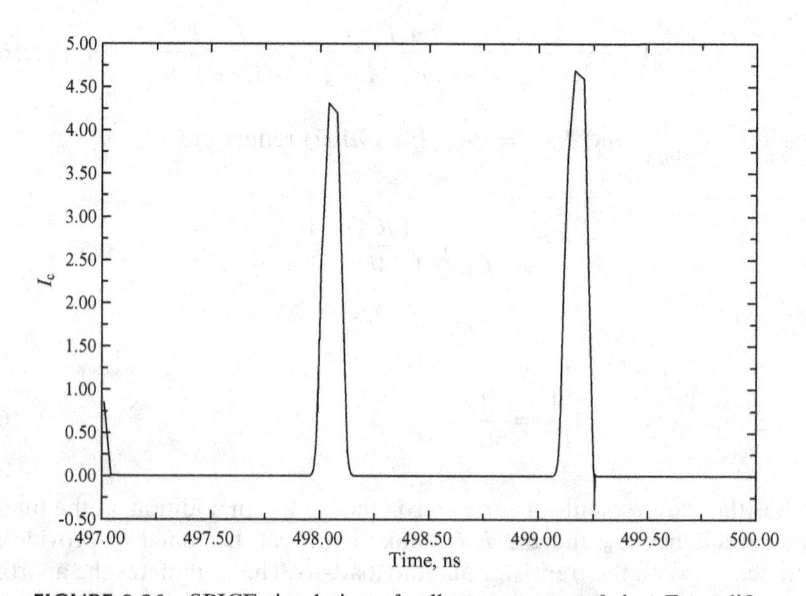

FIGURE 9.26 SPICE simulation of collector current of class F amplifier.

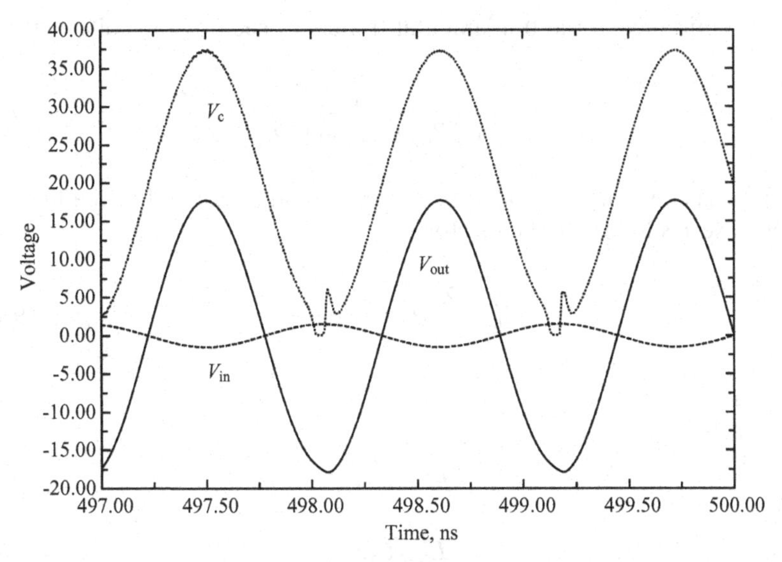

FIGURE 9.27 SPICE simulation of collector and load voltages of class F amplifier.

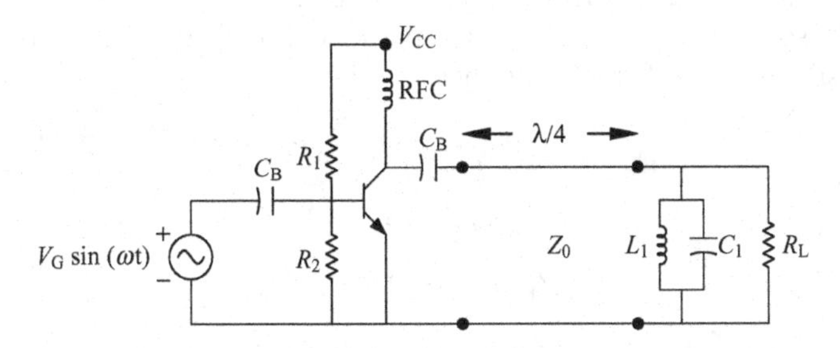

FIGURE 9.28 Class F transmission line power amplifier.

Since actual transistor models are much more complicated than that used here, actual results could be quite different from those shown.

9.7.2 Transmission Line Class F Amplifier

Additional odd harmonics can be controlled by adding additional resonators that make the collector voltage come closer to having a square shape. In effect, an infinite number of odd harmonic resonators can be added if a $\lambda/4$ transmission line at the fundamental frequency replaces the lumped-element L_3C_3 second-harmonic resonator (Fig. 9.28).

The admittance at the fundamental frequency seen by the collector is

$$Y'_L = \frac{Y_0^2}{(1/R_L) + sC_1 + (1/sL_1)} \tag{9.172}$$

The $\lambda/4$ transmission line converts the shunt load at the end of the line to an effective series load at the collector:

$$Z'_L = \frac{Z_0^2}{R_L} + sC_1Z_0^2 + \frac{Z_0^2}{sL_1} \tag{9.173}$$

in which

$$R'_L = \frac{Z_0^2}{R_L}$$

$$L' = C_1Z_0^2$$

$$C' = \frac{Z_0^2}{sL_1}$$

At the first harmonic, the transmission line is $\lambda/2$ and the resonator (L_1, C_1) is a short, so $Z'_L(2\omega_0) = 0$. The effective load for all the harmonics can be found easily at each of the harmonics:

$$Z'_L(2\omega_0) = 0 \qquad \lambda/2$$

$$Z'_L(3\omega_0) = \infty \qquad 3\lambda/4$$

$$Z'_L(4\omega_0) = 0 \qquad \lambda$$

$$Z'_L(5\omega_0) = \infty \qquad 5\lambda/4$$

$$\vdots$$

While this provides open and short circuits to the collector, it is not obvious that these impedances, which act in parallel with the output impedance of the transistor, will provide the necessary amplitude and phase that would produce a square wave at the collector.

There are some practical difficulties in trying to make the transmission line class F amplifier. Most obvious is the physical size of a $\lambda/4$ line. For typical radio frequencies a technique would need to be used to make the line mechanically short while still providing an electrical length of $\pi/2$ at the fundamental operating frequency. Furthermore, the resonator capacitance, C_1, must be sufficiently large to provide approximately zero reactance at $2fo$. At the same time the desired bandwidth also governs the size of C_1 [Eq. (9.165)]. The competing requirements of low second-harmonic impedance and the desired Q may produce an unacceptable compromise. A more extensive harmonic balance analysis of a physics-based model for a metal semiconductor field-effect transistor (MESFET) showed that a power added efficiency of 75% can be achieved at 5 GHz [16].

9.8 FEED-FORWARD AMPLIFIERS

The concept of feed-forward error control was conceived in a patent disclosure by Harold S. Black in 1924 [17]. This was several years prior to his more famous concept of feedback error control. A historical perspective on the feed-forward idea is found in [18]. The feedback approach is an attempt to correct an error after it has occurred. However, a 180° phase difference in the forward and reverse paths in a feedback system can cause unwanted oscillations. In contrast, the feed-forward design is based on cancellation of amplifier errors in the same time frame in which they occur. Signals are handled by wide-band analog circuits, so multiple carriers in a signal can be controlled simultaneously. Feed-forward amplifiers are inherently stable, but this comes at the price of a somewhat more complicated circuit. Consequently, feed-forward circuitry is sensitive to changes in ambient temperature, input power level, and supply voltage variation. Nevertheless feed-forward design offers many advantages that have brought increased interest.

The major source of distortion, such as harmonics, intermodulation distortion, and noise, in a transmitter is the power amplifier. This distortion can be greatly reduced using a feed-forward design. The basic idea is illustrated in Fig. 9.29 where it is seen that the circuit consists basically of two loops. The first one contains the main power amplifier, and the second loop contains the error amplifier. In the first loop, a sample of the input signal is coupled through coup1 reducing the signal by the coupling factor $-C_1$ dB. This goes through the delay line with insertion loss of $-D_1$ dB into the comparator coupler coup3. At the same time the signal passing through the main amplifier with gain G_1 dB

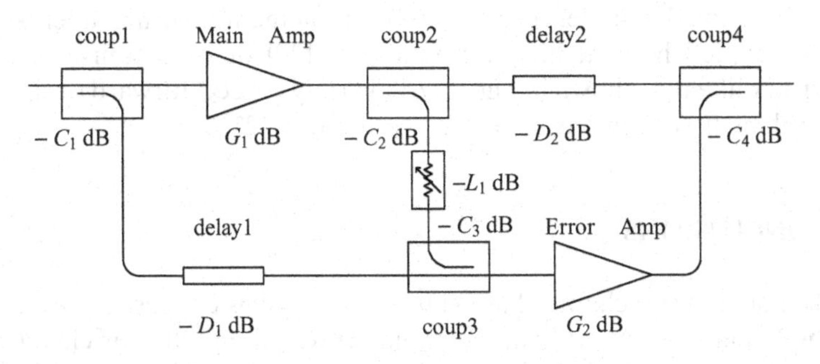

FIGURE 9.29 Linear feed-forward amplifier.

is sampled by coupler coup2, reducing the signal by $-C_2$ dB, the attenuator by $-L_1$ dB, and the coupler coup3 by $-C_3$ dB. The delay line, delay1, is adjusted to compensate for the time delay in the main amplifier as well as the passive components so that two input signals for coup3 are 180° out of phase but synchronized in time. The amplitude of the input signal when it arrives at the error amplifier is

$$-C_1 - D_1 - [G_1 - C_2 - L_1 - C_3] \tag{9.174}$$

which should be adjusted to be zero. What remains is the distortion and noise added by the main amplifier, which is in turn amplified by the error amplifier by G_2 dB. At the same time the signal from the main amplifier with its distortion and noise is attenuated by D_2 dB in the second loop delay line. The second delay line is adjusted to compensate for the time delay in the error amplifier. The relative phase and amplitude of the input signals to coup4 are adjusted so that the distortion terms cancel. The output distortion amplitude

$$-D_2 - [-C_2 - L_1 - C_3 + G_2 - C_4]$$

should be zero for complete cancellation to occur.

The error amplifier will also add distortion and noise to its input signal so that perfect error correction will not occur. Nevertheless, a dramatic improvement is possible since the error amplifier will be operating on a smaller signal (only distortion) that will likely lie in the linear range of the amplifier. Further improvement may be accomplished by treating the entire amplifier in Fig. 9.29 as the main amplifier and adding another error amplifier with its associated circuitry [18].

A typical implementation of a feed-forward system is described in [19] for an amplifier operating in the frequency range of 2.1 to 2.3 GHz with an RF gain of 30 dB and an output power of 1.25 W. This amplifier had intermodulation products at least 50 dB below the carrier level. Their design used a 6-dB coupler for coup1, a 13-dB coupler for coup2, a 10-dB coupler for coup3, and an 8-dB coupler for coup4. In some designs, the comparator coupler, coup3, is replaced by a power combiner.

The directional coupler itself can be implemented using microstrip or stripline coupled lines at higher frequencies [20] or by a transmission line transformer like that shown in Fig. 6.17. A variety of feed-forward designs have been implemented, some using digital techniques [21, 22].

9.9 CONCLUSIONS

The discussion in this chapter has centered on two basic types of power amplifiers: the linear class A, AB, and B amplifiers and the nonlinear class C, D, E, and F amplifiers. The alphabet soup of power amplifiers does continue beyond class F, but these are the most widely used types today. In general, though, higher efficiency comes with the cost of higher distortion. The feed-forward amplifier does attempt to reduce noise and distortion by cancellation, but with the cost of higher complexity and some loss in efficiency. The Doherty power amplifier, though not discussed in this chapter, represents a technique of using two parallel amplifiers where the auxiliary amplifier provides additional current when the main amplifier begins to saturate at high signal level. It is an attempt to provide high efficiency while maintaining signal integrity.

PROBLEMS

9.1. If the crossover discontinuity is neglected, is a class B amplifier considered a linear amplifier or a nonlinear amplifier. Explain your answer.

9.2. A class B amplifier such as that shown in Fig. 9.7 is biased with an 18-V power supply, but the maximum voltage amplitude across each transistor is 16 V. The remaining 2 V is dissipated as loss in the output transformer. If the amplifier is designed to deliver 12 W of RF power, what is

a. The maximum RF collector current?

b. The total dc current from the power supply?

c. The collector efficiency of this amplifier?

9.3. The bipolar class C amplifier equivalent to that shown in Fig. 9.9 has a conduction angle of 60°. It is designed to deliver 75 W of RF output power. The saturated collector–emitter voltage is known to be 1 V and the power supply voltage is 26 V. What is the maximum peak collector current?

9.4. Assume the class C amplifier shown in Fig. 9.9 is excited by the rectangular wave shown in Fig. 9.30. Determine the efficiency of this amplifier as

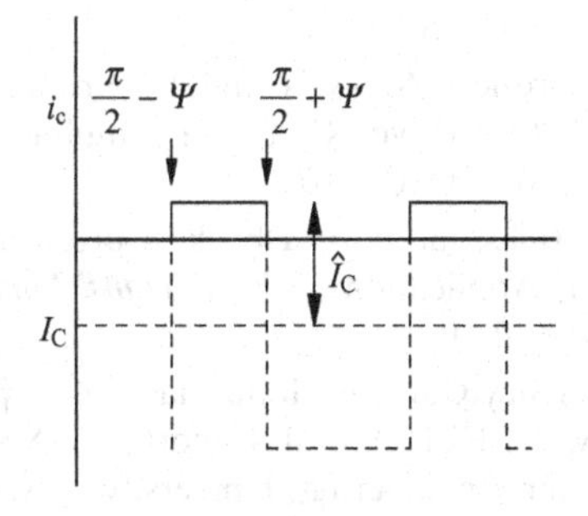

FIGURE 9.30 Square wave pattern for Problem 9.4.

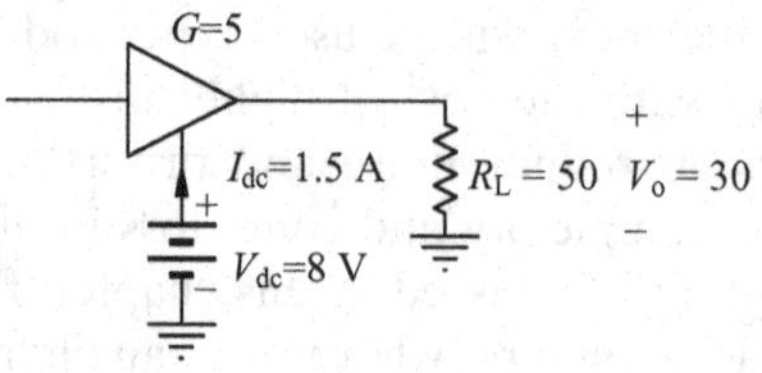

FIGURE 9.31 Amplifier used to determine power efficiency in Problem 9.7.

a function of ψ. If the conduction angle $2\psi = \pi/3$, what is the numerical value for the efficiency?

9.5. A class C amplifier is to be designed using a bipolar transistor to produce a maximum average output power, $P_o = 26\,\text{W}$, at 50 MHz. The transistor being used has a saturation voltage, $V_{CE\text{-sat}} = 2\,\text{V}$. The power supply voltage is $V_{CC} = 28\,\text{V}$. The current wave form shown in Fig. 9.10 can be used where $I_Q = -4\,\text{A}$, $\hat{I}_C = 12\,\text{A}$, and $\psi = 45°$.

a. Determine, r, the ratio of the average power to the peak power.

b. Determine the load resistance needed to realize the required output power.

9.6. A certain power amplifier in the common emitter configuration has a conduction angle of 120°. What is the class type (A, AB, B, C, ...) of the amplifier in Fig. 9.31? The maximum average power delivered to the load by this amplifier is 30 W. What is the peak instantaneous power across the output terminals?

REFERENCES

1. P. R. Gray, P. J. Hurst, S. H. Lewis, and R. G. Meyer, *Analysis and Design of Analog Integrated Circuits*, New York: Wiley, 2001.
2. H. L. Krauss, C. W. Bostian, and F. H. Raab, *Solid State Radio Engineering*, New York: Wiley, 1980.
3. W. L. Everitt, "Optimum Operating Conditions for Class C Amplifiers," *Proc. Inst. Radio Eng.*, **22**, pp. 152–176, Feb. 1934.
4. D. M. Snider, "A Theoretical Analysis and Experimental Confirmation of the Optimum Loaded and Overdrive RF Power Amplifier," *IEEE Trans. Electron Devices*, **ED-14**, pp. 851–857, Dec. 1967.
5. H. Kobayashi, J. M. Hinrichs, and P. M. Asbeck, "Current-Mode Class-D Amplifiers for High-Efficiency RF Applications," *IEEE Trans. Microwave Theory Tech.*, **49**, pp. 2480–2485, December 2001.
6. A. L. Long, High Frequency Current Mode Class-D Amplifiers with High Output Power and Efficiency, ECE Technical Report #03-XX, M.S. Thesis, Dept. of Electrical and Computer Engineering, University of California, Santa Barbara, May 2003.

7. G. D. Ewing, High-Efficiency Radio-Frequency Power Amplifiers, Ph.D. Dissertation, Oregon State University, Corvallis, OR, April 1964.
8. N. O. Sokal and A. D. Sokal, "Class E—A New Class of High-Efficiency Tuned Single-Ended Power Amplifiers," *IEEE J. Solid State Circuits*, **SC-10**(3), pp. 168–176, June 1975.
9. F. H. Raab, "Idealized Operation of the Class-E Tuned Power Amplifier," *IEEE Trans. Circuits Syst.*, **CAS-24**(12), pp. 725–735, Dec. 1977.
10. F. H. Raab, "Effects of Circuit Variations on the Class-E Tuned Power Amplifier," *IEEE J. Solid State Circuits*, **SC-13**(2), pp. 239–247, April 1978.
11. W. H. Cantrell, "Tuning Analysis for the High-Q Class-E Power Amplifier," *IEEE Trans. Microwave Theory Tech.*, **48**(12), pp. 2397–2402, Dec. 2000.
12. W. H. Cantrell, Analysis of High-Q Class-E Power Amplifier, Rectifier and Ampli tude Modulator, Ph.D. Dissertation, University of Texas Arlington, Arlington, TX, December, 2002.
13. T. B. Mader and Z. B. Popovic, "The Transmission-Line High-Efficiency Class-E Amplifier," *IEEE Microwave Guided Wave Lett.*, **5**, pp. 290–292. Sept. 1995.
14. F. H. Raab, "Class-F Power Amplifiers with Maximally Flat Waveforms," *IEEE Trans. Microwave Theory Tech.*, Vol. **45**, pp. 2007–2012, Nov. 1997.
15. C. Trask, "Class-F Amplifier Loading Networks: A Unified Design Approach," *1999 IEEE MTT-S International Symposium Digest*, Piscataway, NJ: IEEE Press, pp. 351–354, 1999.
16. L. C. Hall and R. J. Trew, "Maximum Efficiency Tuning of Microwave Amplifiers," *1991 MTT-S International Symposium Digest*, Piscataway, NJ: IEEE Press, pp. 123–126, 1991.
17. H. S. Black, U.S. Patent 1,686,792, issued Oct. 9, 1929.
18. H. Seidel, H. R. Beurrier, and A. N. Friedman, "Error-Controlled High Power Linear Amplifiers at VHF," *Bell Syst. Tech. J.*, **47**, pp. 651–722, May–June 1968.
19. C. Hsieh and S. Chan, "A Feedforward S-Band MIC Amplifier System," *IEEE J. Solid State Circuits*, **SC-11**, pp. 271–278, April 1976.
20. W. A. Davis, *Microwave Semiconductor Circuit Design*, Chapter 4, New York: Van Nostrand, 1984.
21. S. J. Grant, J. K. Cavers, and P. A. Goud, "A DSP Controlled Adaptive Feed Forward Amplifier Linearizer," *Annual International Conference on Universal Personal Communications*,, pp. 788–792, Cambridge, Ma: Sept. 29–Oct. 2 1996.
22. G. Zhao, F. M. Channouchi, F. Beauregard, and A. B. Kouki, "Digital Implementations of Adaptive Feedforward Amplifier Linearization Techniques," *IEEE Microwave Theory Tech. Symp. Digest*, June 1996.

CHAPTER TEN

Oscillators and Harmonic Generators

10.1 OSCILLATOR FUNDAMENTALS

An oscillator is a circuit that converts energy from a power source (usually a dc power source) to ac energy. In order to produce a self-sustaining oscillation, there necessarily must be feedback from the output to the input, sufficient gain to overcome losses in the feedback path, and a resonator. There are a number of ways to classify oscillator circuits, one of those being the distinction between one-port and two-port oscillators. The one-port oscillator has a load and resonator with a negative resistance at the same port, while the two-port oscillator is loaded in some way at the two ports. In either case there must be a feedback path, although in the case of the one-port circuit this path might be internal to the device itself.

An amplifier with positive feedback is shown in Fig. 10.1. The output voltage of this amplifier is

$$V_\text{o} = aV_\text{i} + a\beta V_\text{o}$$

which gives the closed-loop voltage gain

$$A = \frac{V_\text{o}}{V_\text{i}} = \frac{a}{1 - a\beta} \tag{10.1}$$

Radio Frequency Circuit Design, Second Edition, by W. Alan Davis

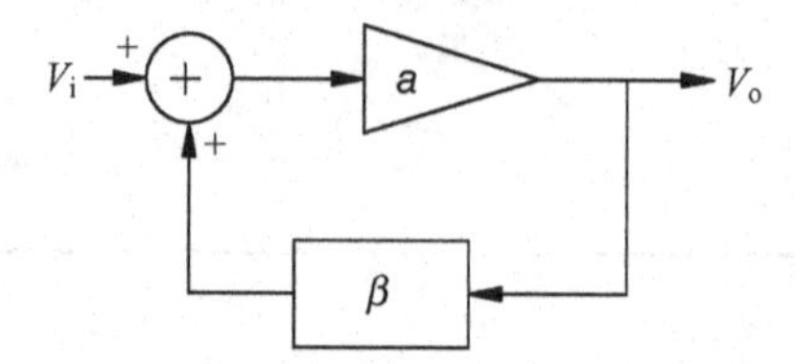

FIGURE 10.1 Circuit with positive feedback.

The positive feedback allows an increasing output voltage to feedback to the input side until the point is reached where

$$a\beta = 1 \tag{10.2}$$

This is called the *Barkhausen criterion* for oscillation and is often described in terms of its magnitude and phase separately. Hence, oscillation can occur when $|a\beta| = 1$ and $\angle a\beta = n \times 360°$, where n is an integer. An alternate way of determining conditions for oscillation is determining when the value $k < 1$ for the stability circle as described in Chapter 8. Still a third way will be considered in Section 10.4.

10.2 FEEDBACK THEORY

The active amplifier part and the passive feedback part of the oscillator can be considered as a pair of two two-port circuits. Usually the connection of these two port circuits occurs in four different ways: series–series, shunt–shunt, series–shunt, and shunt–series (Fig. 10.2). A linear analysis of the combination of these two two-port circuits begins by determining what type of connection exists between them. If, for example, they are connected in series–series, then the best way to describe each of the two-port circuits is in terms of their z parameters. The composite of the two two-port circuits is found by simply adding the z parameters of the two circuits together. Thus, if $[z_a]$ and $[z_f]$ represent the amplifier and feedback circuits connected in series–series, then the composite circuit is described by $[z_c] = [z_a] + [z_f]$. The form of the feedback circuit itself can take a wide range of forms, but, being a linear circuit, it can always be reduced to a set of z, y, h, or g parameters any one of which can be represented by the symbol k for the present. The term that feeds back to the input of the amplifier is k_{12f}. The k_{12f} term, though small, is a significant part of the small incoming signal, so that it cannot be neglected. The open-loop gain, a, of the composite circuit is found by setting $k_{12f} = 0$. Then using the normal circuit analysis, the open-loop gain is determined. The closed-loop gain is found by including k_{12f} in the closed-loop gain given by Eq. (10.1). The Barkhausen criterion for oscillation is satisfied when $ak_{12f} = a\beta = 1$.

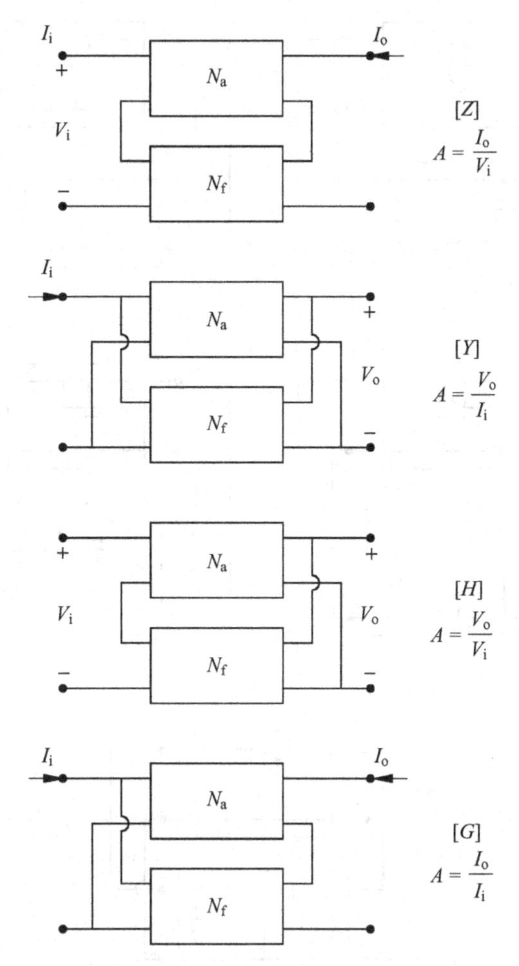

FIGURE 10.2 Four possible ways to connect the amplifier and feedback circuit. Composite circuit is obtained by adding designated two-port parameters. Units for "gain" are as shown.

10.3 TWO-PORT OSCILLATORS WITH EXTERNAL FEEDBACK

There are a wide variety of two-port oscillator circuits that can be designed. The variety of oscillators results from the different ways the feedback circuit is connected to the amplifier and the variety of feedback circuits themselves. Five of these shown in Fig. 10.3 are known as the Colpitts, Hartley, Clapp–Gouriet [1, 2], Armstrong, and Vackar [2, 3] oscillators. The Pierce oscillator is obtained by replacing the inductor in the Colpitts circuit with a crystal that acts like a high-Q inductor. As shown, the first four of these feedback circuits are drawn in a series–series connection while the Vackar is drawn as a series–shunt configuration. Of course, a wide variety of connections and feedback

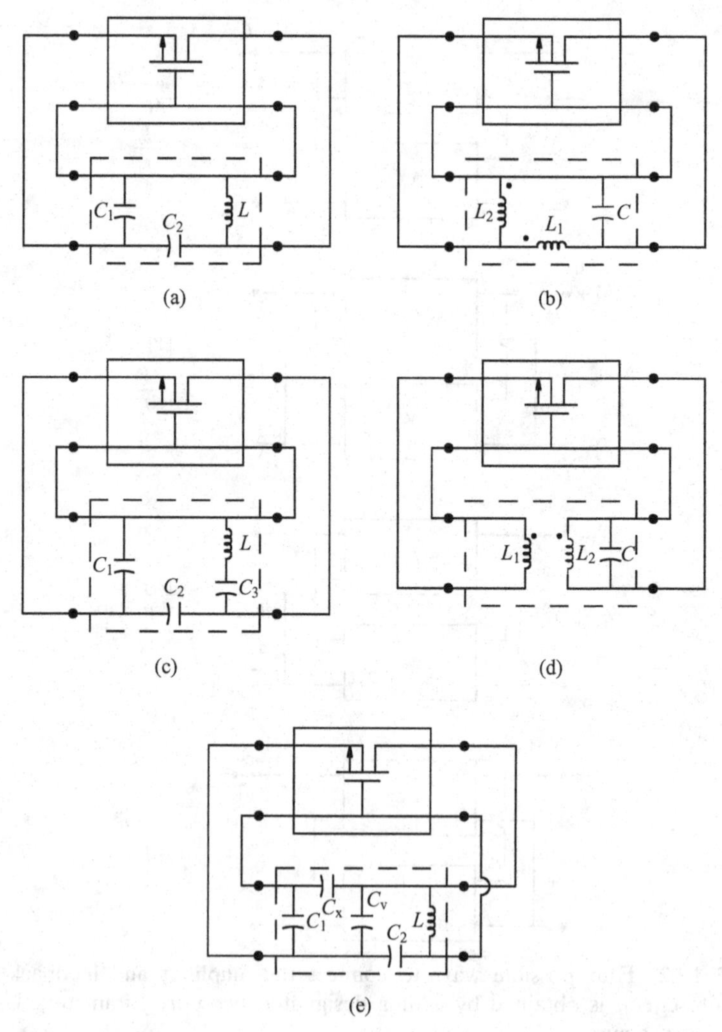

FIGURE 10.3 Oscillator types: (*a*) Colpitts, (*b*) Hartley, (*c*) Clapp–Gouriet, (*d*) Armstrong, and (*e*) Vackar.

circuits are possible. In each of these oscillators, there is a relatively large amount of energy stored in the resonant reactive circuit. If not too much power is dissipated in the load, sustained oscillations are possible.

The Colpitts circuit is generally favored over that of the Hartley because the capacitors in the Colpitts circuit usually have higher Q than inductors at radio frequencies and come in a wider selection of types and sizes. In addition, the inductances in the Hartley circuit can provide a means to generate spurious frequencies because it is possible to resonate the inductors with parasitic device capacitances. Because the first element in the Colpitts circuit is a shunt

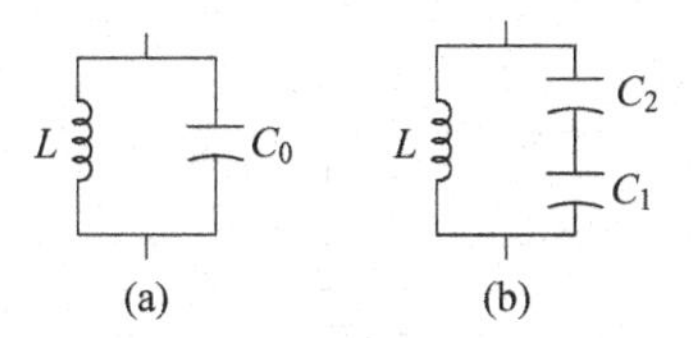

FIGURE 10.4 (*a*) Simple *LC* resonant circuit and (*b*) tapped capacitor *LC* circuit used in Colpitts oscillator.

capacitor, it is a low-pass circuit. For similar reasons, the Hartley oscillator is a high-pass circuit and the Clapp–Gouriet oscillator is a band-pass circuit. There is an improvement in the frequency stability of the tapped capacitor circuit over that of a single *LC* tuned circuit [1]. In a voltage-controlled oscillator application, it is often convenient to vary the capacitance to change the frequency. This can be done using a reverse-biased varactor diode as the capacitor. If the capacitance shown in Fig. 10.4*a* changes because of, say, a temperature shift, the frequency will change by

$$\frac{df_0}{f_0} = -\frac{dC_0}{2C_0} \tag{10.3}$$

In the tapped circuit in Fig. 10.4*b*, C_0 is the series combination of C_1 and C_2. Only C_2 is used for tuning (Colpitts circuit), and it has a frequency stability given by

$$\frac{df}{f} = -\frac{C_0}{2C_2}\frac{dC_2}{C_2} \tag{10.4}$$

This has an improved stability by the factor of C_0/C_2. Furthermore, by increasing C_0 so that C_1 and C_2 are increased by even more while adjusting the inductance to maintain the same resonant frequency, the stability can be further enhanced. The Clapp–Gouriet circuit exhibits even better stability than the Colpitts [2]. In this circuit, C_1 and C_2 are chosen to have large values compared to the tuning capacitor C_3. The minimum transistor transconductance, g_m, required for oscillation for the Clapp–Gouriet circuit increases $\propto \omega^3/Q$. While the Q of a circuit often rises with frequency, it would not be sufficient to overcome the cubic change in frequency. For the Vackar circuit, the required minimum g_m to maintain oscillation is $\propto \omega/Q$. This would tend to provide a slow drop in the amplitude of the oscillations as the frequency rises [2].

The oscillator is clearly a nonlinear circuit, but nonlinear circuits are difficult to treat analytically. In the interest of trying to get an approximate design solution, linear analysis is used. The circuit can be treated by small-signal linear mathematics to just prior to its breaking into oscillation. In going

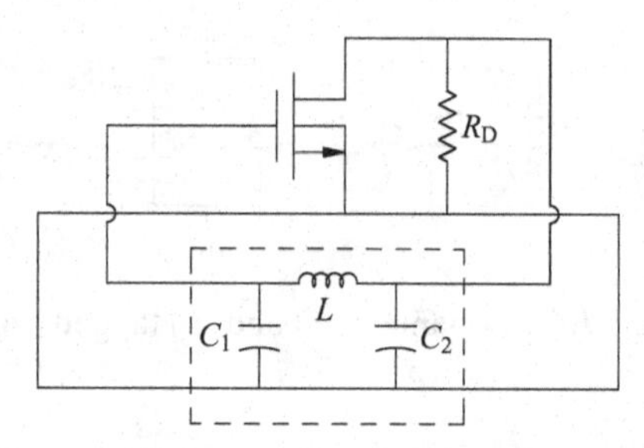

FIGURE 10.5 Colpitts oscillator as shunt–shunt connection.

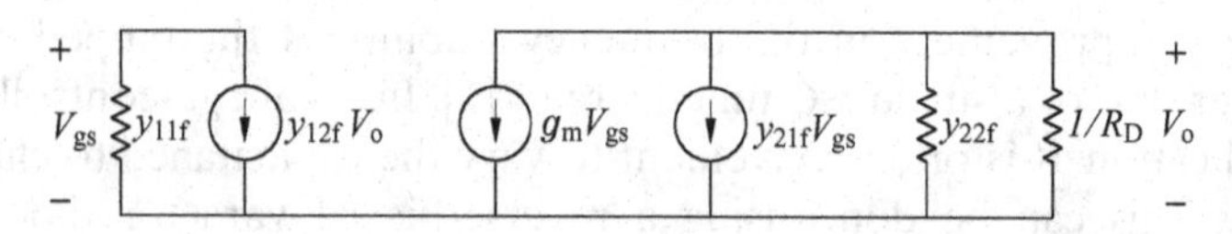

FIGURE 10.6 Equivalent circuit of Colpitts oscillator.

through the transition between oscillation and linear gain, the active part of the circuit does not change appreciably. As a justification for using linear analysis, the previous statement certainly has some flaws. Nevertheless, linear analysis does give remarkably close answers. More advanced computer modeling using methods such as harmonic balance will give more accurate results and in addition provide predictions of output power.

As an example, consider the Colpitts oscillator in Fig. 10.5. Rather than drawing it as shown in Fig. 10.3*a* as a series–series connection, it can be drawn in a shunt–shunt connection by simply rotating the feedback circuit 180° about its x axis. The y parameters for the feedback part are

$$y_{11f} = sC_1 + \frac{1}{sL} \tag{10.5}$$

$$y_{22f} = sC_2 + \frac{1}{sL} \tag{10.6}$$

$$y_{12f} = y_{21f} = \frac{-1}{sL} \tag{10.7}$$

The equivalent circuit for the y parameters now may be combined with the equivalent circuit for the active device (Fig. 10.6). The open-loop gain, a, is found by setting $y_{12f} = 0$:

$$\frac{V_o}{V_{gs}} = -\frac{g_m + y_{21f}}{(1/R_D) + y_{22f}} \tag{10.8}$$

In the usual feedback amplifier theory the y_{21f} term would be considered negligible since the forward gain of the feedback circuit would be very small compared to the amplifier gain. This is not assumed here. The open-loop gain, a, for the shunt–shunt configuration is

$$a = \frac{v_o}{i_i} = \frac{v_o}{-v_{gs} y_{11f}} = \frac{g_m + y_{21f}}{y_{11f}[(1/R_D) + y_{22f}]} \tag{10.9}$$

The negative sign introduced in getting i_i is needed to make the current go north rather than south, as made necessary by the usual sign convention. Finally, using the Barkhausen criterion, oscillation occurs when $\beta a = 1$:

$$1 = a y_{12f} = \frac{y_{12f}(g_n + y_{21f})}{y_{11f}[(1/R_D) + y_{22f}]} \tag{10.10}$$

Making the appropriate substitutions from Eqs. (10.5) through (10.7) results in the following:

$$-\left(g_m - \frac{1}{sL}\right) = sL\left[\left(sC_1 + \frac{1}{sL}\right)\frac{1}{R_D} + \left(sC_1 + \frac{1}{sL}\right)\left(sC_2 + \frac{1}{sL}\right)\right] \tag{10.11}$$

Both the real and imaginary parts of this equation must be equal on both sides. Since $s = j\omega_0$ at the oscillation frequency, all even powers of s are real and all odd powers of s are imaginary. Since g_m in Eq. (10.11) is associated with the real part of the equation, the imaginary part should be considered first:

$$\frac{1}{sL} = sL\left(s^2 C_1 C_2 + \frac{C_1}{L} + \frac{C_2}{L} + \frac{1}{s^2 L^2}\right) \tag{10.12}$$

Solution for the oscillation frequency is

$$\omega_0 = \sqrt{\frac{C_1 + C_2}{LC_1 C_2}} \tag{10.13}$$

Solving the real part of Eq. (10.11) with the now known value for ω_0 gives the required value for g_m:

$$g_m = \frac{C_1}{R_D C_2} \tag{10.14}$$

The value for g_m found in Eq. (10.14) is the minimum transconductance the transistor must have in order to produce oscillations. The small-signal analysis is sufficient to determine conditions for oscillation assuming the frequency of oscillation does not change with current amplitude in the active device. The large-signal nonlinear analysis would be required to determine the precise frequency of oscillation, the output power, the harmonic content of the oscillation, and the conditions for minimum noise.

An alternative way of looking at this example involves simply writing down the node voltage circuit equations and solving them. The determinate for the two nodal equations is zero since there is no input signal:

$$\Delta = \begin{vmatrix} sC_1 + \dfrac{1}{sL} & \dfrac{-1}{sL} \\ \dfrac{-1}{sL} + g_m & sC_2 + \dfrac{1}{sL} + \dfrac{1}{R_D} \end{vmatrix} = 0 \tag{10.15}$$

This gives the same equation as Eq. (10.11) and, of course, the same solution. Solving nodal equations can become complicated when there are several amplifying stages involved or when the feedback circuit is complicated. Advanced theory for feedback amplifiers can be used in a wide variety of circuits.

10.4 PRACTICAL OSCILLATOR EXAMPLE

The oscillator shown in Fig. 10.7 is one of several possible versions of the Hartley circuit. In this circuit, the actual load resistance is $R_L = 50\,\Omega$. Directly loading the transistor with this size resistance would cause the circuit to cease oscillation. Hence, the transformer is used to provide an effective load to the transistor of

$$R = R_L \left(\frac{n_2}{n_3} \right)^2 \tag{10.16}$$

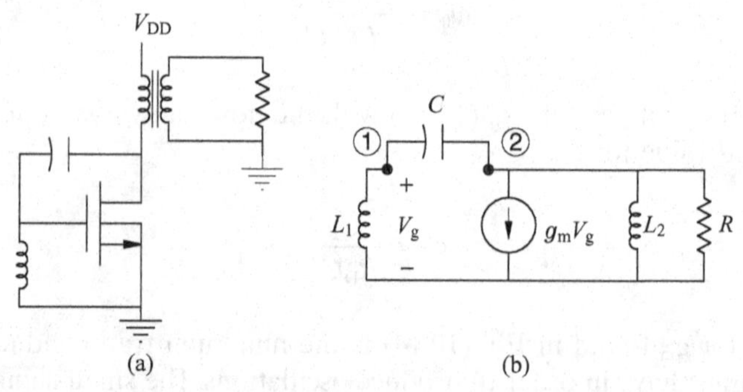

FIGURE 10.7 (*a*) Practical Hartley oscillator and (*b*) equivalent circuit.

and at the same time L_2 acts as one of the inductors required by the Hartley circuit. By solving the network in Fig. 10.7*b* in the same way as described for the Colpitts oscillator, the frequency of oscillation and minimum transconductance can be found:

$$\omega_0 = \frac{1}{\sqrt{C(L_1 + L_2)}} \tag{10.17}$$

$$g_m = \frac{L_2}{L_1 R} \tag{10.18}$$

For a 10-MHz oscillator biased with $V_{DD} = 10\,V$, the inductances L_1 and L_2 are chosen to be both equal to $1\,\mu H$. The capacitance from Eq. (10.17) is 126.6 pF. If the minimum device transconductance for a MOSFET is at least 0.333 mS, then R from Eq. (10.18) is $3000\,\Omega$. This transconductance is considerably smaller than is found in typical BJTs so that the minimum g_m condition for oscillation is much easier to achieve with a BJT. For the resistance R to be $3000\,\Omega$, it will require the transformer turns ratio to be

$$\frac{n_2}{n_3} = \sqrt{\frac{R}{R_L}} = 7.746$$

and

$$L_3 = L_2 \left(\frac{n_3}{n_2} \right)^2 = 1 \left(\frac{1}{7.746} \right)^2 = 0.01667\ \mu H$$

These circuit values can be put into SPICE to check for the oscillation. However, SPICE will give zero output when there is zero input. Somehow, a transient must be used to start the circuit oscillating. If the circuit is designed correctly, oscillations will be self-sustaining after the initial transient. One way to initiate a start-up transient is to prevent SPICE from setting up the dc bias voltages prior to doing a time-domain analysis. This is done by using the UIC (use initial conditions) command in the transient statement. In addition it may be helpful to impose an initial voltage condition on a capacitance or initial current condition on an inductance. A second approach is to use the PWL (piecewise linear) transient voltage somewhere in the circuit to impose a short pulse at $t = 0$, which forever after is turned off. The first approach is illustrated in the SPICE net list for the Hartley oscillator:

```
Hartley Oscillator Example. 10 MHz, RL=50
L1 1 16 1uH
VDC 16 0 dc +.1
C 1 2 126.65pF
L2 3 2 1uh
```

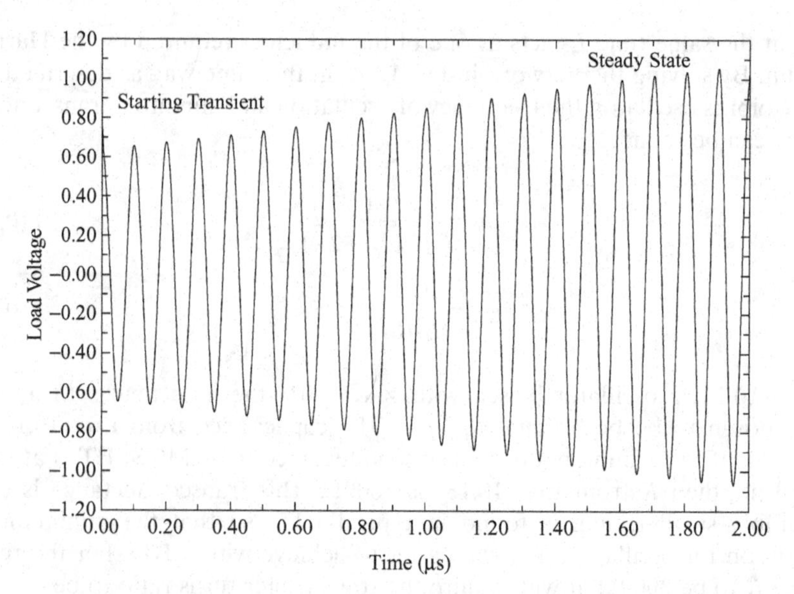

FIGURE 10.8 10-MHz Hartley oscillator time-domain response.

```
L3 4 0 0.1667e-1u
K23 L2 L3 1.
RL 4 0 50.
MOS1 2 1 0 0 MOS-EX L=1.2um W=10um
VDD 3 0 10
.tran 1ns 2us uic
.op
* MOSIS CMOS 1.2um Level 1 version
.MODEL MOS-EX NMOS (LEVEL=1, PHI=0.6, TOX=2.12E-8,
+TPG=1 ,VTO=0.786, LD=1.647E-7,KP=9.6379E-5,
+U0=591.7,RSH=8.5450E1,GAMMA=0.5863,
+NSUB=2.747E16,
+CGDO=4.0241E-10,CGSO=4.0214E-10,
+CGBO=3.6144E-10,CJ=3.8541E-4,MJ=1.1854,CJSW=1.3940E-10,
+MJSW=0.125195,PB=0.8)
.END
```

The result of the circuit analysis in Fig. 10.8 shows the oscillation building up to a steady-state output after many oscillation periods.

10.5 MINIMUM REQUIREMENTS OF THE REFLECTION COEFFICIENT

The two-port oscillator has two basic configurations: (1) a common source FET that uses an external resonator feedback from drain to gate and (2) a common gate FET that produces a negative resistance. In both of these the dc bias and the external circuit determine the oscillation conditions. When a load is connected to an oscillator circuit and the bias voltage is applied, noise in the circuit or start-up transients excites the resonator at a variety of frequencies. However, only the resonant frequency is supported and sent back to the device negative resistance. This in turn is amplified and the oscillation begins building up.

Negative resistance is merely a way of describing a power source. Ohm's law says the resistance of a circuit is the ratio of the voltage applied to the current flowing out of the positive terminal of the voltage source. If the current flows back into the positive terminal of the voltage source, then, of course, it is attached to a negative resistance. The reflection coefficient of a load, Z_L, attached to a lossless transmission line with characteristic impedance, Z_0, is

$$\Gamma = \frac{Z_L - Z_0}{Z_L + Z_0} \tag{10.19}$$

Just like viewing yourself in the mirror, the wave reflected off a positive resistance load would be smaller than the incident wave. It is not expected that an image in the mirror would be brighter than the incident light. However, if the $\Re\{Z_L\} < 0$, then it would be possible for Γ in Eq. (10.19) to be greater than 1. The "mirror" is indeed capable of reflecting a brighter light than was incident on it. Negative resistance produces oscillations when the denominator of Eq. (10.19) approaches 0. The power needed to create the negative resistance must come from an external power source or bias supply.

The conditions for oscillation then for the two-port circuit in Fig. 10.9 are

$$k < 1 \tag{10.20}$$

and

$$Z_G = -Z_i \tag{10.21}$$

where k is the amplifier stability factor from Eq. (8.43) and Z_i is the input impedance of the two-port circuit when it is terminated by Z_L. The expression

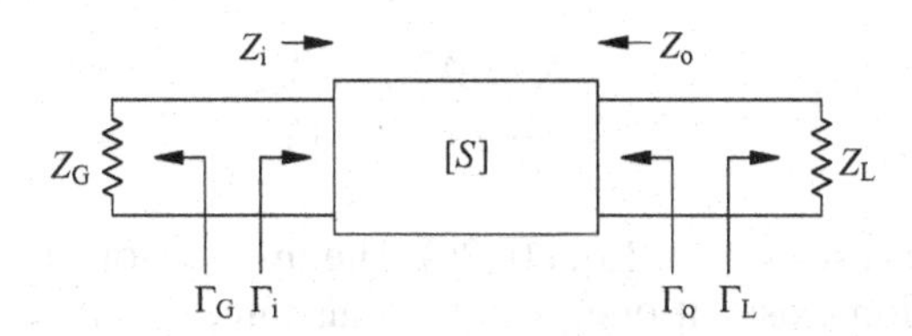

FIGURE 10.9 Doubly terminated two-port circuit.

for oscillation in terms of reflection coefficients is easily found by first determining the expressions for Γ_i and Γ_G:

$$\Gamma_i = \frac{R_i - Z_0 + jX_i}{R_i + Z_0 + jX_i} \tag{10.22}$$

$$\Gamma_G = \frac{R_G - Z_0 + jX_G}{R_G + Z_0 + jX_G} \tag{10.23}$$

If Z_G is now replaced by $-Z_i$ in Eq. (10.23),

$$\Gamma_G = \frac{-R_i - Z_0 - jX_i}{-R_i + Z_0 - jX_i} = \frac{1}{\Gamma_i} \tag{10.24}$$

Thus, Eqs. (10.21) and (10.24) are equivalent conditions for oscillation. In any case the stability factor, k, for the composite circuit with feedback must be less than 1 to make the circuit unstable and thus capable of oscillation.

An equivalent condition for the load port may be found from Eq. (10.24). From Eq. (8.17) in Chapter 8, the input reflection coefficient for a terminated two-port circuit was found to be

$$\begin{aligned}\Gamma_i &= S_{11} + \frac{S_{12}S_{21}\Gamma_L}{1 - S_{22}\Gamma_L} \\ &= \frac{S_{11} - \Delta\Gamma_L}{1 - S_{22}\Gamma_L} = \frac{1}{\Gamma_G}\end{aligned} \tag{10.25}$$

where Δ is the determinate of the S parameter matrix. Solving the right-hand side of Eq. (10.25) for Γ_L gives

$$\Gamma_L = \frac{1 - S_{11}\Gamma_G}{S_{22} - \Delta\Gamma_G} \tag{10.26}$$

But from Eq. (8.18),

$$\begin{aligned}\Gamma_o &= S_{22} + \frac{S_{12}S_{21}\Gamma_G}{1 - S_{11}\Gamma_G} \\ &= \frac{S_{22} - \Delta\Gamma_G}{1 - S_{11}\Gamma_G} = \frac{1}{\Gamma_L}\end{aligned} \tag{10.27}$$

The last equality results from Eq. (10.26). The implication is that if the conditions for oscillation exists at one port, they also necessarily exist at the other port.

10.6 COMMON GATE (BASE) OSCILLATORS

A common gate configuration is often advantageous for oscillators because it has a large intrinsic reverse gain (S_{12g}) that provides the necessary feedback. The subscript g, indicates common gate S parameters. Furthermore, feedback can be enhanced by putting some inductance between the gate and ground. Common gate oscillators often have low spectral purity but wide-band tunability. Consequently, they are often preferred in voltage controlled oscillator (VCO) designs. For a small signal approximate calculation, the scattering parameters of the transistor are typically found from measurements at a variety of bias current levels. Probably the S parameters associated with the largest output power as an amplifier would be those to be chosen for oscillator design. Since common source S parameters, S_{ij}, are usually given, it is necessary to convert them to common gate S parameters, S_{ijg}. Once this is done, the revised S parameters may be used in a direct fashion to check for conditions of oscillation.

The objective at this point is to determine the common gate S parameters with the possibility of having added gate inductance. These are derived from the common source S parameters. The procedure follows:

1. Convert the two-port common source S parameters to two-port common source y parameters (Appendix D).
2. Convert the two-port y parameters to three-port indefinite y parameters (Section 4.10).
3. Convert the three-port y parameters to three-port S parameters (Appendix D).
4. One of the three-port terminals is terminated with a load of known reflection coefficient, r.
5. With one port terminated, the S parameters are converted to two-port S parameters, which could be, among other things, common gate S parameters (Appendix E).

The first step, converting the S parameters to y parameters, can be done using the formulas in Table D.1 or Eq. (D.10) in Appendix D. For example, if the common source S parameters, $[S_s]$, are given, the y parameter matrix is

$$[Y_s] = \frac{Y_0}{D_s}\begin{bmatrix} (1-S_{11s})(1+S_{22s})+S_{12s}S_{21s} & -2S_{12s} \\ -2S_{21s} & (1+S_{11s})(1-S_{22s})+S_{12s}S_{21s} \end{bmatrix} \tag{10.28}$$

where

$$D_s \triangleq (1+S_{11s})(1-S_{22s})-S_{12s}S_{21s} \tag{10.29}$$

Next the y parameters are converted to the 3×3 indefinite admittance matrix using the method given in [4] and Section 4.10. Purely for convenience, the third row and column will be added to the center of the matrix. Then the y_{11} will represent the gate admittance, the y_{22} the source admittance, and the y_{33} the drain admittance. The new elements for the indefinite matrix are then put in between the first and second rows and in between the first and second columns of Eq. (10.28). For example, the new y_{12} is

$$y_{12} = Y_0 \left[\frac{2S_{12s} - (1 - S_{11s})(1 + S_{22s}) + S_{12s}S_{21s} - S_{12s}S_{21s}}{D_s} \right] \tag{10.30}$$

The values for y_{21}, y_{23}, and y_{32} are found similarly. The new y_{22} term is found from $y_{22} = -y_{21} - y_{23}$. The indefinite admittance matrix is then represented as follows:

$$[Y] = \begin{matrix} & g & s & d \\ g \\ s \\ d \end{matrix} \begin{pmatrix} y_{11} & y_{12} & y_{13} \\ y_{21} & y_{22} & y_{23} \\ y_{31} & y_{32} & y_{33} \end{pmatrix} \tag{10.31}$$

The S parameter matrix for 3×3 or higher order can be found from Eq. (D.9) in Appendix D:

$$S = F(1 - G^*Y)(I + GY)^{-1} F^{-1} \tag{10.32}$$

In this equation I is the identity matrix, while G and F are defined in Appendix D. When the reference characteristic impedances, Z_0, are the same in all three ports, the F and the F^{-1} will cancel out. Determining S from Eq. (10.32) is straightforward but lengthy. At this point the common terminal is chosen. To illustrate the process, a common source connection is used in which the the source is terminated by a load with a reflection coefficient, r_s, as shown in Fig. 10.10. If the source is grounded, the reflection coefficient is $r_s = -1$. The relationship between the incident and reflected waves is

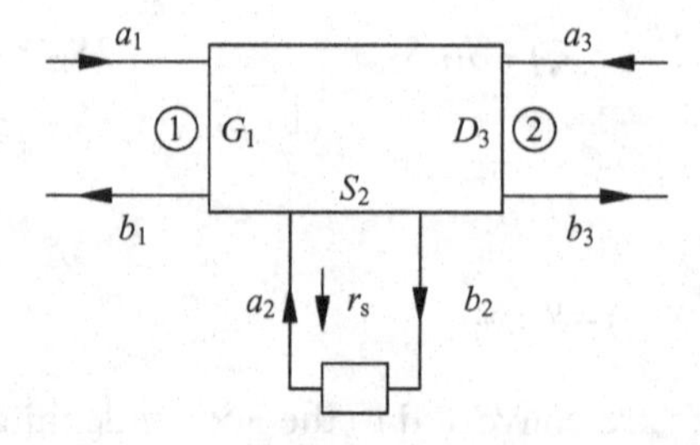

FIGURE 10.10 Three-port with source terminated with r_s.

$$
\begin{aligned}
b_1 &= S_{11}a_1 + S_{12}a_2 + S_{13}a_3 \\
b_2 &= S_{21}a_1 + S_{22}a_2 + S_{23}a_3 \\
b_3 &= S_{31}a_1 + S_{32}a_2 + S_{33}a_3
\end{aligned} \tag{10.33}
$$

Solution for S_{11s} is done by terminating the drain at port 3 with Z_0 so that $a_3 = 0$. The source is terminated with an impedance with reflection coefficient

$$r_s = \frac{a_2}{b_2} \tag{10.34}$$

or for any port

$$r_i = \frac{Z_i - Z_{ref}}{Z_i + Z_{ref}} \tag{10.35}$$

The reflection coefficient is determined relative to the reference impedance, which is the impedance looking back into the transistor. With Eq. (10.34), b_2 can be eliminated in Eq. (10.33) giving a relationship between a_1 and a_2:

$$
\begin{aligned}
\frac{a_2}{r_s} &= S_{21}a_1 + S_{22}a_2 \\
a_2 &= \frac{S_{21}a_1}{1/r_s - S_{22}}
\end{aligned} \tag{10.36}
$$

The ratio between b_1 and a_1 under these conditions is

$$S_{11s} = \frac{b_1}{a_1} = S_{11} + \frac{S_{12}S_{21}}{1/r_s - S_{22}} \tag{10.37}$$

This represents the revised S_{11s} scattering parameter when the source is terminated with an impedance whose reflection coefficient is r_s. In similar fashion the other parameters can be easily found as shown in Appendix E. The numbering system for the common source parameters is set up so that the input port (gate side) is port 1 and the output port (drain) is port 2. Therefore the subscripts of the common source parameters, S_{ijs}, range from 1 to 2. In other words, after the source is terminated, there are only two ports, the input and output. These are written in terms of the three-port scattering parameters, S_{ij}, which of course have subscripts that range from 1 to 3.

The common gate formulas are given in Appendix E. For a particular RF transistor, in which the generator is terminated with a 5-nH inductor, the

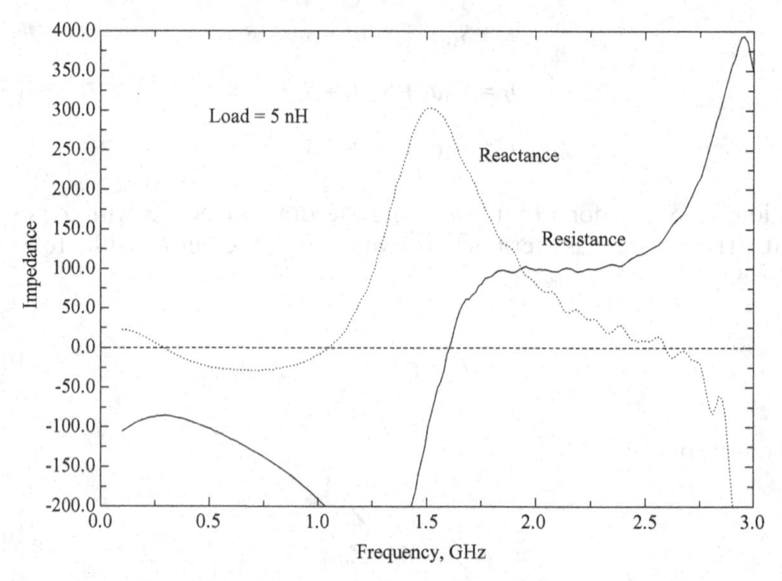

FIGURE 10.11 Plot of load impedance required for oscillation when generator side is terminated with 5-nH inductor.

required load impedance on the drain side to make the circuit oscillate is shown in Fig. 10.11 as obtained from the program SPARC (*S* parameters conversion). Since a passive resistance must be positive, the circuit is capable of oscillation only for those frequencies in which the resistance is above the 0-Ω line. An actual oscillator would still require a resonator to force the oscillator to provide power at a single frequency. A numerical calculation at 2 GHz that illustrates the process is found in Appendix E.

When the real part of the load impedance is less than the magnitude of the negative real part of the device impedance, then oscillations will occur at the frequency where there is resonance between the load and the device. For a one-port oscillator, the negative resistance is a result of feedback, but here the feedback is produced by the device itself rather than by an external path. Specific examples of one-port oscillators use a Gunn or IMPATT diode as the active device. These are normally used at frequencies above the frequency band of interest here. On the surface the one-port oscillator is, in principle, no different than a two-port oscillator whose opposite side is terminated in something that will produce negative resistance at the other end. The negative resistance in the device compensates for positive resistance in the resonator. Noise in the resonator port or a turn on transient starts the oscillation going. The oscillation frequency is determined by the resonant frequency of a high-Q circuit.

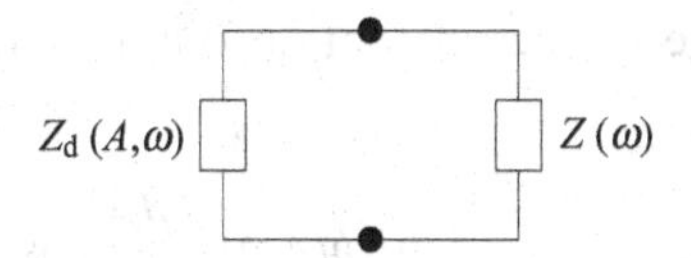

FIGURE 10.12 Oscillator model when passive impedance $Z(\omega)$ is separated from active device $Z_d(A, \omega)$.

10.7 STABILITY OF AN OSCILLATOR

In the previous section, a method has been given to determine whether a circuit will oscillate or not. What is yet to be addressed is whether the oscillation will remain stable in the face of a small current transient in the active device. The simple equivalent circuit shown in Fig. 10.12 can be divided into the part with the active device and the passive part with the high-Q resonator. The current flowing through the circuit is

$$i(t) = A(t)\cos[\omega t + \phi(t)] = \Re\{A(t)e^{j\omega t + \phi(t)}\} \tag{10.38}$$

where A and ϕ are slowly varying functions of time. The part of the circuit with the active device is represented by $Z_d(A, \omega)$ and the passive part by $Z(\omega)$. The condition for oscillation requires the sum of the impedances around the loop to be zero:

$$Z_d(A, \omega) + Z(\omega) = 0 \tag{10.39}$$

Ordinarily, the passive circuit selects the frequency of oscillation by means of a high-Q resonator. The relative variation of the impedance of the active device with frequency is small, so Eq. (10.39) can be approximated by

$$Z_d(A) + Z(\omega) = 0 \tag{10.40}$$

In phaser notation the current is

$$I = Ae^{j\phi} \tag{10.41}$$

and

$$Z(\omega) = R(\omega) + jX(\omega) \tag{10.42}$$

so that the voltage drop around the closed loop in Fig. 10.12 is

$$\begin{aligned} 0 &= \Re\{[Z(\omega) + Z_d(A)]I\} \\ &= [R(\omega) + R_d(A)]A\cos(\omega t + \phi) - [X(\omega) + X_d(A)]A\sin(\omega t + \phi) \end{aligned} \tag{10.43}$$

The time rate of change of the current is found by taking the derivative of Eq. (10.38):

$$\begin{aligned}\frac{di}{dt} &= -A\left(\omega+\frac{d\phi}{dt}\right)\sin(\omega t+\phi)+\frac{dA}{dt}\cos(\omega t+\phi)\\ &= \Re\left\{\left[j\left(\omega+\frac{d\phi}{dt}\right)+\frac{1}{A}\frac{dA}{dt}\right]Ae^{j\omega t+\phi}\right\}\end{aligned} \tag{10.44}$$

Ordinarily, in ac circuit analysis, d/dt is equivalent to $j\omega$ in the frequency domain. Now with variation in the amplitude and phase, the time derivative is equivalent to

$$\frac{d}{dt} \to j\omega' = j\left[\omega+\frac{d\phi}{dt}-j\frac{1}{A}\frac{dA}{dt}\right] \tag{10.45}$$

The Taylor series expansion of $Z(\omega')$ about ω_0 is

$$Z\left(\omega+\frac{d\phi}{dt}-j\frac{1}{A}\frac{dA}{dt}\right) \approx Z(\omega_0)+\frac{dZ}{d\omega}\left(\frac{d\phi}{dt}-j\frac{1}{A}\frac{dA}{dt}\right) \tag{10.46}$$

Consequently, an expression for the voltage around the closed loop can be found:

$$\begin{aligned}\Re\{(Z+Z_d)I\} &= \left[R(\omega_0)+R_d(A)+\frac{dR}{d\omega}\frac{d\phi}{dt}+\frac{dX}{d\omega}\frac{1}{A}\frac{dA}{dt}\right]A\cos(\omega t+\phi)\\ &\quad -\left[X(\omega_0)+X_d(A)+\frac{dX}{d\omega}\frac{d\phi}{dt}-\frac{dR}{d\omega}\frac{1}{A}\frac{dA}{dt}\right]A\sin(\omega t+\phi)\end{aligned} \tag{10.47}$$

Multiplying Eq. (10.47) by $\cos(\omega t + \phi)$ and integrating will produce Eq. (10.48) by the orthogonality property of sine and cosine. Similarly, multiplying Eq. (10.47) by $\sin(\omega t + \phi)$ will produce Eq. (10.49):

$$0 = R(\omega)+R_d(A)+\frac{dR}{d\omega}\frac{d\phi}{dt}+\frac{dX}{d\omega}\frac{1}{A}\frac{dA}{dt} \tag{10.48}$$

$$0 = -X(\omega)+X_d(A)-\frac{dX}{d\omega}\frac{d\phi}{dt}+\frac{dR}{d\omega}\frac{1}{A}\frac{dA}{dt} \tag{10.49}$$

Multiplying Eq. (10.48) by $dX/d\omega$ and Eq. (10.49) by $dR/d\omega$ and adding will eliminate the $d\phi/dt$ term. A similar procedure will eliminate dA/dt. The results are

$$0 = [R(\omega)+R_d(A)]\frac{dX}{d\omega}-[X(\omega)+X_d(A)]\frac{dR}{d\omega}+\left|\frac{dZ(\omega)}{d\omega}\right|^2\frac{1}{A}\frac{dA}{dt} \tag{10.50}$$

$$0 = [X(\omega) + X_d(A)]\frac{dX}{d\omega} + [R(\omega) + R_d(A)]\frac{dR}{d\omega} + \left|\frac{dZ(\omega)}{d\omega}\right|^2 \frac{d\phi}{dt} \tag{10.51}$$

Under steady-state conditions, the time derivatives are zero. Combination of Eqs. (10.50) and (10.51) gives

$$\frac{dR/d\omega}{dX/d\omega} = \frac{R(\omega) + R_d(A)}{X(\omega) + X_d(A)} = -\frac{X(\omega) + X_d(A)}{R(\omega) + R_d(A)} \tag{10.52}$$

The only way for this equation to be satisfied is that it satisfies Eq. (10.40). However, suppose there is a small disturbance in the current amplitude of δA from the steady-state value of A_0. Based on Eq. (10.40) the resistive and reactive components would become

$$\begin{aligned} R(\omega_0) + R_d(A) &= R(\omega_0) + R_d(A_0) + \delta A \frac{\partial R_d(A)}{\partial A} \\ &= \delta A \frac{\partial R_d(A)}{\partial A} \end{aligned} \tag{10.53}$$

$$X(\omega_0) + X_d(A) = \delta A \frac{\partial X_d(A)}{\partial A} \tag{10.54}$$

The derivatives are, of course, assumed to be evaluated at $A = A_0$. Substituting these into Eq. (10.50) gives the following differential equation with respect to time:

$$0 = \delta A \frac{\partial R_d(A)}{\partial A}\frac{dX(\omega)}{d\omega} - \delta A \frac{\partial X_d(A)}{\partial A}\frac{dR(\omega)}{d\omega} + \left|\frac{dZ(\omega)}{d\omega}\right|^2 \frac{1}{A_0}\frac{d\delta A}{dt} \tag{10.55}$$

or

$$0 = \delta A S + \alpha \frac{d\delta A}{dt} \tag{10.56}$$

where

$$S \triangleq \frac{\partial R_d(A)}{\partial A}\frac{dX(\omega)}{d\omega} - \frac{\partial X_d(A)}{\partial A}\frac{dR(\omega)}{d\omega} \tag{10.57}$$

and

$$\alpha \triangleq \left|\frac{dZ(\omega)}{d\omega}\right|^2 \frac{1}{A_0} \tag{10.58}$$

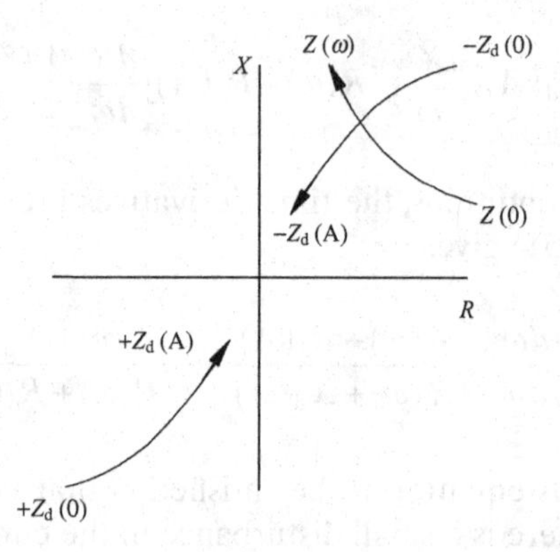

FIGURE 10.13 Locus of points for oscillator passive and active impedances.

The solution of Eq. (10.56) is

$$\delta A = Ce^{-St/\alpha}$$

which is stable if $S > 0$. The Kurokawa stability condition for small changes in the current amplitude occurs when Eq. (10.57) is positive [5]. As an example, consider the stability of a circuit whose passive circuit impedance changes with frequency as shown in Fig. 10.13 and device impedance that changes with current amplitude shown in the third quadrant of Fig. 10.13. As the current amplitude increases, $R_d(A)$ and $X_d(A)$ both increase:

$$\frac{\partial R_d(A)}{\partial A} > 0$$

$$\frac{\partial X_d(A)}{\partial A} > 0$$

As frequency increases, the passive circuit resistance, $R(\omega)$, decreases and the circuit reactance, $X(\omega)$, increases:

$$\frac{dR(\omega)}{d\omega} < 0$$

$$\frac{dX(\omega)}{d\omega} > 0$$

From Eq. (10.57) this would provide stable oscillations at the point where $Z(\omega)$ and $-Z_d(A)$ intersect. If there is a small change in the current amplitude, the circuit tends to return back to the A_0, ω_0 oscillation point.

If there is a small perturbation in the current phase rather than the current amplitude, the stability criterion can be found in similar fashion as above. In this case the equations similar to (10.53) and (10.54) are

$$R(\omega)+R_{\mathrm{d}}(\phi)=\delta\phi\frac{\partial R_{\mathrm{d}}(\phi)}{\partial\phi} \tag{10.59}$$

$$X(\omega_0)+X_{\mathrm{d}}(\phi)=\delta\phi\frac{\partial X_{\mathrm{d}}}{\partial\phi} \tag{10.60}$$

This is substituted into Eq. (10.51) in which the device impedance is given as a function of ϕ rather than A. The terms $R(\omega_0) + R_{\mathrm{d}}(\phi)$ and $X(\omega_0) + X_{\mathrm{d}}(\phi)$ are replaced by Eqs. (10.59) and (10.60), respectively:

$$0=\delta\phi\frac{\partial X_{\mathrm{d}}(\phi)}{\partial\phi}\frac{dX(\omega)}{d\omega}+\delta\phi\frac{\partial R_{\mathrm{d}}}{\partial\phi}\frac{dR(\omega)}{d\omega}+\left|\frac{dZ(\omega)}{d\omega}\right|^2\frac{d\delta\phi}{dt} \tag{10.61}$$

$$0=S'\delta\phi+\alpha'\frac{d\phi}{dt} \tag{10.62}$$

where

$$S\triangleq\frac{\partial X_{\mathrm{d}}(\phi)}{\partial\phi}\frac{dX(\omega)}{d\omega}+\frac{\partial R_{\mathrm{d}}(\phi)}{\partial\phi}\frac{dR(\omega)}{d\omega} \tag{10.63}$$

and

$$\alpha'\triangleq\left|\frac{dZ(\omega)}{d\omega}\right|^2 \tag{10.64}$$

Since

$$\delta\phi=Ce^{-S't/\alpha'}$$

the oscillator is stable with respect to small changes in phase if $S' > 0$.

10.8 INJECTION-LOCKED OSCILLATORS

A free-running oscillator frequency can be modified by applying an external frequency source to the oscillator. Such injection-locked oscillators can be used as high-power FM amplifiers when the circuit Q is sufficiently low to accommodate the frequency bandwidth of the signal. If the injection signal voltage, $V(\omega_{\mathrm{in}})$, is at a frequency close to but not necessarily identical to the free-running frequency of the oscillator and is placed in series with the passive impedance, $Z(\omega_{in})$, in Fig. 10.12, then the loop voltage is

$$[Z(\omega_{in})+Z_d(A)]I=V \tag{10.65}$$

The amplitude of the current at the free-running point is A_0 and the relative phase between the voltage and current is ϕ. Hence,

$$Z(\omega_{in})=-Z_d(A)+\frac{|V|}{A_0}e^{-j\phi} \tag{10.66}$$

Up to this point, the passive impedance has been left rather general. As a specific example, the circuit can be considered to be a high-Q series resonant circuit determined by its inductance and capacitance together with some cavity losses, R_c, and a load resistance, R_L:

$$Z(\omega_{in})=j\left(\omega_{in}L-\frac{1}{\omega_{in}C}\right)+R_c+R_L \tag{10.67}$$

Since ω_{in} is close to the circuit free-running oscillator frequency ω_0,

$$\begin{aligned} Z(\omega_{in}) &= j\frac{L}{\omega_{in}}(\omega_{in}^2-\omega_0^2)+R_c+R_L \\ &\approx j2L\Delta\omega_m+R_c+R_L \end{aligned} \tag{10.68}$$

where $\Delta\omega_m=\omega_0-\omega_{in}$ comes from the Taylor series expansion of $Z(\omega_{in})$.

Equation (10.66) represented in Fig. 10.14 is a modification of that shown in Fig. 10.13 for the free-running oscillator case. If the magnitude of the injection voltage, V, remains constant, then the constant magnitude vector, $|V|/A_0$, which must stay in contact with both the device and circuit impedance lines, will change its orientation as the injection frequency changes [thereby chang-

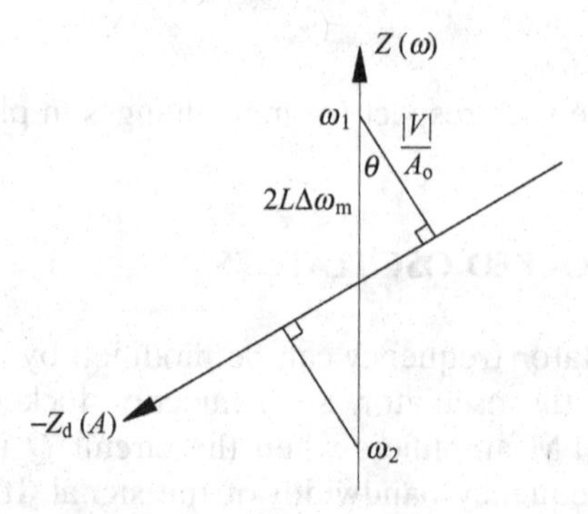

FIGURE 10.14 Injection-locked frequency range.

ing $Z(\omega_{in})$]. However, there is a limit to how much the $|V|/A_0$ vector can move because circuit and device impedances grow too far apart. In that case injection lock ceases. The example in Fig. 10.14 illustrates the simple series resonant cavity where the circuit resistance is independent of frequency. Furthermore, the $|V|/A_0$ vector is drawn at the point of maximum frequency excursion from ω_0. Here $|V|/A_0$ is orthogonal to the $Z_d(A)$ line. If the frequency moves beyond ω_1 or ω_2, the oscillator loses lock with the injected signal. At the maximum locking frequency,

$$|2\Delta\omega_m L\cos\theta| = \frac{|V|}{A_0} \tag{10.69}$$

The expressions for the oscillator power delivered to the load, P_o, the available injected power, and the external circuit Q_{ext} are

$$P_o = \tfrac{1}{2} R_L A_0^2 \tag{10.70}$$

$$P_i = \frac{|V|^2}{8R_L} \tag{10.71}$$

$$Q_{ext} \approx \frac{\omega_0 L}{R_L} \tag{10.72}$$

When these are substituted into Eq. (10.69) the well-known injection locking range is found [6]:

$$\Delta\omega_m = \frac{\omega_0}{Q_{ext}}\sqrt{\frac{P_i}{P_o}}\,\frac{1}{\cos\theta} \tag{10.73}$$

The total locking range is from $\omega_0 - \Delta\omega_m$ to $\omega_0 + \Delta\omega_m$. The expression originally given by Adler [6] did not included the $\cos\theta$ term. However, high-frequency devices often exhibit a phase delay of the RF current with respect to the voltage. This led to Eq. (10.73) where the device and circuit impedance lines are not necessarily orthogonal [7]. In the absence of information about the value of θ, a conservative approximation for the injection range can be made by choosing $\cos\theta = 1$. The frequency range over which the oscillator frequency can be pulled from its free-running frequency is proportional to the square root of the injected power and inversely proportional to the circuit Q as might be expected intuitively.

10.9 OSCILLATOR PHASE NOISE

The fluctuations in the amplitude and especially the phase of an oscillator is an important limitation on the quality of an oscillator. In a receiver, the noise

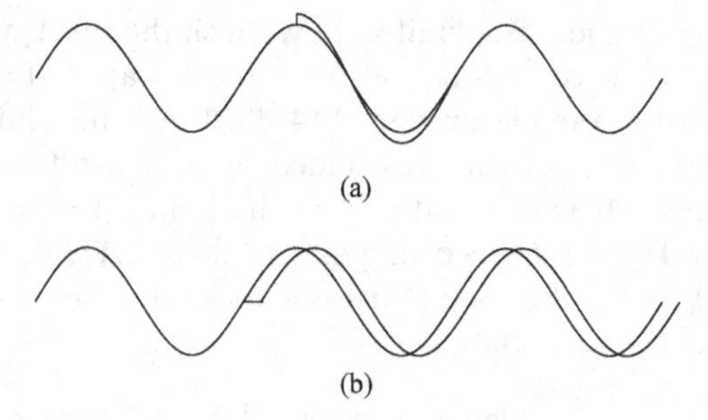

FIGURE 10.15 Effect of noise (*a*) injected at peak and (*b*) at zero crossing.

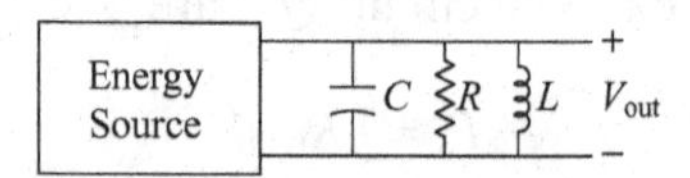

FIGURE 10.16 Circuit model for phase noise calculation.

in the local oscillator of a mixer translates to noise in the intermediate frequency (IF) output. This implies the channel bandwidth must be larger than that required by the signal to accommodate the added phase noise. In a digital system using a clock, phase noise produces timing jitter. A noise current spike will primarily affect the amplitude of the oscillation if it occurs at one of the two extrema of the oscillation waveform. It will primarily affect the phase of the oscillation if it occurs during the zero crossing of the waveform (Fig. 10.15). When the noise fluctuation occurs at the waveform extrema in a stable oscillator, the amplitude will be quickly restored to its equilibrium value, and there will be no long-term effects. When the noise fluctuation occurs at the zero crossing, the phase change is permanent. Phase noise, $\mathcal{L}\{\Delta\omega\}$, is defined as the ratio of the noise power in a certain bandwidth (usually 1 Hz) at a certain offset frequency, $\Delta\omega$, away from the main carrier frequency to the signal power. The units for phase noise are typically given in dBc/Hz although the hertz part is inside the logarithm. This will be clarified later in this section.

The analysis of phase noise is done with a simple *RLC* resonator excited by an ideal negative resistance energy source (Fig. 10.16). A variety of models for phase noise have been proposed, but the *linear time-varying* theory developed by Hajimiri and Lee provides both a reasonably tractable and accurate model [8–10]. One of the assumptions they make is that a small increase in the input signal from a noise perturbation will produce a proportional output phase response. While the large-signal oscillator is clearly nonlinear, the small-signal perturbation is assumed to be linear. The second assumption is that the low-frequency 1/*f* noise can be folded up to the oscillator output band by the periodic and therefore time-varying signal.

The linearity assumption allows defining an impulse response function that relates the input noise impulse to the output phase response. This is modeled as the unit step function, $u(t)$, in the response function, h_ϕ. Since phase noise is much more important than amplitude noise, only the phase impulse response is useful:

$$h_\phi(t,\tau)=\frac{\Gamma(\omega_0\tau)}{q_{max}}u(t-\tau) \tag{10.74}$$

Hajimiri [8] defines $\Gamma(x)$ as the *impulse sensitivity function*. This function is periodic (though not necessarily sinusoidal) with a period equal to that of the oscillator. The maximum charge displacement on the tank capacitor, q_{max}, normalizes $\Gamma(x)$ so that it is independent of signal level. This function is maximum at the signal zero crossings and zero at the extrema of the oscillation. Hajimiri [8] shows how values for this function might be obtained by simulation methods or approximate analytical methods for special cases. The phase response to a noise current is

$$\phi(t)=\int_{-\infty}^{\infty}h_\phi(t,\tau)i(\tau)d\tau=\frac{1}{q_{max}}\int_{-\infty}^{t}\Gamma(\omega_0\tau)i(\tau)d\tau \tag{10.75}$$

Since $\Gamma(x)$ is a periodic function, it can be expanded into a Fourier series:

$$\Gamma(\omega_0\tau)=\frac{c_0}{2}+\sum_{n=1}^{\infty}c_n\cos(n\omega_0\tau+\theta_n) \tag{10.76}$$

Since θ_n represents the phase of the uncorrelated noise, it plays no significant role and is set to zero. The injected noise current, $i(t)$, is represented as a sine wave at a multiple, m, of the oscillation frequency, ω_0,

$$i(t)=I_m\cos[(m\omega_0+\Delta\omega)t] \tag{10.77}$$

where $\pm\Delta\omega$ is the frequency offset above and below $m\omega_0$ where there is "significant" noise. Equations (10.76) and (10.77) are substituted into Eq. (10.75). The orthogonality of the cosine functions demands that only the case where $m = n$ survives. Also the product of the dc terms is dropped out for now in order to focus on the frequency terms:

$$\phi(t)=\frac{c_nI_m}{q_{max}}\int_{-\infty}^{t}\cos[(m\omega_0+\Delta\omega)\tau]\cos(n\omega_0\tau)d\tau \tag{10.78}$$

From the trigonometric double-angle identity

$$\phi(t) = \frac{c_m I_m}{2q_{\max}} \int_{-\infty}^{t} \cos[(2m\omega_0 + \Delta\omega)\tau] + \cos(\Delta\omega\tau)\, d\tau \tag{10.79}$$

$$= \frac{c_m I_m}{2q_{\max}} \left[\frac{\sin[(2m\omega_0 + \Delta\omega)\tau]}{2m\omega_0 + \Delta\omega} + \frac{\sin(\Delta\omega\tau)}{\Delta\omega} \right]_{-\infty}^{t} \tag{10.80}$$

$$\approx \frac{c_m I_m \sin(\Delta\omega t)}{2q_{\max}\Delta\omega} \tag{10.81}$$

It is assumed that $\Delta\omega << \omega_0$ so that the first term is negligible. Also, the evaluation at $t = -\infty$ is assumed to be zero. This shows two sideband offsets, $\pm\Delta\omega$, of the phase spectrum around the oscillation frequency even if the noise is injected at some integer multiple, m, above ω_0. As pointed out in [8, 10], the noise voltage and subsequently the noise power in the two sidebands are found using the phase-to-voltage converter:

$$v_{\text{out}}(t) = \cos[\omega_0 t + \phi(t)] \tag{10.82}$$

$$= \cos[\omega_0 t + K_m \sin \Delta\omega t] \tag{10.83}$$

where K_m is the prefix to $\sin \Delta\omega t$ in Eq. (10.81). Thus

$$v_{\text{out}}(t) = \cos(\omega_0 t)\cos(K_m \sin \Delta\omega t) - \sin(\omega_0 t)\sin(K_m \sin \Delta\omega t) \tag{10.84}$$

$$\approx \cos(\omega_0 t) - \sin(\omega_0 t)\sin(K_m \Delta\omega t) \tag{10.85}$$

Since for small argument $\sin(K_m\Delta\omega t) \approx K_m \sin(\Delta\omega t)$,

$$v_{\text{out}}(t) \approx \cos(\omega_0 t) - \frac{K_m}{2}[\cos(\omega_0 - \Delta\omega)t - \cos(\omega_0 + \Delta\omega)t] \tag{10.86}$$

The ratio of the noise power in the sidebands at $\Delta\omega$ to the carrier power is proportional to $K_m/2$ or in log form:

$$P_{\text{SBC}} = 10 \log \left[\frac{I_m c_m}{4q_{\max}\Delta\omega} \right]^2 \tag{10.87}$$

For white noise with a wide band of frequencies, all the I_m in Eq. (10.87) are equal to each other, and the sum of the I_m^2 is $\langle i^2 \rangle / \Delta f$. This is the equivalent rms noise current. The total noise spectral density relative to the carrier is

$$\mathcal{L}\{\Delta\omega\} = 10 \log \left(\frac{1}{4q_{\max}^2 \Delta\omega^2} \frac{\langle i^2 \rangle}{\Delta f} \sum_{m=0}^{\infty} c_m^2 \right) \tag{10.88}$$

Experimentally, it is easier to find the rms value of Γ^2 than the individual c_m^2 components. Parseval's theorem gives the link between these:

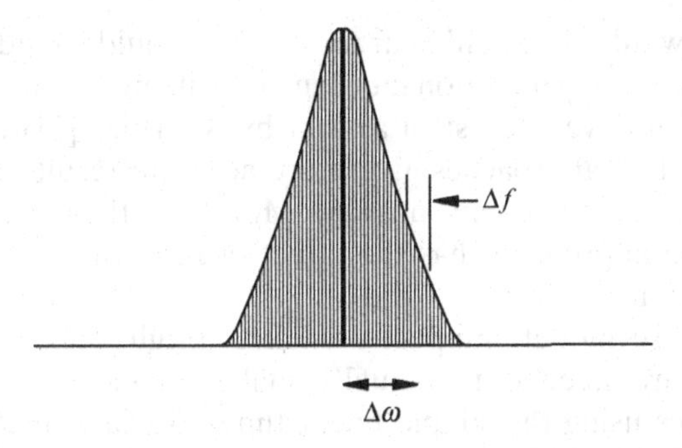

FIGURE 10.17 Phase noise over band of $\pm\Delta\omega$ within Δf.

$$\sum_{m=0}^{\infty} c_m^2 = \frac{1}{\pi}\int_0^{2\pi} |\Gamma(x)|^2\, dx = 2\Gamma_{\text{rms}}^2 \tag{10.89}$$

so that Eq. (10.88) is

$$\mathcal{L}\{\Delta\omega\} = 10\log\left(\frac{\Gamma_{\text{rms}}}{q_{\max}^2}\frac{\langle i^2\rangle/\Delta f}{2\Delta\omega^2}\right) \tag{10.90}$$

Equation (10.90) has two frequency terms: $\Delta\omega$ is the offset frequency from the carrier where the noise is measured, and Δf is the band over which the noise is measured, typically 1 Hz. Figure 10.17 illustrates the distinction between $\Delta\omega$ and Δf.

The sideband power given in Eq. (10.90) represents first a translation of noise generated at $m\omega_0$ to near dc by the Fourier coefficients. The $1/f$ noise is turned into $1/f^2$ noise near dc. This is a result of the implied integration of the impulse response function. The phase-to-voltage transformation upconverts the near dc $1/f^2$ noise back up to ω_0. Thus, the oscillator will have both the original $1/f$ noise as well as a $1/f^2$ noise component. Minimization of the effects of noise implies using a high-Q resonator and a large signal. Also, the $\Gamma(x)$ function should have a small value of Γ_{dc} so as to minimize upconversion of noise at low frequencies to ω_0. Often the device has in itself a flicker noise component that is proportional to $1/f$. When this is upconverted back up to ω_0, the phase noise then also includes a $1/f^3$ frequency dependence as has been experimentally observed.

The Hajimiri model gives design insights on how noise might be minimized and provides a physical mechanism for observed phenomena. Its practical weakness lies in determining the impulse sensitivity function, $\Gamma(x)$. Hajimiri [8] describes three methods for finding Γ that involve SPICE simulation or approximate analytical methods.

However, the assumption of linearity in an oscillator and the approximations that it brings is troublesome. The actual determination of $\Gamma(x)$ is lengthy.

The alternative would be a nonlinear model that would include both AM and PM as part of the noise generation mechanism. Still more general and accurate models have been developed such as that by Kaertner [11] and Demiri [12, 13]. While these two approaches give more accurate results, they do require simulation techniques to provide answers. Moreover, the increased rigor associated with these approaches has the effect of reducing the physical insight that the Hajimiri model has.

The phase noise models support oscillator insight and analysis. Practical design criteria are needed to actually make an oscillator. Optimization, however, requires using the simple linear time-invariant model. What is lost in accuracy is gained in finding an optimum design. The linear model assumes the existence of the parallel resonant tank circuit shown in Fig. 10.16. The tank admittance near resonance at $\omega_o + \Delta\omega$ is

$$
\begin{aligned}
Y(\omega) &= G + \frac{j}{\omega L}(\omega^2 LC - 1) \\
&\approx G\left[1 + \frac{jR}{\omega_0 L}\left(\frac{\omega_0^2 + 2\omega_0\Delta\omega + (\Delta\omega)^2}{\Delta\omega_0^2} - 1\right)\right] \\
&\approx G\left[1 + jQ\left(\frac{2\Delta\omega}{\omega_0}\right)\right]
\end{aligned}
\tag{10.91}
$$

where $R = 1/G$. The magnitude squared of the impedance is, therefore,

$$
|Z|^2 \approx R^2\left(\frac{\omega_0}{2Q\Delta\omega}\right)^2 \tag{10.92}
$$

The mean-square noise voltage per hertz is determined using Eq. (10.92):

$$
\frac{\langle v^2\rangle}{\Delta f} = \frac{\langle i^2\rangle}{\Delta f}|Z|^2 \tag{10.93}
$$

$$
= 4kTGR^2\left(\frac{\omega_0}{2Q\Delta\omega}\right)^2 \tag{10.94}
$$

$$
= 4kTG\left(\frac{\omega_0}{2\Delta\omega}\right)^2(\omega_0 L)^2 \tag{10.95}
$$

The phase (noise-to-signal ratio) is

$$
\mathcal{L}\{\Delta\omega\} = \frac{4kTG}{V_o^2}\left(\frac{\omega_0}{2\Delta\omega}\right)^2(\omega_0 L)^2 \tag{10.96}
$$

Zhu [14] started with this simple formula to arrive at a procedure for minimizing phase noise in an LC voltage-controlled oscillator. In integrated circuit

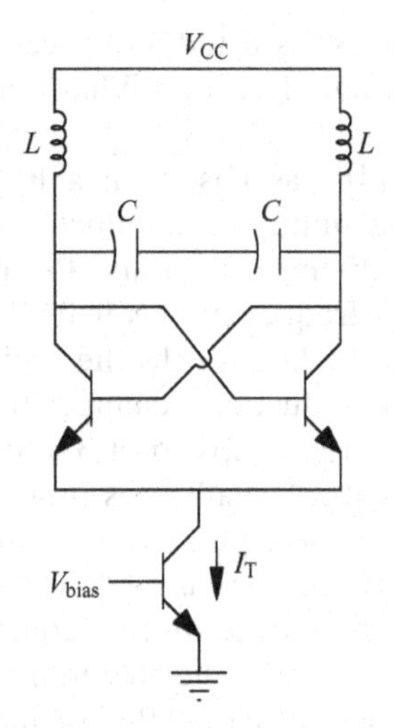

FIGURE 10.18 The *LC* oscillator.

design, the inductor takes up the major part of the real estate. Design of the oscillator is focused on controlling the inductor size. In the subsequent analysis the inductor area and the line spacing between turns is maintained constant, while the width and number of turns are varied to achieve various values of inductance.

The focus here will be on the *LC* oscillator circuit in Fig. 10.18 since it provides lower phase noise than the popular ring oscillator. The *LC* oscillator is biased by a tail current source, I_T. The tank circuit current waveform is a square wave that provides voltage harmonics. Only the fundamental of the voltage wave is supported so that its amplitude is $V_0'(4/\pi)$. Since half of the current flows through the left side and half through the right side, this voltage is

$$V_0 = 2I_T R/\pi \tag{10.97}$$

To emphasize that the G in Eq. (10.96) is the parallel tank circuit conductance, G is replaced by the symbol, g_p. In terms of the tail current, I_T, Eq. (10.96) is

$$\mathcal{L}\{\Delta\omega\} = \frac{\pi^2 kTg_p^3}{I_T^2}\left(\frac{\omega_0}{2\Delta\omega}\right)^2 (\omega_0 L)^2 \tag{10.98}$$

The output waveform is proportional to the tail current, so, as expected, the phase noise decreases with increasing signal power. Similar calculations were carried out in [14] for noise resulting from the base spreading resistance and the shot noise of the transistor. In each case the phase noise was found to be proportional to $g_p^3 L^2$. The tail current can be increased to the point where the oscillation signal becomes distorted. However, once this is set to a constant, it is not part of the design decisions. What remains for design analysis is minimizing $g_p^3 L^2$.

The analysis done in [14] was based on a fixed spiral inductor size of $400 \times 400\,\mu\text{m}^2$, which had a minimum conductor spacing of $2\,\mu$m. The line width was varied between 25 and $55\,\mu$m and the number of turns, n, varied from 2 to 5. The operating frequency was 900 MHz. Zhu's [14] simulations show that by varying n and the line widths the optimum Q and $L^2 g_p^3$ occur at different values of L. As the inductance changes from 2 to 9 nH by varying n and line width, the maximum Q drops from 3.8 to 3.2, or 16%. At the same time, the $L^2 g_p^3$ decreases 35.5%. Usually, it is thought that high $Q(= 1/g_p \omega_o L)$ would imply small L to achieve low-phase noise. Zhu [14] has shown that large L gives lower phase noise. While increasing L^2 does increase the noise, the g_p decreases the noise even more. Furthermore, a comparison of two designs, one optimized for maximum Q and the other for minimum $L^2 g_p^3$ showed that the latter had a 3.6-dB lower phase noise than the oscillator designed for maximum Q.

10.10 HARMONIC GENERATORS

The previous sections have been concerned with fundamental frequency oscillators. It is also possible to use a highly stable low-frequency oscillator and use a frequency multiplier to obtain the desired radio frequency. The nonlinearity of a resistance in a diode can be used in mixers to produce a sum and difference of two input frequencies (see Chapter 11). If a large signal is applied to a diode, the nonlinear resistance can produce harmonics of the input voltage. However, the efficiency of the nonlinear resistance can be no greater than $1/n$, where n is the order of the harmonic. However, a nonlinear susceptance (reciprocal of reactance) as found in a reverse-biased diode can provide efficient frequency upconversion:

$$\frac{dv}{dq} = S = S_0 \left(1 - \frac{v}{\phi}\right)^{\gamma} \tag{10.99}$$

where ϕ is the built-in voltage and typically is between +0.5 and +1 V. The applied voltage v is considered positive when the diode is forward biased. The exponent γ for a varactor diode typically ranges from 0 for a step recovery diode to $\frac{1}{3}$ for a graded junction diode to $\frac{1}{2}$ for an abrupt junction diode. Using the nonlinear capacitance of a diode theoretically allows for generation of harmonics with an efficiency of 100% with a loss-free diode. This assertion is supported by the Manley–Rowe relations [15, 16], which describe the power balance when two frequencies, f_1 and f_2, along with their harmonics are present in a lossless circuit:

$$\sum_{m=0}^{\infty}\sum_{n=-\infty}^{\infty}\frac{mP_{m,n}}{mf_1+nf_2}=0 \tag{10.100}$$

$$\sum_{n=0}^{\infty}\sum_{m=-\infty}^{\infty}\frac{nP_{m,n}}{mf_1+nf_2}=0 \tag{10.101}$$

These equations are basically an expression of the conservation of energy. From Eq. (10.100)

$$P_1=-\sum_{m=2}^{\infty}P_{m,0} \qquad n=0 \tag{10.102}$$

The depletion elastance given by Eq. (10.99) is valid for forward voltages up to about $v/\phi=\frac{1}{2}$. Under forward bias, the diode will exhibit diffusion capacitance that tends to be more lossy in varactor diodes than the depletion capacitance associated with reverse-bias diodes. Notwithstanding, an analysis of harmonic generators will be based on Eq. (10.99) for all applied voltages up to $v=\phi$. This is a reasonably good approximation when the minority carrier life time is long relative to the period of the oscillation. The maximum elastance (minimum capacitance) will occur at the reverse breakdown voltage, V_B. The simplified model for the diode then is defined by two voltage ranges:

$$\frac{S}{S_{\max}}=\left(\frac{\phi-v}{\phi-V_B}\right)^{\gamma} \qquad v\le\phi \tag{10.103}$$

$$\frac{S}{S_{\max}}=0 \qquad v>\phi \tag{10.104}$$

Integration of Eq. (10.99) gives

$$-\int_{v}^{\phi}\frac{\phi\,d(1-v/\phi)}{(1-v/\phi)}=S_0\int_{q}^{q_\phi}dq \tag{10.105}$$

$$\frac{(\phi-v)^{(1-\gamma)}}{1-\gamma}=S_0(q_\phi-q) \tag{10.106}$$

This can be evaluated at the breakdown point where $v=V_B$ and $q=Q_B$. Since V_B and Q_B are negative quantities, their signs in Eq. (10.107) and following will be effectively reversed. Taking the ratio of this with Eq. (10.106) gives the voltage and charge relative to that at the breakdown point:

$$\frac{\phi-v}{\phi-V_B}=\left(\frac{q_\phi-q}{q_\phi-Q_B}\right)^{1/(1-\gamma)} \tag{10.107}$$

For the abrupt junction diode where $\gamma=\frac{1}{2}$, it is possible to produce power at mf_1 when the input frequency is f_1 except for $m=2$ [17]. Higher order terms require that the circuit support intermediate frequencies called *idlers*. While the circuit allows energy storage at the idler frequencies, no external currents can flow at these idler frequencies. Thus, multiple lossless mixing can produce output power at mf_1 with high efficiency when idler circuits are available.

Design of a varactor multiplier consists in predicting the input and output load impedances for maximum efficiency, the value of the efficiency, and the output power. A quantity called the *drive*, D, may be defined where q_{max} represents the maximum stored charge during the forward swing of the applied voltage:

$$D = \frac{q_{max} - Q_B}{q_\phi - Q_B} \tag{10.108}$$

If $q_{max} = q_\phi$, then $D = 1$. An important quality factor for a varactor diode is the cutoff frequency. This is related to the series loss, R_s, in the diode:

$$f_c = \frac{S_{max} - S_{min}}{2\pi R_s} \tag{10.109}$$

When $D \geq 1$, $S_{min} = 0$. When $f_c/(nf_1) > 50$, the tabulated values* given in [18] provide the necessary circuit parameters. These tables have been coded in the program MULTIPLY. The efficiency given by [18] assumes loss only in the diode where $f_{out} = mf_1$:

$$\eta = \exp(\alpha f_{out}/f_c) \tag{10.110}$$

The output power at mf_1 is found to be

$$P_m = \beta \frac{\omega_1(\phi - V_B)^2}{S_{max}} \tag{10.111}$$

The values of α and β are given in [17, 18]. If the varactor has a dc bias voltage, V_o, then the normalized voltage is

$$V_{o,norm} = \frac{\phi - V_o}{\phi - V_B} \tag{10.112}$$

This value corresponds to the selected drive level. Finally, the input and load resistances are found from the tabulated values. The elastances at all supported harmonic frequencies up to and including m are also given. These values are useful for knowing how to reactively terminate the diode at the

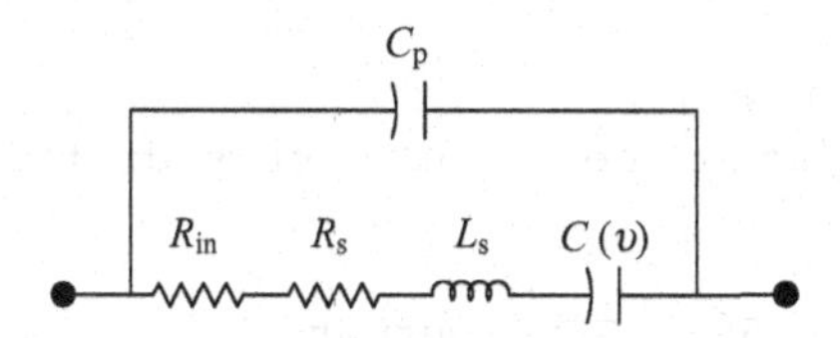

FIGURE 10.19 Intrinsic varactor diode with package.

idler and output frequencies. A packaged diode will have package parasitic circuit elements as shown in Fig. 10.19 that must be considered in the design of a matching circuit. When given these package elements, the program MULTIPLY will find the appropriate matching impedances required external to the package. Following is an example run of MULTIPLY in the design of a 1–2–3–4 (idlers at each of these harmonics) varactor quadrupler with an output frequency of 2 GHz. The **bold** numbers are user input values:

```
Input frequency, GHz.  =
0.5
Diode Parameters
Breakdown Voltage =
60
Built-in Potential phi =
0.5
Specify series resistance or cutoff frequency, Rs OR fc. <R/F>
f
Zero Bias cutoff frequency (GHz), fc =
50.
Junction capacitance at 0 volts (pF), Co =
0.5
Package capacitance (pF), Series inductance (nH) =
0.1, 0.2
For a Doubler Type A
For a 1-2-3 Tripler Type B
For a 1-2-4 Quadrupler Type C
For a 1-2-3-4 Quadrupler Type D
For a 1-2-4-5 Quintupler Type E
For a 1-2-4-6 Sextupler Type F
For a 1-2-4-8 Octupler Type G
For a 1-4 Quadrupler using a SRD, Type H
For a 1-6 Sextupler using a SRD, Type I
For a 1-8 Octupler using a SRD, Type J
Ctrl C to end
d
Type G for Graded junction (Gamma = .3333)
Type A Abrupt Junction (Gamma = .5)
Choose G or A
```

```
g
Drive is  1.0< D < 1.6.
Linear extrapolation done for D outside this range.
Choose drive.
2.0
Input Freq =   0.5000 GHz, Output Freq =   2.0000 GHz,
fc =   50.0000 GHz, Rs = 31.4878 Ohms.
Pout =   78.50312 mWatt, Efficiency =   75.47767%
At Drive  2.00, DC Bias Voltage = -7.76833
Harmonic elastance values
S0( 1) =  0.197844E+13
S0( 2) =  0.313252E+13
S0( 3) =  0.296765E+13
S0( 4) =  0.263791E+13
Total Capacitance with package cap.
CT0( 1) = 0.605450E-12
CT0( 2) = 0.419232E-12
CT0( 3) = 0.436967E-12
CT0( 4) = 0.479087E-12
Inside package, Rin =   643.400     RL =  346.470
Diode model Series Ls, Rin+Rs, S(v) shunted by Cp
Required impedances outside package.
Zin =   456.218     +j -606.069
Zout =   208.267     + j -242.991
Match these impedances with their complex conjugate
Match idler 2 with conjugate of 0 + j  -379.181
Match idler 3 with conjugate of 0 + j  -242.125
```

PROBLEMS

10.1. Verify Eqs. (10.3) and (10.4).

10.2. The crystal-controlled oscillator in Fig. 10.20 uses a tank circuit on the output side to achieve high effective reactance to help stabilize the oscillator. The narrow-band crystal is inductive when this circuit oscillates.

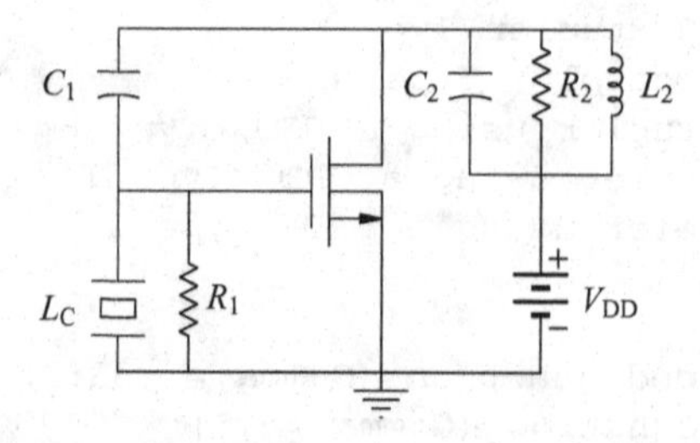

FIGURE 10.20 Crystal-controlled oscillator for Problem 10.2.

a. Write down the small-signal equivalent circuit for this oscillator.

b. Write down the equations needed to determine the frequency of oscillation and the minimum transistor g_m oscillation to occur.

10.3. In Appendix D derive Eq. (D.9) from (D.10).

10.4. In Appendix E derive the common gate S parameters from the presumably known three-port S parameters.

10.5. Prove the stability factor S' that is given in Eq. (10.63).

10.6. The measurements of a certain active device as a function of current give Z_d (10 mA) = $-20 + j30\,\Omega$ and Z_d (50 mA) = $-10 + j15\,\Omega$. The passive circuit to which this is connected is measured at two frequencies: Z (800 MHz) = $12 - j10\,\Omega$ and Z (1000 MHz) = $18 - j40\,\Omega$. Determine whether the oscillator will be stable in the given ranges of frequency and current amplitude. Assume linear interpolation between the given values is justified.

REFERENCES

1. J. K. Clapp, "An Inductance-Capacitance Oscillator of Unusual Frequency Stability," *Proc. IRE*, **36**, pp. 356–358, March 1948.
2. J. K. Clapp, "Frequency Stable LC Oscillators," *Proc IRE*, **42**, pp. 1295–1300, Aug. 1954.
3. J. Vackar, "LC Oscillators and Their Frequency Stability," *Telsa Tech. Reports*, Czechoslovakia, pp. 1–9, 1949.
4. W. K. Chen, *Active Network and Feedback Amplifier Theory*, New York: McGraw Hill, 1980.
5. K. Kurokawa, "Some Basic Characteristics of Broadband Negative Resistance Oscillator Circuits," *Bell Syst. Tech. J.*, **48**, pp. 1937–1955, July–Aug. 1969.
6. R. Adler, "A Study of Locking Phenomena in Oscillators," *Proc. IRE.*, **22**, pp. 351–357, June 1946.
7. K. Kurokawa, "Injection Locking of Microwave Solid-State Oscillators," *Proc IEEE*, **61**, pp. 1386–1410, Oct. 1973.
8. A. Hajimiri and T. H. Lee, "A General Theory of Phase Noise in Electrical Oscillators," *IEEE J. Solid-State Circuits*, **33**, pp. 179–194, Feb. 1998.
9. A. Hajimiri and T. H. Lee, "Corrections to 'A General Theory of Phase Noise in Electrical Oscillators'," *IEEE J. Solid-State Circuits*, **33**, p. 928, June 1998.
10. T. H. Lee and A. Hajimiri, "Oscillator Phase Noise: A Tutorial," *IEEE J. Solid-State Circuits*, **35**, pp. 326–336, March 2000.
11. F. X. Kaertner, "Determination of the Correlation Spectrum of Oscillators with Low Noise," *IEEE Trans. Microwave Theory Tech.*, **37**, pp. 90–101, Jan. 1989.
12. A. Demir, A. Mehrotra, and J. Roychowdhury, "Phase Noise in Oscillators: A Unifying Theory and Numerical Methods for Characterization," *IEEE Trans. Circuits Syst. I: Fund. Theory Appl.*, **47**, pp. 655–674, May 2000.

13. A. Demir, "Phase Noise and Timing Jitter in Oscillators with Colored-Noise Sources," *IEEE Trans. Circuits Syst. I: Fund. Theory Appl.*, **49**, pp. 1782–1791, Dec. 2002.
14. Z. Zhu, *Low Phase Noise Voltage Controlled Oscillator Design*, Ph.D. Dissertation, University of Texas at Arlington, Texas, 2005.
15. J. M. Manley and H. E. Rowe, "Some General Properties of Nonlinear Elements: Part I—General Energy Relations," *Proc. IRE*, **44**, July 1956, pp. 904–913.
16. H. E. Rowe, "Some General Properties of Nonlinear Elements: Part II—Small Signal Theory," *Proc. IRE*, **46**, pp. 850–860, May 1958.
17. M. Uenohara and J. W. Gewartowski, "Varactor Applications," in H. A. Watson, ed., *Microwave Semiconductor Devices and Their Circuit Applications*, New York: McGraw-Hill, pp. 194–270, 1969.
18. C. B. Burckhardt, "Analysis of Varactor Frequency Multipliers for Arbitrary Capacitance Variation and Drive Level," *Bell Syst Tech. J*, **44**, pp. 675–692, April 1965.

CHAPTER ELEVEN

RF Mixers

11.1 NONLINEAR DEVICE CHARACTERISTICS

A typical mixer is a three-port circuit that accepts two signals at two different frequencies and produces at the third port a signal that is the sum or difference of the two input frequencies. Production of a new frequency or frequencies requires a nonlinear resistance device. The two most common semiconductor nonlinear characteristics are of the form $e^{qV(t)/kT}$ as found in *pn* junction diodes or bipolar junction transistors and of the form $I_{DSS}[1 - V(t)/V_t]^2$ as found in field-effect transistors. Schottky barrier diodes are not described here since they are mostly used out of necessity for low-noise high-microwave-frequency applications.

When a *pn* junction is excited by two signals (plus a dc term):

$$V(t) = V_{dc} + V_p \cos \omega_p t + V_1 \cos \omega_t t \tag{11.1}$$

The device current is of the form

$$I(t) = I_S \exp\left(\frac{V_{dc}}{V_T}\right)\left[\exp\left(\frac{V_p}{V_T}\right)\cos \omega_p t \exp\left(\frac{V_1}{V_T}\right)\cos \omega_1 t\right] \tag{11.2}$$

where the thermal voltage, V_T, is defined as kT/q, k is Boltzmann's constant, T is the absolute temperature, and q is the magnitude of the electronic charge. It is known, however, that this can be simplified by expressing it in terms of modified Bessel functions because

$$e^{z\cos\theta} = I_0(z) + 2\sum_{n=1}^{\infty} I_n(z)\cos n\theta \tag{11.3}$$

Radio Frequency Circuit Design, Second Edition, by W. Alan Davis

where $I_n(z)$ is the modified Bessel function of order n and argument z [1]. The Bessel function has the property that as n increases and z decreases the function itself decreases. The two exponentials in Eq. (11.2) indicate there are two infinite series of the form shown in Eq. (11.3): the first with summation index n and the second with index m. If $\bar{V}_{dc} = V_{dc}/V_T$, $\bar{V}_p = V_p/V_T$, and $\bar{V}_1 = V_1/V_T$, the current given in Eq. (11.2) can be found by the appropriate substitution:

$$\begin{aligned} I(t) &= I_s e^{\bar{V}_{dc}} \left[I_0(\bar{V}_p) + 2\sum_{n=1}^{\infty} I_n(\bar{V}_p)\cos(n\omega_p t) \right] \times \left[I_0(\bar{V}_1) + 2\sum_{m=1}^{\infty} I_n(\bar{V}_1)\cos(m\omega_1 t) \right] \\ &= I_{dc} e^{\bar{V}_{dc}} I_0(\bar{V}_p) I_0(\bar{V}_1) \\ &\quad + 2I_{dc} e^{\bar{V}_{dc}} \left[I_0(\bar{V}_1) \sum_{n=1}^{\infty} I_n(\bar{V}_p)\cos(n\omega_p t) + I_0(\bar{V}_p) \sum_{m=1}^{\infty} I_m(\bar{V}_1)\cos(m\omega_1 t) \right] \\ &\quad + 4I_{dc} e^{\bar{V}_{dc}} \left[\sum_{n=1}^{\infty} I_n(\bar{V}_p)\cos(n\omega_p t) \right] \left[\sum_{m=1}^{\infty} I_m(\bar{V}_1)\cos(m\omega_1 t) \right] \end{aligned} \tag{11.4}$$

The basic result is a set of frequencies $n\omega_p + m\omega_1$, where n and m can take on any integer. The actual current values at any given frequency would be greatly modified by circuit impedances at these frequencies as well as variations in the device itself. In the usual mixer application shown in Fig. 11.1, the amplitude of the local oscillator voltage, V_p, is typically 40 dB greater than the RF signal voltage V_1. Consequently, the number of frequencies drops to

$$\omega_n = n\omega_p + \omega_0 \tag{11.5}$$

The higher order mixing products are reduced in amplitude by approximately $1/n$. The usual desired output for a receiver is the intermediate frequency (IF), ω_0. The frequencies of primary interest are given the following names:

Local oscillator (pump) frequency (LO): ω_p
Intermediate frequency (IF): $\omega_0 = \omega_1 - \omega_p$
Input signal frequency (RF): ω_1

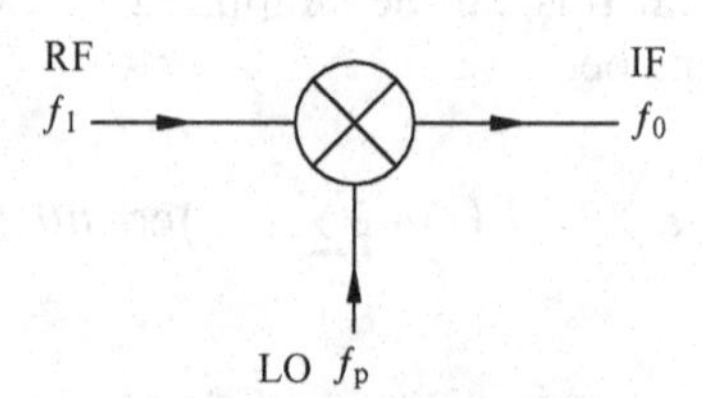

FIGURE 11.1 Schematic diagram for mixer.

Image frequency: $\omega_{-1} = -\omega_p + \omega_0$
Sum frequency: $\omega_2 = 2\omega_p + \omega_0$

In the FET type of nonlinearity, the current as a result of excitation given by Eq. (11.1) is

$$\frac{I(t)}{I_{DSS}} = \left(1 - \frac{V_p}{V_t}\cos\omega_p t - \frac{V_1}{V_t}\cos\omega_1 t\right)^2 \tag{11.6}$$

$$= 1 - 2\left(\frac{V_p}{V_t}\cos\omega_p t + \frac{V_1}{V_t}\cos\omega_1 t\right) + \frac{V_p^2}{2V_t^2}(1 + \cos 2\omega_p t) + \frac{V_1^2}{2V_t^2}(1 + \cos 2\omega_1 t)$$
$$+ \frac{V_p V_1}{V_t^2}[\cos(\omega_p + \omega_1)t + \cos(\omega_p - \omega_1)t] \tag{11.7}$$

While it may appear that the FET is "less nonlinear" than the *pn* junction type of nonlinearity, it should be remembered that the circuit into which the device is embedded will reflect back into the nonlinear device and create multiple mixing products. However, ultimately the frequencies will follow at least potentially the values shown in Fig. 11.2.

Readily apparent from the foregoing, a measure of patience is necessary to unravel all the frequency terms and their relative amplitudes. The nonlinear device is sometimes modeled as a power series of the applied voltages:

$$I(t) = I_{dc} + aV(t) + bV^2(t) + cV^3(t)\ldots \tag{11.8}$$

Rather than directly determining the mixing products by multiplication, it is more convenient to determine these in the frequency domain by employing the Fourier transform [2]. The most convenient way of writing the pair is symmetrically where f is used rather than ω:

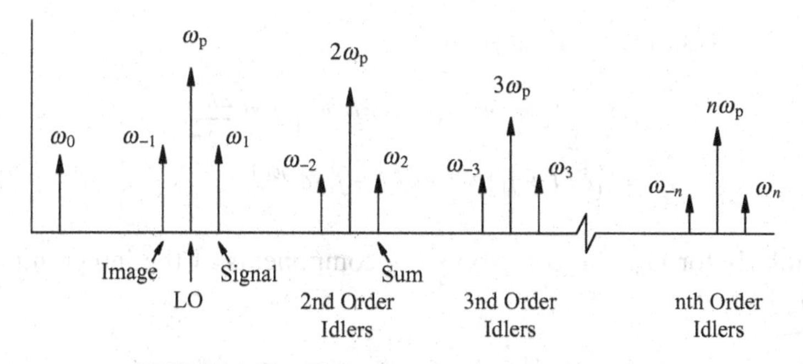

FIGURE 11.2 Frequency components in mixer.

$$g(t) = \int_{-\infty}^{\infty} G(f) e^{-j2\pi ft} df \tag{11.9}$$

$$G(f) = \int_{-\infty}^{\infty} g(t) e^{j2\pi ft} dt \tag{11.10}$$

The Fourier transform of an exponential function is a Dirac δ function:

$$\mathcal{F}(e^{-jf_a}) = \delta(f - f_a) \tag{11.11}$$

If two voltages are represented as $V_a = \cos(\omega_a t + \theta_a)$ and $V_b = \cos(\omega_b t + \theta_b)$, then these are to be multiplied together. Before doing this, they can each be converted into the frequency domain by Eq. (11.10) and substituted into the convolution theorem:

$$G_o(f) = \int_{-\infty}^{\infty} G_a(\lambda) G_b(f - \lambda) d\lambda \tag{11.12}$$

For sinusoidal voltages, the integral is simply a series of δ functions that are trivial to integrate. The process proceeds most easily graphically. The function $G_a(\lambda)$ is fixed and the $G_b(f - \lambda)$ is allowed to slide from right to left. The nonzero parts of the integration occurs when δ functions coincide.

As an example, consider the frequencies that would result from the product of two voltages:

$$V_o(t) = V_a(t) V_b(t) \tag{11.13}$$

where

$$V_a(t) = \cos(2\pi f_1 t + \theta_1)$$

and

$$V_b(t) = \cos(2\pi f_2 t + \theta_2)$$

The Fourier transform of $V_a(t)$ is

$$G_a(f) = \int_{-\infty}^{\infty} V_a(t) e^{j2\pi ft} dt \tag{11.14}$$

$$= \int_{-\infty}^{\infty} \frac{1}{2} \left[e^{j(2\pi f_1 t + \theta_1)} + e^{-j(2\pi f_1 t + \theta_1)} \right] e^{j2\pi ft} \frac{2\pi\, dt}{2\pi}$$

$$= \tfrac{1}{2} \left[\delta(f + f_1) e^{j\theta_1} + \delta(f - f_1) e^{-j\theta_1} \right] \tag{11.15}$$

and similarly for G_b. The corresponding components of the integrand of Eq. (11.12) are

$$G_a(\lambda) = \tfrac{1}{2} \left[\delta(\lambda + f_1) e^{j\theta_1} + \delta(\lambda - f_1) e^{-j\theta_1} \right] \tag{11.16}$$

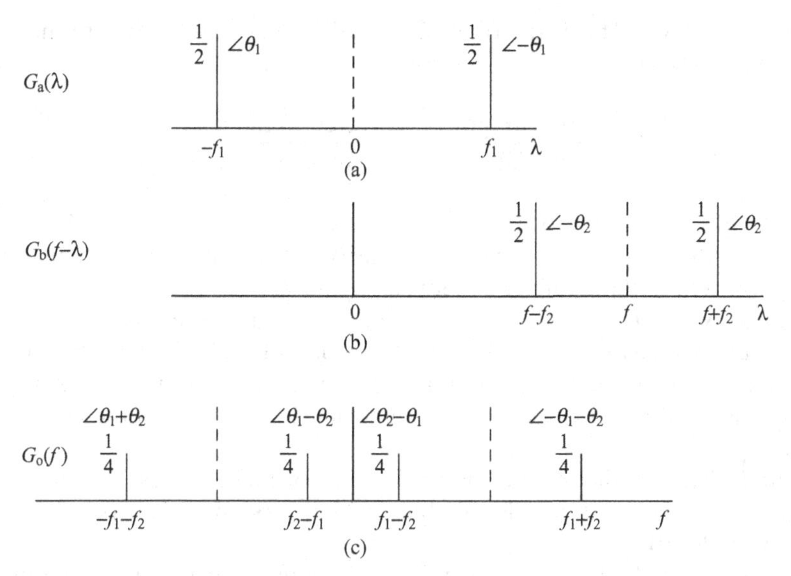

FIGURE 11.3 Graphical integration of convolution integral where (*a*) is $G_a(\lambda)$, (*b*) is $G_b(f-\lambda)$, and (*c*) is result of integration.

$$G_b(f-\lambda)=\tfrac{1}{2}\left[\delta(f-\lambda+f_2)e^{j\theta_2}+\delta(f-\lambda-f_2)e^{-j\theta_2}\right] \tag{11.17}$$

The Fourier transforms, $G_a(\lambda)$ and $G_b(f-\lambda)$ are shown in Figs. 11.3*a* and 11.3*b*, which display both the magnitude and the phase of the terms. As λ increases, $G_b(f-\lambda)$ moves from right to left. No contribution to the convolution integral occurs until a δ function in Fig. 11.3*a* coincides with a δ in Fig. 11.3*b*. The first of these occurs at $\lambda=f_1=f-f_2$ or $f=f_1+f_2$. The amplitude is $\frac{1}{4}$ and the phase is $\angle-\theta_1-\theta_2$. This is the rightmost line shown in Fig. 11.3*c*. As λ continues to increase, all four intercepts between $G_a(\lambda)$ and $G_b(f-\lambda)$ are found. While the amount of effort in using the frequency-domain approach described here and the time-domain approach of multiplying sines and cosines in this example is about the same, adding a third frequency quickly tilts the ease of calculation toward the frequency-domain approach.

11.2 FIGURES OF MERIT FOR MIXERS

The quality of a mixer rests on a number of different mixer parameters, which of course must fit the application under consideration. The first of these is *conversion loss L*. This is the ratio of the delivered output power to the input available power:

$$L = \frac{\text{output IF power delivered to the load, } P_0}{\text{available RF input signal power, } P_1}$$

Clearly, the conversion loss is dependent on the load of the input RF circuit as well as the output impedance of the mixer at the IF port. The conversion loss for a typical diode mixer is between 6 and 7 dB.

The *noise figure* is a measure of the noise added by the mixer itself to the RF input signal as it gets converted to the output IF. It specifically excludes the noise figure of the following IF amplifier and neglects the $1/f$ flicker noise. In practice, the mixer noise figure is very nearly the same as the conversion loss.

The *isolation* is the amount of local oscillator power that leaks into either the IF or the RF ports. For double-balanced mixers this value typically lies in the 15- to 20-dB range.

A single- or double-balanced mixer will convert energy in the upper or lower sidebands with equal efficiency. Consequently, noise in the sideband with no signal will be added to the IF output, which of course will increase the noise figure by 3 dB in the IF port. *Image rejection* mixers will block this unwanted noise from the IF port.

The *conversion compression* is the RF input power above which the RF input versus the IF output deviates from linearity by a given amount. For example, the 1-dB compression point occurs when the conversion loss increases by 1 dB above the conversion loss in the low-power linear range. A typical value of 1.0-dB compression occurs when the RF power is +20 dBm and the LO is +7 dBm.

The *LO drive power* is the required LO power level needed to make the mixer operate in optimal fashion. For a double-balanced mixer, this is typically +6 to +20 dBm.

The *dynamic range* is the maximum RF input power range for the mixer. The maximum amplitude is limited by the conversion compression, and the minimum amplitude is limited by the noise figure.

The *input intercept point* is the RF input power at which the output power levels of the undesired intermodulation products and the desired IF output would be equal. In defining the input intercept point, it is assumed that the IF output power does not compress. It is therefore a theoretical value and is obtained by extrapolating from low power levels. The higher this power level, the better is the mixer. Sometimes, an output intercept point is used. This is the input intercept point minus the conversion loss. The idea of intercept points is described in greater detail in Section 11.7.

The *two-tone third-order intermodulation point* is a measure of how the mixer reacts when two equal amplitude radio frequencies excite the RF input port of the mixer.

11.3 SINGLE-ENDED MIXERS

Mixers are usually classed as either single ended, single balanced, or double balanced. The technical advantages of the double-balanced mixer over the other two usually precludes using the slightly lower cost of the single-ended or single-balanced types in RF circuits. They are used though in millimeter-wave circuits where geometrical constraints and other complexities favor using the simpler single-ended mixer.

The single-ended mixer in Fig. 11.4 shows that the RF input signal and the local oscillator signal enter the mixer at the same point. Some degree of isolation between the two is achieved by using a directional coupler in which the RF signal enters the direct port and the local oscillator enters through the coupled port. The amplitude of the local oscillator, even after passing through the coupler, is large enough to turn the diode on and off during each cycle. Indeed, the LO power is so large that it causes clipping of the LO voltage, thereby approximating a square wave. The small RF signal is then presented with alternately a short or open circuit at the LO rate. It is this turning on and off of the radio frequency that produces the $|nf_p \pm f_1|$ set of frequencies. The one of most interest in the standard receiver is the IF frequency, $f_0 = f_p - f_1$. Among the disadvantages of the single-ended mixer are a high noise figure, a large number of frequencies generated because of the nonlinear diode, a lack of isolation between the RF and LO signals, and large LO currents in the IF circuit. The RF to LO isolation problem can be very important since the LO can leak back out the RF port and be radiated through the receiver antenna. The LO currents in the IF circuit would have to be filtered out with a low-pass filter that has sufficient attenuation at the LO frequency to meet system specifications. It does have the advantage of requiring lower LO power than the other types of mixers.

Rather than using a switching diode, an FET can be switched at the LO rate. One such design is when the LO and RF signal both enter the FET gate and the output IF signal is developed in the drain circuit. The nonlinearity of the FET implies that fewer spurious signals are generated than the "more" nonlinear diode. Furthermore, it is possible to achieve conversion gain between the input RF and output IF signals. A second alternative would be to excite the gate with the RF signal and the source with the LO; then the output IF is

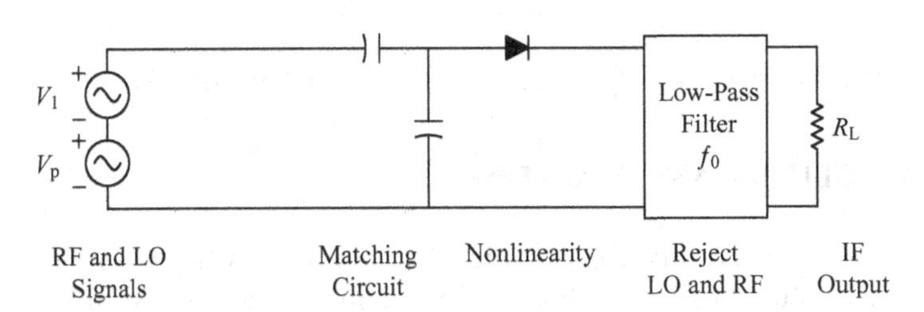

FIGURE 11.4 Single-ended mixer.

developed in the drain. This circuit offers improved isolation between the RF and LO signals but at the cost of higher LO power requirements. The dual-gate FET can be used where one gate is excited with the LO and the other with the RF. The IF is again developed in the drain circuit. This circuit offers even better isolation between RF and LO, but its gain is somewhat lower.

11.4 SINGLE-BALANCED MIXERS

The single-balanced (or simply balanced) mixer has either two or four diodes as shown in the examples of Fig. 11.5. In all these cases, when the LO voltage has a large positive value, all the diodes are shorted. When the LO voltage has a large negative value, all the diodes are open. The LO power cannot reach the IF load nor the RF load because of circuit symmetry. However, the incoming RF voltage sees alternately a path to the IF load and a blockage to the IF load. The block may be either an open circuit to the IF load or a short circuit to ground.

It is assumed the LO voltage is much greater than the RF voltage ($V_p \gg V_1$) as in Eq. (11.5). The LO voltage can be approximated as a square wave with period $T = 1/f_p$ that modulates the incoming RF signal (Fig. 11.6). A Fourier analysis of the square wave results in a switching function designated by $S(t)$:

$$S(t) = \frac{1}{2} + \sum_{n=1}^{\infty} \frac{\sin(n\pi/2)}{n\pi/2} \cos n\omega_p t \tag{11.18}$$

If the input RF signal is expressed as $V_1 \cos \omega_1 t$, then the output voltage is thus multiplied by the switching function:

$$V_0 = V_1 \cos \omega_1 t \cdot S(t) \tag{11.19}$$

$$= V_1 \cos \omega_1 t \left(\frac{1}{2} + \sum_{n=1}^{\infty} \frac{\sin(n\pi/2)}{n\pi/2} \cos n\omega_p t \right) \tag{11.20}$$

Clearly, the RF input signal voltage will be present in the IF circuit. However, only the odd harmonics of the local oscillator voltage will effect the IF load. Thus, the spurious voltages appearing in the IF circuit are

$$f_1, f_p + f_1, 3f_p \pm f_1, 5f_p \pm f_1, \ldots$$

and all even harmonics of f_p are suppressed (or balanced out).

11.5 DOUBLE-BALANCED MIXERS

The double-balanced mixer is capable of isolating both the RF input voltage and the LO voltage from the IF load. The slight additional cost of some extra diodes and a balun is usually outweighed by the improved intermodulation

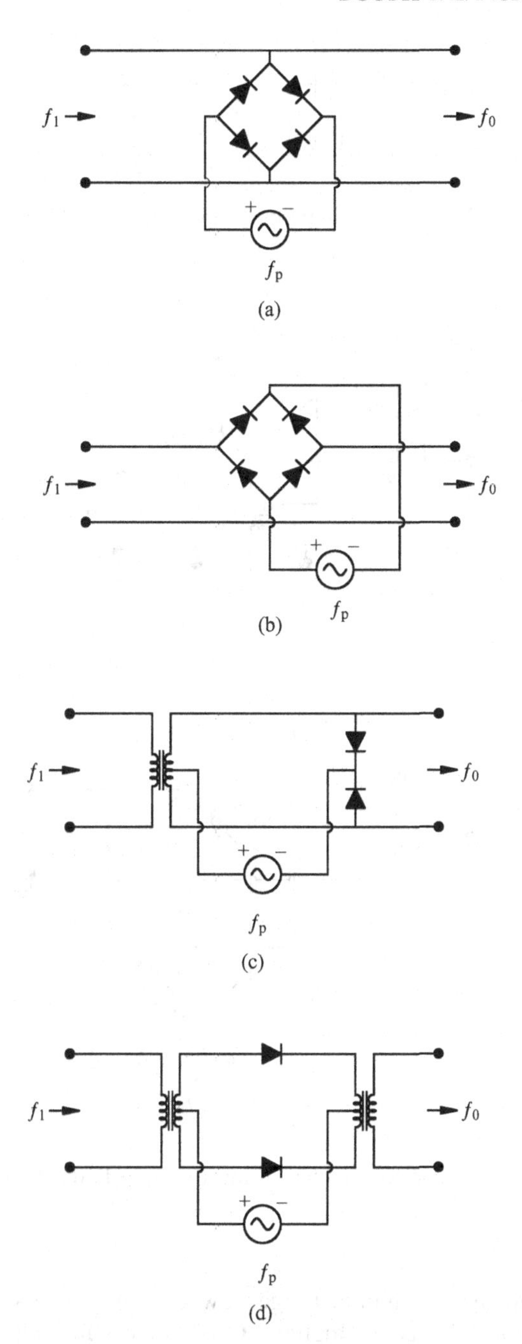

FIGURE 11.5 Four possible single-balanced mixers.

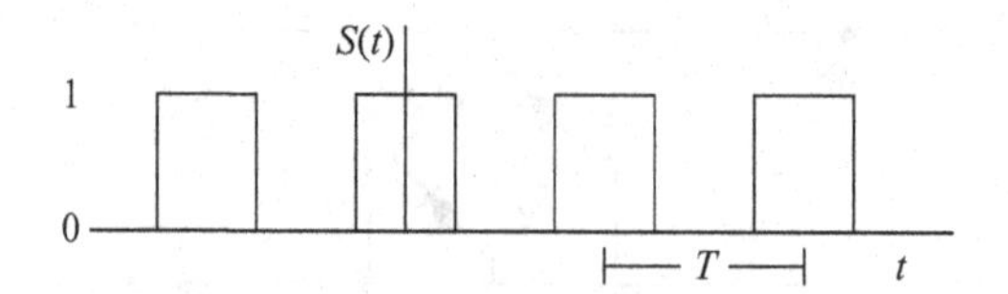

FIGURE 11.6 Single-balanced mixer waveform.

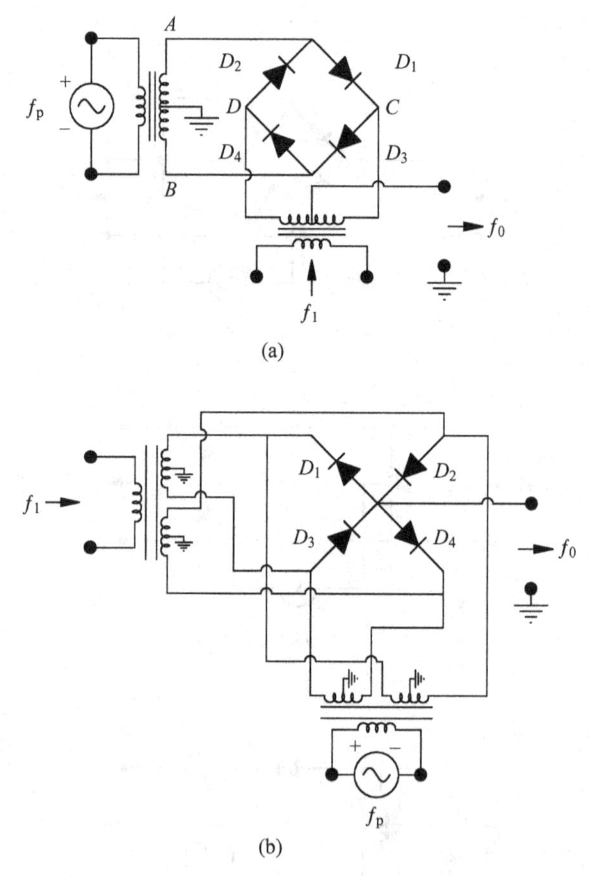

FIGURE 11.7 Double-balanced mixers using (*a*) ring diode design and (*b*) diode star design.

suppression, improved dynamic range, low conversion loss, and AM noise cancellation at the expense of higher LO power requirement. The two most widely used double-balanced mixers for the RF and microwave band are the "ring" mixer and the "star" mixer depicted in Fig. 11.7. In the single-balanced mixer, all the diodes were either turned on or turned off, depending on the instantaneous polarity of the local oscillator voltage. The distinguishing feature of the double-balanced mixer is that half the diodes are on and half are off at any given time. The diode pairs are switched on or off according to the local oscillator polarity. Thus, the path from the signal port with frequency f_1 to the intermediate frequency port, $f0$, reverses polarity at the rate of $1/f_p$.

In Fig. 11.7*a*, when the LO is positive at the upper terminal so that node *A* is positive and node *B* is negative, diodes D_1 and D_3 are shorted while diodes D_2 and D_4 are open. When the LO is negative, diodes D_2 and D_4 are shorted. Thus, nodes *C* and *D* alternately become virtual ground to the RF signal. The RF signal, f_1, is sent to the IF port at alternate polarities governed by the LO frequency, f_p. The switching of the polarity at the LO frequency, f_p, of the current in the IF circuit produces the difference frequency, f_0.

An analysis of this mixer can be done using SPICE with diodes and ideal center-tapped transformers. An ideal transformer could be modeled in SPICE with mutual inductances having a coupling coefficient of 1 and very high inductance. However, this would result in very long time constants when doing a transient analysis. An alternative is to model the transformer with voltage-controlled voltage sources and current-controlled current sources with a multiplication factor equal to the transformer turns ratio, *n*. A variety of sources for this circuit can be found on the Internet, and the one used here is based on [3]. Figure 11.8*a* shows a center-tapped transformer with an equivalent representation. This second form is modeled by the circuit in Fig. 11.8*b*, which uses ideal controlled sources. The very large primary resistance, R_p, and the very small secondary resistance, R_s, are included to help SPICE converge. The net list for this transformer model is given in the form of a subcircuit. The local oscillator in this circuit is set at 900 MHz and the RF signal at 800 MHz. The resulting time-domain output shown in Fig. 11.9 is not easily interpreted. The Fourier transform in Fig. 11.10 clearly shows the resulting IF output frequency at 100 MHz along with other frequencies generated by the mixer.

```
Double Balanced Diode Mixer
* Local Oscillator
vp      10    0    sin(0    6.    900e6)
rp       10    1    0.01
xpump   1    0    2    0    3    ctap
* rf signal
vrf      20    0    sin(0    0.2    800e6)
rrf      20    4    0.01
xrf      4    0    5    6    7    ctap
rdummy       7    0    100meg
* Diode ring
d1 2   7    diodem
d2      5    2    diodem
d3      7    3    diodem
d4      3    5    diodem
rlif     6    0    5000
.model    diodem    d  (rs=0.)
*.op
*      print step    final    begin prt    ceiling    UIC
```

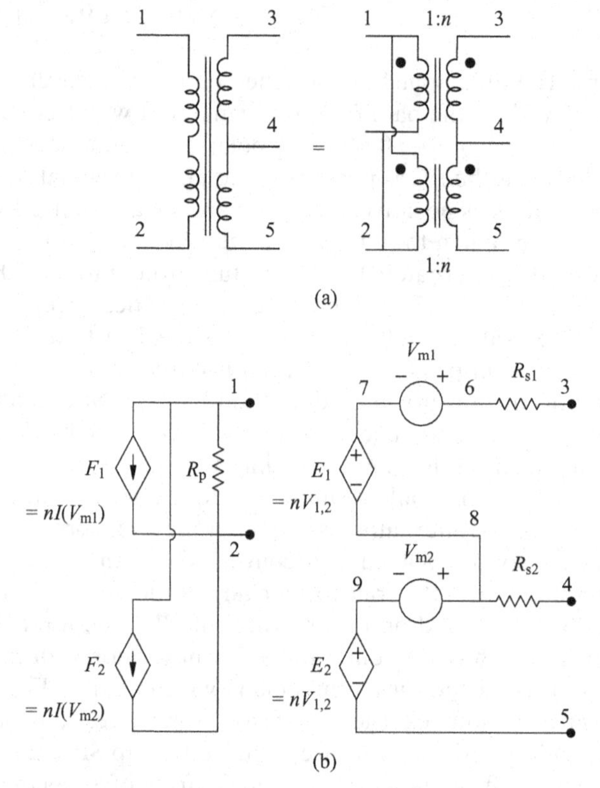

FIGURE 11.8 (*a*) Center-tapped transformer used in mixer and (*b*) SPICE model for ideal transformer.

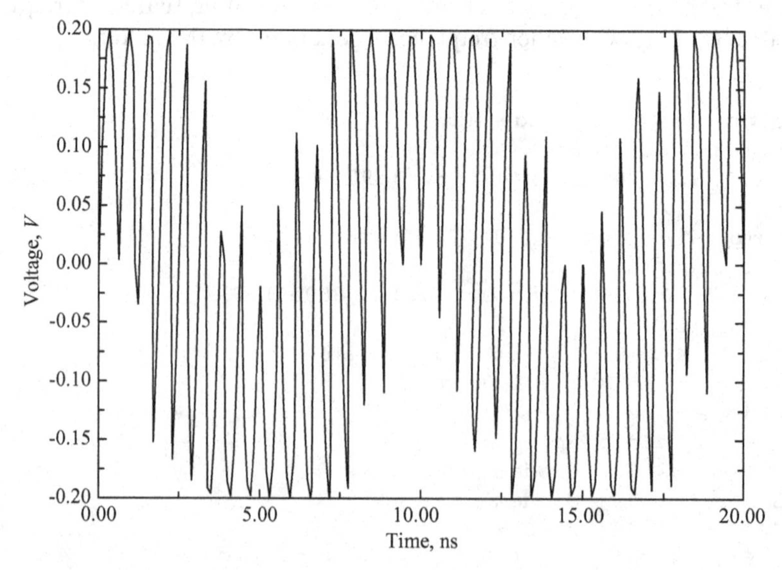

FIGURE 11.9 Time-domain voltage at IF load of double-balanced mixer using ideal switches.

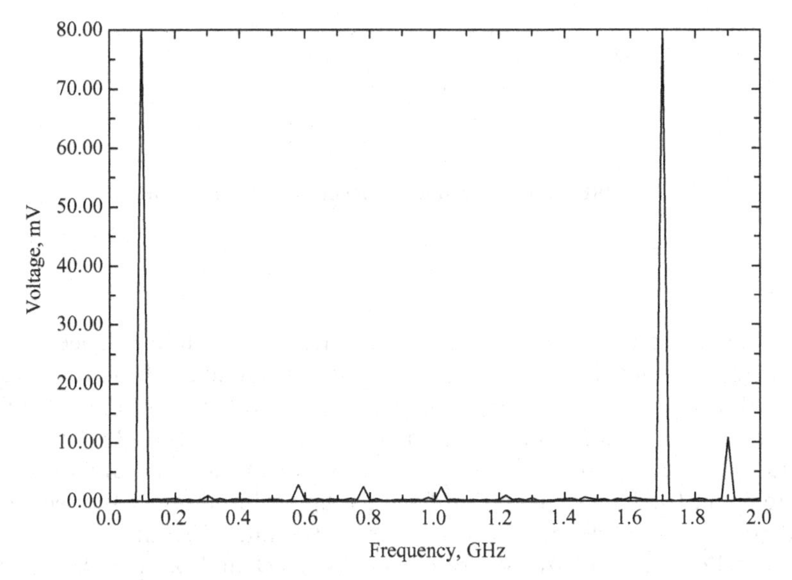

FIGURE 11.10 Fast Fourier transform of time function clearly shows frequency components off double-balanced mixer.

```
*.tran    0.1ns    200ns    180ns    10ps
.tran    0.1ns    200ns    180ns
* if output is v(6)
.end
.subckt    ctap 1    2    3    4    5
* node 1 positive primary
* node 2 negative primary
* node 3 positive first secondary
* node 4 secondary center tap
* node 5 negative second secondary
* Each transformer part is a 1:n turns ratio (here given n=2)
* fp is the primary cccs, and es is the secondary vcvs.
*es1    7    9    1    2    "n"
es1    7    8    1    2    1
* Current Controlled Current Source
*fp1    1    2    vmeas1    "n"
fp1    1    2    vmeas1    1
rp    1    2    1meg
rs1    6    3    1u
vmeas1 7    6    dc    0
* Current Controlled Current Source
fp2    1    2    vmeas2    1
es2    9    5    1    2    1
rs2    8    4    1u
vmeas2  9    8    dc    0
.ends
```

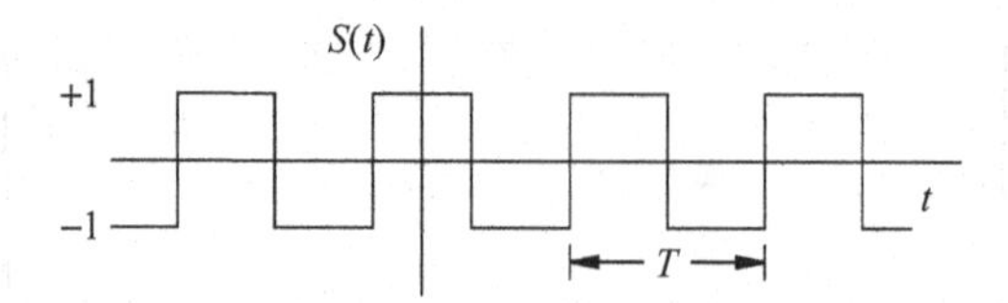

FIGURE 11.11 Double-balanced mixer waveform.

The star circuit shown in Fig. 11.7*b* also acts as a double-balanced mixer. An advantage over the ring mixer is that the central node of the four diodes allows direct connection to the IF circuit. On the other hand the star mixer requires a more complicated transformer in the RF signal and LO ports. When the LO voltage is positive, diodes D_1 and D_2 are shorted and diodes D_3 and D_4 are open. The RF signal current from the upper terminals of the secondary winding flows to the IF port at the center of the star. When the LO voltage is negative, diodes D_3 and D_4 are shorted and diodes D_1 and D_2 open. The current then flows from the lower terminals of the RF signal transformer secondary. The RF signal current in the IF circuit has switched polarity. The switching rate produces an output at the difference frequency, f_0. In both these cases the switching function is shown in Fig. 11.11. Fourier analysis provides the following time-domain representation of the switching function, which differs from Eq. (11.18) by a lack of a dc term:

$$S(t) = 2\sum_{n=1}^{\infty} \frac{\sin(n\pi/2)}{n\pi/2} \cos n\omega_p t \tag{11.21}$$

The IF voltage is found by multiplying the signal voltage by the switching function:

$$\begin{aligned} V_0 &= V_1 \cos\omega_1 t \cdot S(t) \\ &= 2V_1 \cos\omega_1 t \left(\sum_{n=1}^{\infty} \frac{\sin(n\pi/2)}{n\pi/2} \cos n\omega_p t \right) \end{aligned} \tag{11.22}$$

Clearly, there is no RF signal nor LO voltage seen in the IF circuit, nor any even harmonics of the LO voltage.

The above description of mixers has assumed the use of ideal diodes. The diodes are in fact either *pn* or Schottky barrier (metal–semiconductor) junctions with a nonzero forward voltage drop and nonzero leakage current in the reverse-bias condition. The Schottky barrier devices are particularly useful when low noise is required at high microwave frequencies. The device and package parasitic elements limit mixer frequency response, although designs

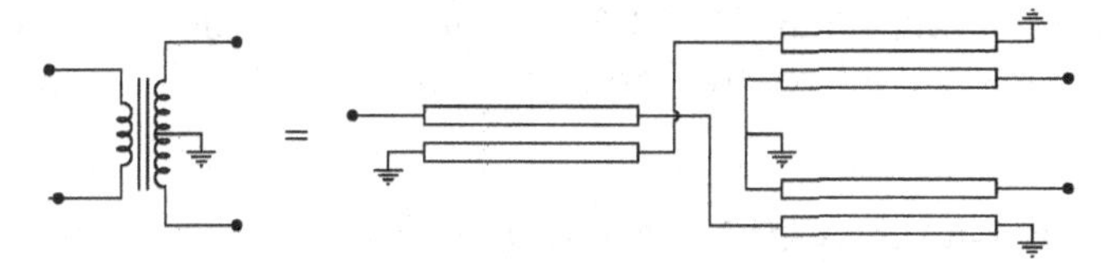

FIGURE 11.12 Transmission line transformer equivalent to center-tapped transformer.

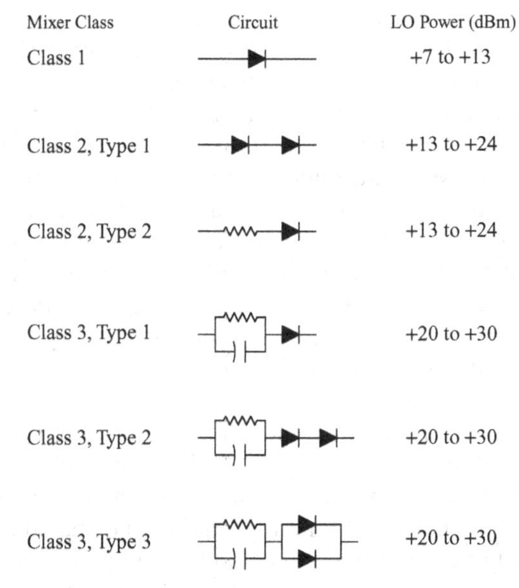

Mixer Class
Circuit
LO Power (dBm)
Class 1
+7 to +13
Class 2, Type 1
+13 to +24
Class 2, Type 2
+13 to +24
Class 3, Type 1
+20 to +30
Class 3, Type 2
+20 to +30
Class 3, Type 3
+20 to +30

FIGURE 11.13 Double-balanced mixer classes based on elements in each branch. Required LO power levels increase with circuit complexity [6].

based on the above analysis have been made to work at frequencies exceeding 26 GHz.

This analysis was also based on the availability of ideal center-tapped transformers. At RF frequencies, these can be realized using transmission line transformers as shown in Fig. 11.12.

The double-balanced ring mixer described above used a single diode in each arm of the ring. Such a mixer is termed a class 1 mixer. Class 2 mixers are obtained by replacing the single diode in each arm of the ring with two diodes in series or with a diode and resistor in series (Fig. 11.13). The precision resistor in the later case can be adjusted to improve the ring balance and thus the intermodulation distortion. More complex ring elements can be used to further improve intermodulation distortion with the added cost of increasing the amount of LO power required to drive the diodes. More detailed information on design of RF and microwave mixers is available in [4, 5].

11.6 DOUBLE-BALANCED TRANSISTOR MIXERS

Transistors can also be used as the mixing element in all three types of mixers described above, though only the double-balanced configuration is described here. These are called *active mixers* because they provide the possibility of conversion gain, which the diode mixers are not capable of doing. They produce approximately the same values of port isolation and suppression of even harmonic distortion as the diode mixers. One example of such a circuit is a transistor ring of enhancement mode *n*-channel MOSFETs in which the gate voltage must exceed V_t in order for the transistor to turn on (Fig. 11.14). When the LO voltage is positive as indicated, the pair of transistors on the right-hand side is turned on and the left-hand pair is turned off. When the LO voltage is negative, the two pairs of transistors switch roles. In this process, the path from the RF signal switches back and forth between the positive and negative IF ports at the LO switching rate. While the balance of the polarity of the RF signal voltage precludes it from being seen at the IF port, the difference frequency generated by the switching action does appear across the IF terminals.

An alternative design is based on the Gilbert cell multiplier [7]. An analysis of the elementary Gilbert cell in Fig. 11.15 is most easily accomplished by assuming the base and reverse-bias saturation currents are negligible, that the output resistances of the transistors are infinite, and that the bias source is ideal. Current continuity of, transistors Q_1, Q_2, and Q_5 demands:

$$I_{C5} = I_{C1} + I_{C2} \tag{11.23}$$

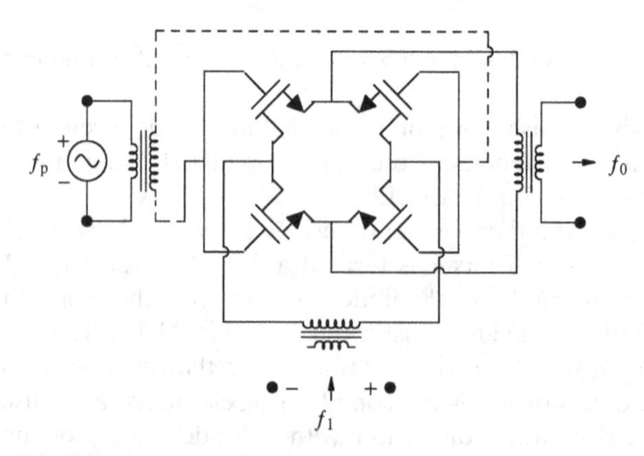

FIGURE 11.14 Double-balanced mixer using MOSFETs.

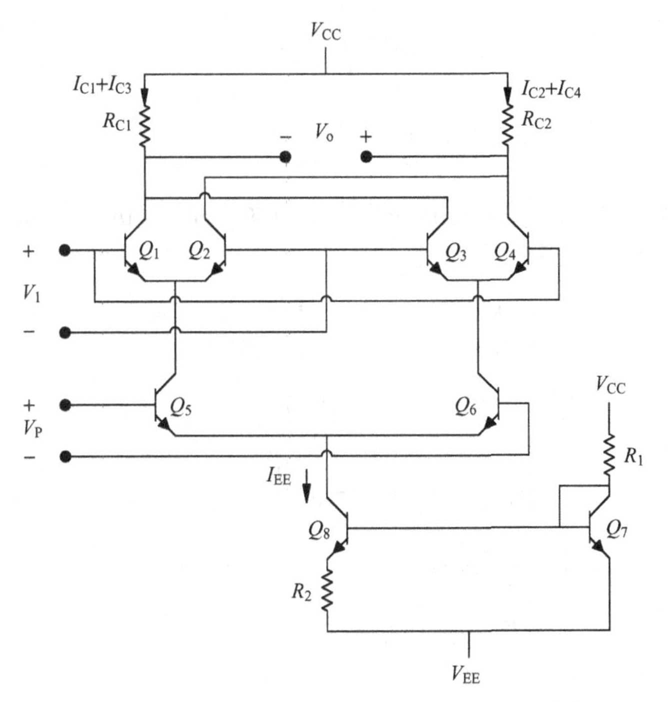

FIGURE 11.15 Gilbert cell can be used as a modulator.

The ratio of the Shockley diode equations with negligible saturation current gives a second relationship:

$$\frac{I_{C1}}{I_{C2}} = \frac{e^{V_{BE1}}/V_T}{e^{V_{BE2}}/V_T} = e^{V_1/V_T} \tag{11.24}$$

where $V_{BE1} - V_{BE2} = V_1$ and an expression for I_{C1} can be found. In like manner the currents for Q_2, Q_3, and Q_4 are found:

$$I_{C1} = \frac{I_{C5}}{1 + e^{-V_1/V_T}} \tag{11.25}$$

$$I_{C2} = \frac{I_{C5}}{1 + e^{V_1/V_T}} \tag{11.26}$$

$$I_{C3} = \frac{I_{C6}}{1 + e^{V_1/V_T}} \tag{11.27}$$

$$I_{C4} = \frac{I_{C6}}{1 + e^{-V_1/V_T}} \tag{11.28}$$

For Q_5 and Q_6 the collector currents are

$$I_{C5} = \frac{I_{EE}}{1+e^{-V_2/V_T}} \tag{11.29}$$

$$I_{C6} = \frac{I_{EE}}{1+e^{V_2/V_T}} \tag{11.30}$$

The output voltage is proportional to the difference of the currents through the collector resistors:

$$\begin{aligned}
V_o &= [(I_{C1}+I_{C3})-(I_{C2}+I_{C4})]R \qquad (11.31)\\
&= [(I_{C1}-I_{C4})-(I_{C2}-I_{C3})]R\\
&= \frac{R(I_{C5}-I_{C6})}{1+e^{-V_1/V_T}} - \frac{R(I_{C5}-I_{C6})}{1+e^{V_1/V_T}}\\
&= \frac{I_{EE}R}{1+e^{-V_1/V_T}}\left(\frac{1}{1+e^{-V_2/V_T}} - \frac{1}{1+e^{V_2/V_T}}\right)\\
&\quad - \frac{I_{EE}R}{1+e^{V_1/V_T}}\left(\frac{1}{1+e^{-V_2/V_T}} - \frac{1}{1+e^{V_2/V_T}}\right)\\
&= \frac{I_{EE}R}{1+e^{-V_1/V_T}}\left(\frac{e^{V_2/2V_T}}{e^{V_2/2V_T}+e^{-V_2/2V_T}} - \frac{e^{-V_2/2V_T}}{e^{-V_2/2V_T}+e^{V_2/2V_T}}\right)\\
&\quad - \frac{I_{EE}R}{1+e^{V_1/V_T}}\left(\frac{e^{V_2/2V_T}}{e^{V_2/2V_T}+e^{-V_2/2V_T}} - \frac{e^{-V_2/2V_T}}{e^{-V_2/2V_T}+e^{V_2/2V_T}}\right)\\
&= I_{EE}R\tanh\left(\frac{V_2}{2V_T}\right)\tanh\left(\frac{V_1}{2V_T}\right) \qquad (11.32)
\end{aligned}$$

Since $\tanh x \approx x$ for $x \ll 1$, monotonically increasing output occurs when $V_i \ll 2V_T$ where $i = 1, 2$. At the other extreme $\tanh x \approx 1$ when $x \gg 1$.

The modulator application typically has one large input voltage (LO) and one small one (RF signal). A positive value of the LO voltage, shown as V_1 in Fig. 11.15, will then turn on Q_1 and Q_4, while Q_2 and Q_3 are turned off. As in the previous double-balanced mixers, the LO switches the RF signal voltage path to the IF port at the frequency, f_p, so that the difference frequency is generated. A SPICE analysis of the Gilbert cell shown below again demonstrates the production of an IF output between the collectors of Q_1 and Q_4.

```
Gilbert Cell
vrf   1   4    sin  (0   .2    800meg ) dc   0
vp    8   9    sin  (0   2     900meg ) dc   0
vcc   7   0    dc   15
vee   0   12   dc   15
q1    2   1    3    device
q2    6   4    3    device
q3    2   4    5    device
```

```
q4      6    1    5     device
q5      3    8    10    device
q6      5    9    10    device
q7      11   11   12    device
q8      10   11   13    device
r1      7    11   15
r2      13   12   100
rc1     7    2    30k
rc2     7    6    30k
.model    device    npn
.dc vrf    -100m 10m vp -100m 100m 20m
* print step, final time, print start, step ceiling
.tran    1ns      100ns     0
* if output is v(2,6)
* dc analysis
.tf    v(6) vrf
*.tf   v(6) vp
.end
```

This same circuit can be realized using field-effect transistors. In either case, a large RF signal input can cause the mixer to operate outside of its linear region. The mixer dynamic range can be improved by adding emitter (source) degeneracy. This is a small resistor (usually in the hundreds of ohms) in the emitter circuit. Another scheme is to include a filter between the lower two transistors and the upper ones [8]. Distortion products produced in Q_5 and Q_6 are thus filtered out before the RF signal reaches the transistors being switched by the LO. A 20-dB improvement in dynamic range over the conventional Gilbert cell is reported using this filtering technique.

11.7 SPURIOUS RESPONSE

The previous sections considered some representative mixer circuits. Here, some of the primary performance criteria for mixers are described. The first of these are the spurious frequencies generated when the mixer is excited by a single-tone RF signal. A second measurement of mixer performance results from exciting it with two tones near to each other that produces two IF terms. The latter is termed *two-tone intermodulation distortion.*

Single-tone intermodulation is an effect that is a result of the imbalance in the transformers or the diodes used in the mixer. A distinction is made between the inherent nonlinear current–voltage curve of a diode and the nonlinearity associated with the switching action of the diode [9]. Fitting a polynomial function to an ideal diode characteristic whose current is zero when off and whose $i - v$ slope is a straight line when the diode is on, would yield a polynomial fitting function with many powers of the independent variable. Indeed the switching of the diode appears to be the predominant effect in a mixer.

Analytical estimates of intermodulation distortion suppression can be made solely on the basis of the switching action of the diodes in the mixer, rather than any curvature of individual diode curves. Such an expression is presented in Appendix H. That equation has also been coded in the program IMSUP. Basically, the intermodulation suppression is measured in dBc (dB below the carrier) for a set of frequencies $nf_p \pm mf_1$.

Two-tone intermodulation distortion is best explained by following a simple experimental procedure. Exciting the RF port of the mixer with two RF signals, f_{1a} and f_{1b}, spaced close together within the pass band of the mixer input will produce the following frequencies:

$$(\pm m_1 f_{1a} \pm m_2 f_{1b}) \pm nf_p \tag{11.33}$$

The order of the mixing product is $m_1 + m_2$. It would be nice if the IF output were only $|f_{1a} - f_p|$ and $|f_{1b} - f_p|$ since that would represent the down-converted signal to the IF output. Those terms containing harmonics of f_p would be far outside the band of interest and could be filtered out. There are essentially two possibilities for the second-order intermodulation products:

$$(\pm f_{1a} \pm 1 f_{1b}) \pm f_p$$

$$(\pm 1 f_{1a} \mp 1 f_{1b}) \pm f_p$$

In the first case, the output is near $3f_p$, and is therefore well outside the IF pass band. The second case presents an output frequency slightly above or below the local oscillator frequency, f_p, which again is well outside the IF pass band. However, the third-order intermodulation products are prolematic:

$$(\pm 2 f_{1a} \mp 1 f_{1b}) \pm f_p$$

$$(\pm 1 f_{1a} \mp 2 f_{1b}) \pm f_p$$

A numerical example illustrates what occurs with the third-order intermodulation products. If $f_p = 500\,\text{MHz}$, the desired RF input signal is $f_{1a} = 410\,\text{MHz}$, and a second signal of the same amplitude is at $f_{1b} = 400\,\text{MHz}$. The first-order products would give the desired output IF frequencies and a high frequency that could be easily filtered out:

$$|f_{1a} \pm f_p| = 90{,}910 \text{ MHz}$$

$$|f_{1b} \pm f_p| = 100{,}900 \text{ MHz}$$

The third-order intermodulation products would be

$$|2 f_{1a} - f_{1b} \pm f_p| = |820 - 400 \pm 500| = 80{,}920 \text{ MHz}$$

$$|2 f_{1b} - f_{1a} \pm f_p| = |800 - 410 \pm 500| = 110{,}890 \text{ MHz}$$

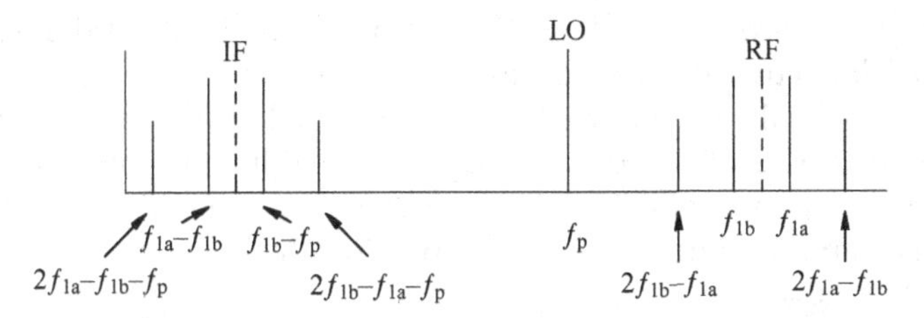

FIGURE 11.16 Third-order intermodulation distortion.

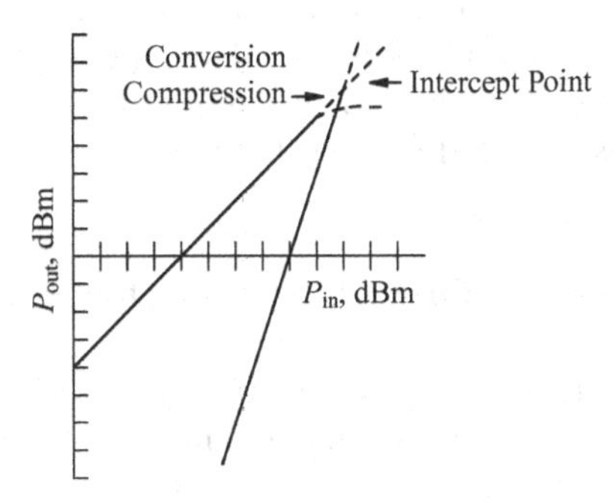

FIGURE 11.17 Two-tone third-order intermodulation intercept point.

As shown if Fig. 11.16, the undesired 80- and 110-MHz third-order intermodulation products could lie inside the IF pass band and thus distort the signal. The surest defense against this is to keep the amplitudes of the third-order intermodulation products small.

The measure of the size of the third-order intermodulation product is the intersection of third-order term with the desired first-order term, $f_0 = f_p - f_1$ (Fig. 11.16). The second-order intermodulation product is a result of having two RF signals that are multiplied together because of a quadratic nonlinearity:

$$[A\cos\omega_{1a}t \cdot B\cos\omega_{1b}t]\cos\omega_p t$$

The resulting amplitude proportional to AB will increase 2 dB when A and B each increase by 1 dB. The third-order intermodulation product is a result of a cubic nonlinearity:

$$[A^2\cos^2\omega_{1a}t \cdot B\cos\omega_{1b}t]\cos\omega_p t$$

The resulting amplitude proportional to A^2B will increase by 3 dB for every 1-dB rise in A and B. Thus, when the RF signal rises by 1 dB, the desired IF term will rise by 1 dB, but the undesired third-order intermodulation term rises by 3 dB (Fig. 11.17). The interception of the extrapolation of these two lines in the output power versus input power is called the *third-order intercept point.*

The input power level where this intersection occurs is called the *input intercept point*. The actual third-order intermodulation point is not directly measured, but it is usually found by extrapolation from lower power levels. It gives a single-valued criterion for determining the upper limit of the dynamic range of a mixer (or power amplifier). The conversion compression is the point where the desired IF output drops by 1 dB below the linear extrapolation from the low-level values.

The range of mixer LO frequencies and RF signal frequencies should be chosen so as to reduce to a minimum the possibility of producing intermodulation products that will end up in the IF bandwidth. When dealing with multiple bands of frequencies, keeping track of all the possibilities that may cause problems is often done with the aid of computer software available on the Internet or can be easily coded (see Problem 11.5).

11.8 SINGLE-SIDEBAND NOISE FACTOR AND NOISE TEMPERATURE

The frequency-independent noise power from a resistor is to a good approximation kT where k is Boltzmann's constant and T is the absolute temperature. In a two-port circuit shown in Fig. 11.18, a generator resistance, R_G, produces the equivalent noise temperature of T_G. The network itself is characterized as having a certain transducer power gain, G_T, and noise temperature. When describing the noise temperature of a two-port circuit, it must be decided if the noise is measured at the input or the output. The noise power at the output is presumably

$$T_{out} = G_T T_{in} \tag{11.34}$$

where T_{in} is the noise temperature referred to the input port and G_T is the transducer power gain. For mixers, this is the conversion gain or loss between the signal and IF ports. In the land where amplifiers are broadband, linear, and have wide dynamic range, Eq. (11.34) is accurate. It is assumed low-level random noise voltages are amplified the same way a clean sinusoid is. So the noise power delivered to the load, Z_L, is

$$N_L = k(G_T T_G + T_{out}) \tag{11.35}$$

or

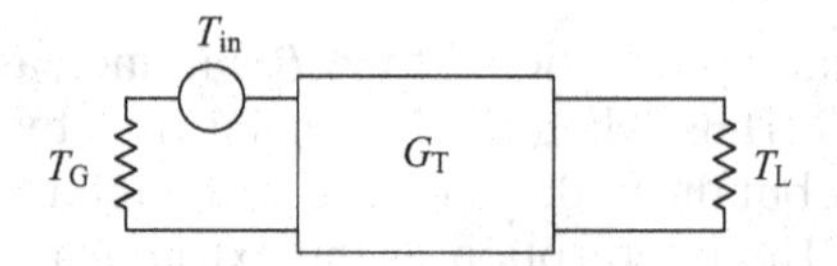

FIGURE 11.18 Noise within circuit referred to input side.

$$\begin{aligned} T_L &= G_T T_G + T_{out} \\ &= G_T(T_G + T_{in}) \end{aligned} \tag{11.36}$$

While the load will generate its own noise, this load noise is defined out of the equation since this is the noise *delivered* to the load.

The noise factor from Eq. (7.30) is

$$\begin{aligned} F &= \frac{S_{in}/N_{in}}{S_{out}/N_{out}} = \frac{S_{in}}{S_{out}}\left(\frac{N_{out}}{N_{in}}\right) \\ &= \frac{1}{G_T}\frac{G_T(T_G + T_{in})}{T_G} \\ &= \left(1 + \frac{T_{in}}{T_G}\right) \end{aligned} \tag{11.37}$$

The noise factor depends on the temperature of the generator, which by convention is set to room temperature, $T_G = 290\,\text{K} \triangleq T_0$. Thus, the noise characteristics of a two-port circuit such as a mixer (the LO port being conceptually ignored) can be characterized with either noise factor or noise temperature. Because of the greater expansion of the temperature scale over that of noise figure in decibels, noise temperature is preferred when describing very low noise systems and noise figure for higher noise systems. However, the concept of noise temperature becomes increasingly convenient when describing mixers with their multiple frequency bands.

The noise figure of a mixer can be described in terms of single-sideband (SSB) noise figure or double-sideband (DSB) noise figure. If the IF term, ω_0 in Fig. 11.2 comes solely from the signal, ω_1, and the image frequency, ω_{-1}, is entirely noise free, then the system is described in terms of its single-sideband noise figure, F_{SSB} (Fig. 11.19*a*). Double-sideband noise figure comes from considering both the noise contributions of the signal and the image frequencies

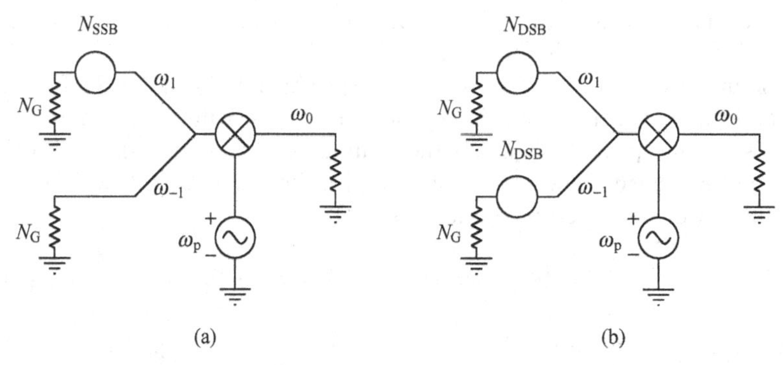

FIGURE 11.19 Mixer noise specification using (*a*) single-sideband noise and (*b*) double-sideband noise.

(Fig. 11.19*b*). In general, the output noise of the mixer will be the sum of the noise generated within the mixer itself and the noise power coming into the mixer multiplied by the mixer conversion gain. The noise power from inside the mixer itself can be referred to either the output port or the input port as described by Eq. (11.34). If all the internal mixer noise is referred back to the input RF signal port, then this is designated as N_{SSB}. The total noise power delivered to the load is found by multiplying N_{SSB} by the RF port conversion gain, G_{rf}, and adding to this the power entering from the signal source, N_G, at both the RF signal and image frequencies:

$$N_L = (N_{SSB} + N_G)G_{rf} + N_G G_{im} \tag{11.38}$$

The gains at the RF signal and image frequencies, G_{rf} and G_{im}, are typically very close to being the same since these two frequencies are close together. The terms in this definition are readily measurable, but Eq. (11.38) is at variance with the way the IEEE standards define single-sideband noise figure [4]. The single-sideband noise factor is conventionally defined as the ratio of the total noise power delivered to the load to the noise power entering at the RF signal frequency from a generator whose temperature is T_0 and when the mixer itself is considered to be noise free:

$$F_{SSB} = \frac{N_L}{N_G G_{rf}} \tag{11.39}$$

Assuming that $G_{rf} = G_{im}$,

$$\begin{aligned} F_{SSB} &= \frac{N_{SSB} G_{rf} + 2G_{rf} N_G}{G_{rf} N_G} \\ &= \frac{T_{SSB}}{T_0} + 2 \end{aligned} \tag{11.40}$$

Since N_{SSB} is referred to the mixer input, so its associated noise temperature, T_{SSB}, is also referred to the input side.

If the internal mixer noise power is referred back to both the RF frequency band and the image frequency band, then this power will be designated as the double-sideband power, N_{DSB}. For the double-sideband analysis, both the RF signal and image frequencies are considered as inputs to the mixer. In this case the total power delivered to the load is

$$N_L = (N_G + N_{DSB})(G_{rf} + G_{im}) \tag{11.41}$$

The double-sideband noise factor is determined by taking the ratio of the power delivered to the load from both of these frequency bands if the mixer were considered noise free:

$$F_{\mathrm{DSB}} = \frac{N_{\mathrm{L}}}{(G_{\mathrm{rf}} + G_{\mathrm{im}})N_{\mathrm{G}}} \tag{11.42}$$

Substituting Eq. (11.41) into Eq. (11.42) and again assuming $G_{\mathrm{rf}} = G_{\mathrm{im}}$,

$$F_{\mathrm{DSB}} = \frac{T_{\mathrm{DSB}}}{T_0} + 1 \tag{11.43}$$

In the single-sideband case, all mixer noise power is referred to the mixer input at the RF signal frequency. In the double-sideband case, all the mixer noise is referred to the mixer input at both the RF signal and image frequencies. Since the internal mixer power is split between the two frequency bands,

$$T_{\mathrm{SSB}} = 2T_{\mathrm{DSB}} \tag{11.44}$$

so that

$$F_{\mathrm{SSB}} = \frac{T_{\mathrm{SSB}}}{T_0} + 2 = \frac{2T_{\mathrm{DSB}}}{T_0} + 2 = 2F_{\mathrm{DSB}} \tag{11.45}$$

This illustrates the often stated difference between single- and double-sideband noise figures. Noise figure specification of a mixer should always state which of these two is being used.

11.9 SPECIAL MIXER APPLICATIONS

A single-sideband mixer can be obtained using the mixers discussed thus far with 90° and 180° couplers such as those described in Chapter 6. This circuit is shown in Fig. 11.20 where the signal and local oscillator are given as:

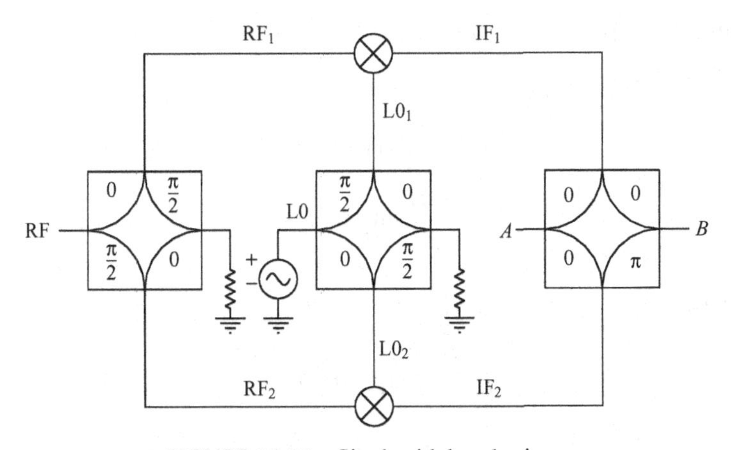

FIGURE 11.20 Single-sideband mixer.

$$\text{RF} = C\cos(\omega_1 t) \tag{11.46}$$

$$\text{LO} = D\cos(\omega_\text{p} t) \tag{11.47}$$

The plan is to find the signals delivered at ports A and B. The input RF signal as well as the LO signal are split and phase shifted by the couplers into RF_1, RF_2, LO_1, and LO_2.

$$\text{RF}_1 = \frac{C}{\sqrt{2}}\cos(\omega_1 t) \tag{11.48}$$

$$\begin{aligned}\text{RF}_2 &= \frac{C}{\sqrt{2}}\cos\left(\omega_1 t - \frac{\pi}{2}\right)\\ &= \frac{C}{\sqrt{2}}\sin(\omega_1 t)\end{aligned} \tag{11.49}$$

$$\text{LO}_1 = \frac{D}{\sqrt{2}}\sin(\omega_\text{p} t) \tag{11.50}$$

$$\text{LO}_2 = \frac{D}{\sqrt{2}}\cos(\omega_\text{p} t) \tag{11.51}$$

The two intermediate frequencies are obtained by the mixing of the RF_1 with LO_1 and the mixing of RF_2 with LO_2:

$$\text{IF}_1 = \frac{CD}{2}[\sin(\omega_\text{p} t)\cos(\omega_1 t)] \tag{11.52}$$

$$\text{IF}_2 = \frac{CD}{2}[\cos(\omega_\text{p} t)\sin(\omega_1 t)] \tag{11.53}$$

The output at port A is the sum of the two IF terms with $0°$ phase difference:

$$\begin{aligned}A &= \frac{CD}{2}[\text{IF}_1 + \text{IF}_2]\\ &= \frac{CD}{2}\sin[(\omega_\text{p} + \omega_1)t]\end{aligned} \tag{11.54}$$

At port B,

$$\begin{aligned}B &= \frac{CD}{2}[\text{IF}_1 - \text{IF}_2]\\ &= \frac{CD}{2}\sin[(\omega_\text{p} - \omega_1)t]\end{aligned} \tag{11.55}$$

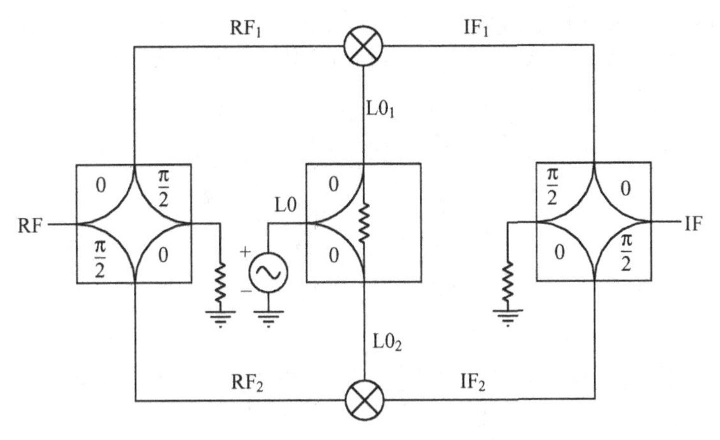

FIGURE 11.21 Image rejection mixer.

The output at port *A* is the upper sideband and the output at port *B* is the lower sideband. Each sideband is separated out to different ports.

An image rejection mixer can be obtained using the circuit in Fig. 11.21. Two frequencies enter the receiver on either side of the local oscillator frequency, one being the image of the other. The lower frequency is $\omega_{RFL} = \omega_p - \omega_0$ and the higher frequency is $\omega_{RFH} = \omega_p + \omega_0$. For ω_{RFL}:

$$\mathrm{RFL} = \frac{C}{\sqrt{2}}\cos[(\omega_p - \omega_0)t] \tag{11.56}$$

then

$$\mathrm{RF}_1 = \mathrm{RFL}$$

This is the same as the SSB case above.

$$\begin{aligned}\mathrm{IF}_1 &= \frac{CD}{2}\cos(\omega_p t)\cos[(\omega_p - \omega_0)t] \\ &= \frac{CD}{4}\cos(\omega_0 t)\end{aligned} \tag{11.57}$$

where the $2\omega_p$ term is filtered out. For the other branch the intermediate frequency is offset by $-\pi/2$:

$$\begin{aligned}\mathrm{IF}_2 &= \frac{CD}{2}\cos(\omega_p t)\cos\left[(\omega_p - \omega_1) - \frac{\pi}{2}\right] \\ &= -\frac{CD}{4}\sin(\omega_0 t)\end{aligned} \tag{11.58}$$

Again it is assumed the high-frequency term is filtered out. At the output end, the IF_2 term is shifted in phase, so the total IF is given below:

$$\begin{aligned} \mathrm{IF} &= \mathrm{IF}_1 = \mathrm{IF}_2\angle\left(-\frac{\pi}{2}\right) \\ &= \frac{CD}{4}\left\{\cos(\omega_0 t) - \sin\left[\left(\omega_0 t - \frac{\pi}{2}\right)\right]\right\} \\ &= \frac{CD}{2}\cos(\omega_0 t) \end{aligned} \tag{11.59}$$

Consequently, the lower RF input image frequency passes on through to the IF circuit. The destination of the upper image frequency, ω_{RFH} needs to be found:

$$\mathrm{RF}_1 = \frac{C}{\sqrt{2}} = \cos[(\omega_p + \omega_0)t] \tag{11.60}$$

$$\begin{aligned} \mathrm{IF}_1 &= \frac{CD}{2}\{\cos_p t \cos[(\omega_p + \omega_0)t]\} \\ &= \frac{CD}{4}\cos(\omega_0 t) \end{aligned} \tag{11.61}$$

Similarly for RF_2

$$\begin{aligned} \mathrm{RF}_2 &= \frac{C}{\sqrt{2}}\cos\left[(\omega_p + \omega_0)t - \frac{\pi}{2}\right] \\ &= \frac{C}{\sqrt{2}}\sin[(\omega_p + \omega_0)t] \end{aligned} \tag{11.62}$$

$$\begin{aligned} \mathrm{IF}_2 &= \frac{CD}{2}\cos(\omega_p t)\sin[(\omega_p + \omega_0)t] \\ &= \frac{CD}{4}\sin(\omega_0 t) \end{aligned} \tag{11.63}$$

The IF_1 and IF_2 terms are combined through the coupler, which shifts one of the signals by –90°:

$$\begin{aligned} \mathrm{IF} &= \mathrm{IF}_1 + \mathrm{IF}_2\angle\left(-\frac{\pi}{2}\right) \\ &= \frac{CD}{4}[\cos(\omega_0 t) - \cos(\omega_0 t)] \\ &= 0 \end{aligned} \tag{11.64}$$

Hence, the upper image frequency, ω_{RFH} does not get through this mixer circuit. Only the lower sideband is received.

11.10 CONCLUSIONS

The mixer component in the transceiver can take a wide variety of forms. Their design is usually based on efficiently controlling the unwanted frequencies that are generated by the device nonlinearity. The single mixer can be combined with other mixers to provide a wide variety of modulation and demodulation schemes.

PROBLEMS

11.1. Using the Fourier transform pair, show that $\mathcal{F}(e^{-j\omega t}) = \delta(\omega - \omega_0)$.

11.2. Two closely separated frequencies are delivered to the input signal port of a mixer of a receiver. The center frequency of the receiver is 400 MHz, and the two input frequencies are at 399.5 and 400.5 MHz. The mixer has a conversion loss of 6 dB and the local oscillator is at 350 MHz. The power level of these two input frequencies are –14 dBm (dB below a milliwatt). At this input power, the third-order modulation products are at –70 dBm.

a. What are the numerical values for the output frequencies of most concern to the receiver designer?

b. What is the output third-order intercept point?

11.3. A nonlinear device is modeled by the following equation:

$$i_o(t) = I_0 + av_i(t) + bv_i^2(t) + cv_i^3(t) + \cdots$$

where $I_0 = 10^{-2}$, $a = 5 \times 10^{-3}$, $b = 2 \times 10^{-3}$, and $c = 10^{-3}$. The applied input voltage is

$$v_i(t) = 1\cos 2\pi f_1 t + \frac{1}{2}\cos 2\pi f_2 t$$

Find the series representation of the output current in terms of sines and cosines of the various frequencies by means of the convolution theorem. Finding the $v_i^3(t)$ is done by first finding $v_i^2(t)$ and then multiplying that by $v_i(t)$ by means of the convolution theorem.

11.4. For the mixer circuit in Fig. 11.22, f_p is 10 MHz and f_1 is 8 MHz. Assume the diode model is

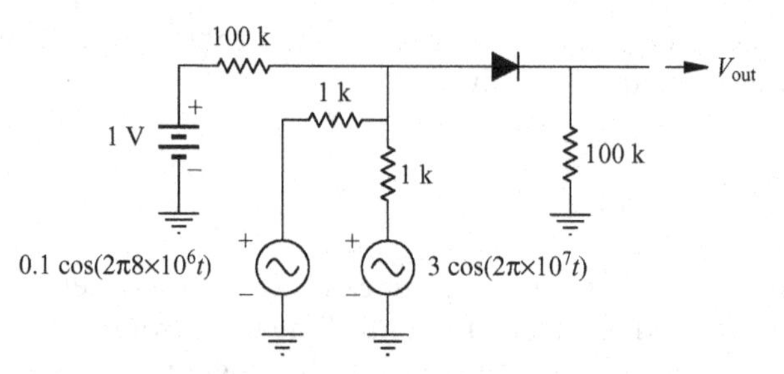

FIGURE 11.22 Single end mixer for Problem 11.4.

```
.model d1n4148 d(is=2.682n n=1.836 rs=.5664 ikf=44.17m xti=3
+ eg=1.11 cjo=4p m=.3333 vj=.5 fc=.5 bv=100 ibv=100u tt=11.54n)
```

Using SPICE, determine the amplitudes of the mixing products at 2, 8, 10, and 18 MHz. A suggested step size for the transient analysis is 1 ns, and the time duration is 1 μs.

11.5. Write a program that determines the spurious responses of a mixer in a receiver where frequencies other than the desired radio frequency is inside the band of the amplifier. You are asked to not only provide the answers to the given specific problem, but a listing of the source code. The nonlinear mixer produces sum and difference frequencies of the RF and LO inputs as well as with smaller levels of intermodulation products. These frequencies are given by

$$f_s = |nf_{LO} - mf_{RF}|$$

where m and n are positive integers. Some of these products may fall inside the IF pass band and thus interfere with the desired signal. The spurious responses may be determined by the following procedure:

a. Divide the RF tuning range of the receiver into a number of frequencies, each spaced by the IF bandwidth.

b. For each RF frequency, compute the required LO frequency from

$$f_{LO} = f_{RF} \pm f_{IF}$$

c. Compute the intermodulation frequency from (1) for $1 \le m \le M$ and $1 \le n \le N$. The values M and N represent the order of the mixer nonlinearity at each of these frequencies.

d. A spurious response lies within the IF pass band if $f = |mf_1 - nf_p| \le$ the IF pass band.

Determine the spurious responses for a receiver with an input radio frequency range from 840 to 850 MHz, an IF frequency of 168 MHz, and an IF bandwidth of 10 MHz for orders up to 10 for M and N. Consider both the high side and low side for the f_{LO} frequency.

REFERENCES

1. M. Abramowitz and I. A. Stegun, *Handbook of Mathematical Functions with Formulas, Graphs, and Mathematical Tables*, New York: Wiley, Equations 9.6.33–9.6.35, 1972.
2. H. L. Krauss, C. W. Bostian, and F. H. Raab, *Solid State Radio Engineering*, New York: Wiley, 1980.
3. L. G. Meares and C. Hymowitz, "SPICE Models for Power Electronics," http://www.intusoft.com/articles/satcore.pdf, April 2002.
4. S. A. Maas, *Microwave Mixers*, 2nd. ed., Norwood MA: Artech, 1993.
5. G. D. Vendelin, A. M. Pavio, and U. L. Rohde, *Microwave Circuit Design*, New York: Wiley, Chapter 7, 1990.
6. Watkins-Johnson Co., *Microwave Designer's Handbook*, 97–98 Catalog SanJose, CA.
7. B. Gilbert, "A Precise Four-Quadrant Multiplier with Subnanosecond Response," IEEE J. Solid State Circuits, pp. 365–373, Dec. 1968.
8. J. M. Moniz and B. Maoz, "Improving the Dynamic Range of Si MMIC Gilbert Cell Mixers for Homodyne Receivers," *IEEE J. 94 Microwave and Millimeter-Wave Monolithic Circuits Symposium*, pp. 103–106, 1994.
9. D. G. Tucker, "Intermodulation Distortion in Rectifier Modulators," *Wireless Engr.*, 145–152, June 1954.

CHAPTER TWELVE

Phase-Lock Loops

12.1 INTRODUCTION

Phase-lock loops (PLL) became widespread with the availability of high-quality integrated circuit operational amplifiers (op-amps) in the 1960s. Since then this versatile circuit has found applications across the frequency spectrum in consumer, commercial, deep space, and military projects. Understanding the operation of the PLL incorporates RF techniques, oscillator design, closed-loop control theory, analog circuit design, and digital circuit design. The references listed at the end of this chapter provide insight beyond the present scope [1–4]. This chapter begins with the basic concepts, which are later expanded into practical designs.

12.2 PLL DESIGN BACKGROUND

The PLL uses negative feedback to tailor its performance. The primary design parameters are response time, transient performances, bandwidth, damping ratio, and phase margin. The type and order of a closed-loop system define the complexity and response to a stimulus. At least two of the components of a PLL, the voltage-controlled oscillator (VCO) and the phase detector, are high-frequency components. There may also be amplifiers, mixers, frequency multipliers, and other oscillators that require RF techniques.

Many integrated circuits are presently available that combine many of the PLL functions on a single chip. Operational amplifiers often are used in filter circuits, which may be either inverting or noninverting designs as required by the phase-lock loop design. Integrators, dc amplifiers, Schmitt triggers, and offset circuits are used to set the loop operation. Resistor/capacitor circuits provide phase shift for stability. The VCO keeps the loop locked.

12.3 PLL APPLICATIONS

A phase-lock loop can be used to multiply, divide, or filter different frequencies. The latter is illustrated in a space probe rapidly moving away from Earth. To recover data from the probe, the transmitter frequency must be known. The signal is very weak because of the distance, and the low signal-to-noise ratio requires a very small receiver filter bandwidth to recover the data. However, because of the relative motion, there is a significant and changing Doppler shift to the transmit frequency. The system requires a filter that may be only a few hertz wide, operating at a varying frequency that is centered at several gigahertz. The PLL helps keep the filter centered on the correct frequency.

An electronic PLL is one form of a closed-loop system. The cruise control in an automobile is another. A switching power supply, a camera's light meter, a radio's automatic gain control, the temperature control in a building, a car's emission system controls, and a touch-tone dialing system are examples of closed-loop systems. A broadcast receiver changes frequency with a button push or electronically. Each time the station is accurately centered with no manual adjustment required. Physically, these PLLs are all very different, but they all have common characteristics and must be made stable.

The concept of feedback control is illustrated by the simple action of controlling the speed of a car. If the desired speed is 60 mph, then this becomes the reference speed. Any deviation from this speed is an error. The accelerator pedal is the control element. On level terrain, a constant pressure on the pedal will maintain constant speed. As the car goes up a hill, it will slow down, and the difference between the actual speed and the reference value generates an error. This error generates a command to push the accelerator pedal to increase the speed, but there will continue to be a slight error. As the car crests the hill and starts down, the speed will increase. Releasing pedal pressure will slow the acceleration, but an error will remain until a steady-state condition is again reached. For this example, the driver's brain is the feedback path. The driver controls the sense of the feedback by knowing when to push and when to release the pedal. By his reaction time, he controls how close to the reference he maintains the car's speed. He may decide to rapidly change the correction to tightly match the desired speed, or he may choose to compensate slowly so his speed averages out to the correct value. His actions coupled with the car's controls form a system closely analogous to a phase-lock loop. Replace the human with an electrical circuit that senses the speed error, include another circuit that tempers the response time, and couple it to the accelerator controls. This is the typical cruise control system.

12.4 PLL BASICS

A PLL is a closed-loop system used for frequency control. Three and sometimes four building blocks are common to most PLL designs:

1. The phase detector.
2. The loop filter.
3. The voltage-controlled oscillator.
4. In addition a frequency divider is used when the output frequency is to be a multiple of the reference input frequency.

Figure 12.1 illustrates the connection of these blocks to make a complete phase-lock loop. The phase detector has two inputs and one output. This block can be realized by a specialized mixer based on those described in Chapter 11 where the IF port pass band goes down to dc. If the two input signals are very close in frequency, then the output will contain a term at twice the input frequency and a term that is almost zero frequency. The loop error signal in the PLL is the near-zero term. This error signal goes to the loop filter, which may be as simple as a capacitor and a resistor, or it can be one or more operational amplifiers with many resistors and capacitors. The VCO is the control element of this loop. The input is a control voltage from the loop filter, while the output is the required frequency. The time integral of this frequency is the phase that when compared with the input phase gives the error voltage to the loop filter.

When the loop is first turned on, the VCO frequency is not controlled by an error voltage. The loop filter output voltage can be anywhere between the high and low limits set by the power supply. However, the phase detector produces an error voltage that is the difference between the actual VCO frequency and the reference frequency. Like the cruise control example, this signal tells the loop filter whether the VCO frequency is too high or too low relative to the reference frequency. If the error signal indicates that the VCO frequency is less than the reference, the loop filter adjusts the control voltage to raise the VCO frequency. If the VCO frequency is too high, the loop filter changes the voltage and lowers the VCO frequency. The loop filter sets how fast the error is corrected. Some loops may be designed for a fast bumpy ride,

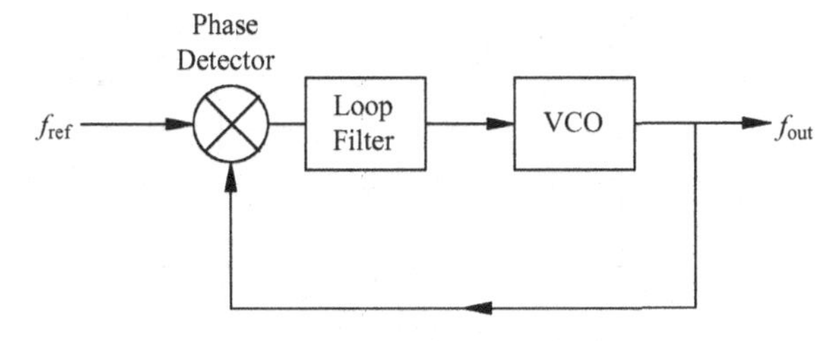

FIGURE 12.1 Basic phase-lock loop.

while others may require a slow response, resulting in a smooth ride. When the loop filter has done its job, the VCO frequency will exactly match the reference frequency, and the two inputs will have a constant phase difference. This match in frequency and constant phase difference will be maintained even if the reference frequency changes. With each change, the PLL again goes through the settling out process. If the reference is noisy, the PLL is in a continual state of change, working hard to follow the input.

12.5 LOOP DESIGN PRINCIPLES

The important top-level PLL parameters are the input and output frequency, the response time, the loop bandwidth, and the loop damping ratio. The PLL block diagram in Fig. 12.2 includes a frequency divider in the feedback path. Both frequencies coming into the phase detector must be locked together at the same frequency. This will force the output frequency, f_{out}, to be N times the reference frequency, f_{ref}. The actual design process can be summarized in three steps. First, each component should be thoroughly understood and tested individually. Second, when the components are assembled together, they should give the required phase margin and bandwidth for stability. Third, closed-loop analysis should show that the final connection matches the system level goals both in theory and experimentally. The basic building blocks, except in exotic applications, are those shown in Fig. 12.2. This section describes in greater depth each of these functions.

12.5.1 Phase Detectors

Phase detectors come in many configurations. These include those with logic level inputs, passive and active analog designs, and sampling versions used for high-frequency multiplication. In addition, there are phase detectors with automatic frequency search features to aid in initial frequency acquisition. In its simplest form, a phase detector is a frequency mixer. As described in Chapter 11, when two signals come into the mixer, the output consists

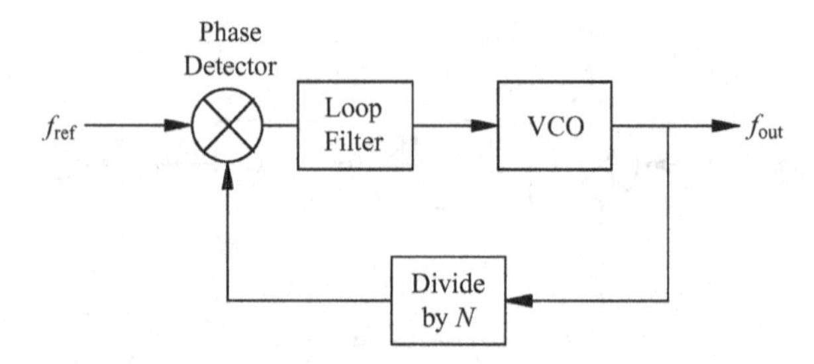

FIGURE 12.2 Phase-lock loop with frequency counter.

primarily in the sum and difference frequencies. The sum frequency is filtered out by the loop filter. The difference frequency, historically called the *beat note*, is typically a few kilohertz or less in a PLL. If the two input frequencies are exactly the same, the phase detector output is the phase difference between the two inputs. This loop error signal is filtered and used to control the VCO frequency. The two input signals can be represented by sine waves:

$$V_1 = V_a \sin(\omega_1 t + \phi_1) \tag{12.1}$$

$$V_2 = V_b \sin(\omega_2 t + \phi_2) \tag{12.2}$$

The difference frequency term is the error voltage given as

$$V_e = K_m V_1 V_2 = \frac{K_m V_a V_b}{2} \cos[(\omega_1 - \omega_2)t + (\phi_1 - \phi_2)] \tag{12.3}$$

where K_m is a constant describing the conversion loss of the mixer. Equation (12.3) is a time-varying cosine waveform at the beat note frequency. When the two frequencies are identical, the output voltage is a function of the phase difference, $\Delta\phi = \phi_1 - \phi_2$:

$$V_e = \frac{K_m V_a V_b}{2} \cos(\Delta\phi) \tag{12.4}$$

This is maximum when $\Delta\phi = 0°$, a minimum when $\Delta\phi = 180°$, and zero when $\Delta\phi = 90°$ or $270°$ (Fig. 12.3).

In the frequency domain, the phase detector can be modeled as

$$K_{pd} \frac{a}{a + s} \tag{12.5}$$

which at low frequencies is simply the slope of the voltage versus phase curve. The units for K_{pd} is V/rad:

$$K_{pd} = \frac{dV_e}{d\Delta\phi} = -\frac{K_m V_a V_b}{2} \sin(\Delta\phi) \tag{12.6}$$

Thus, $K_{pd} = 0$ when $\Delta\phi = 0°$ or $180°$ and is at its extreme values at $\Delta\phi = 90°$ or $270°$.

12.5.2 Voltage-Controlled Oscillator

The voltage-controlled oscillator is the PLL control element in which the output frequency changes monotonically with its input tuning voltage. A linear

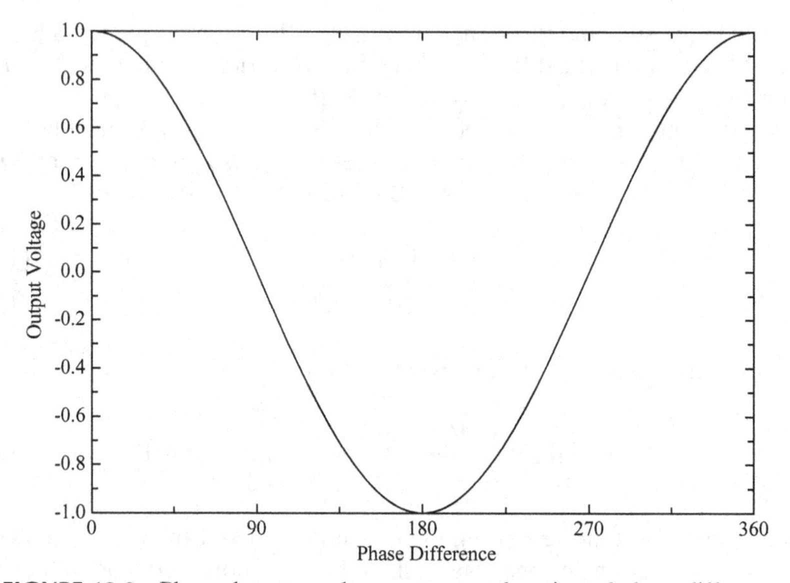

FIGURE 12.3 Phase detector voltage output as function of phase difference.

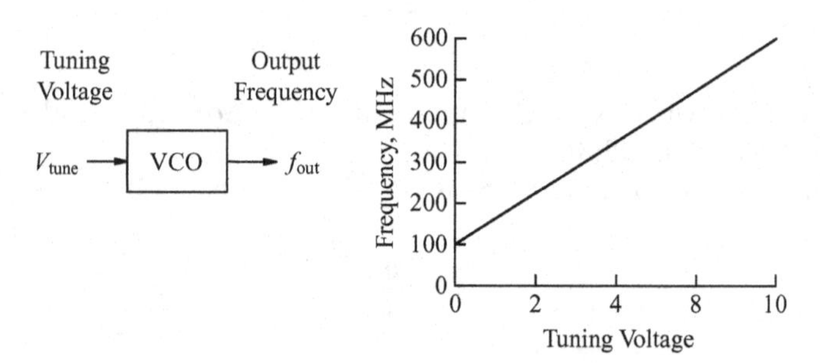

FIGURE 12.4 Voltage-controlled oscillator tuning.

frequency versus tuning voltage is an adequate model for understanding its operation (Fig. 12.4):

$$\omega_{\text{out}} = K_{\text{vco}} V_{\text{tune}} + \omega_0 \tag{12.7}$$

In a PLL, the ideal VCO output phase may be expressed as

$$\phi_2(t) = \omega_0 t + \int_0^t K_{\text{vco}} V_{\text{tune}} dt + \phi_0 \tag{12.8}$$

where $\phi_2(t)$ is the output phase of the VCO, ω_0 is the free-running VCO frequency when the tuning voltage is zero, and K_{vco} is the tuning rate with the dimension of rad/s-V.

The error voltage from the phase detector first steers the frequency of the VCO to exactly match the reference frequency, then holds it there with a constant phase difference. It is modeled as having a low-frequency gain K_{vco} and one or more poles of the following form:

$$\frac{K_{\text{vco}}}{s(s+a)} \tag{12.9}$$

12.5.3 Loop Filters

A loop filter is a low-frequency circuit that filters the phase detector error voltage, which in turn controls the VCO frequency. The filter may be either active or passive, but it is usually a simple analog design. In extreme cases it might be an entire microprocessor. This discussion will be limited to analog loop filters such as the representative topologies shown in Fig. 12.5. Figure 12.5*a* shows an op-amp integrator with nearly infinite dc gain. This is the loop

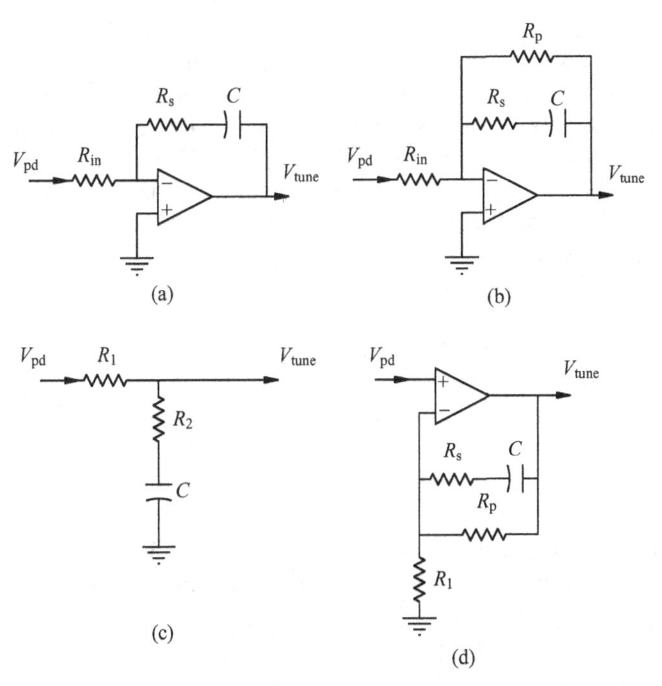

FIGURE 12.5 Loop filter used in (*a*) type 2 second-order PLL, (*b*) type 2 third-order PLL, (*c*) type 1 second-order PLL, and (*d*) type 1 second-order inverting PLL.

filter often associated with the type 2 PLL. The order and type of a PLL is defined in Section 12.9. Figure 12.5*b* shows an operational amplifier loop filter with a finite gain and is associated with a type 1 PLL. Figure 12.5*c* is a passive filter used with a phase detector whose output is current rather than voltage. This type of detector is frequently found in synthesizer ICs and is associated with a type 2 PLL. While the loop filter is a simple circuit, its characteristic is important in determining the final closed-loop operation. The wrong design may make the loop unstable, causing oscillation or have an undesirable response. The loop filters shown in Fig. 12.5 are of the form

$$F(s) = \pm k \frac{s+a}{s+b} \tag{12.10}$$

where $-a$ is a zero and $-b$ is a pole.

12.5.4 Frequency Dividers

When the output frequency must be a multiple of the input frequency, frequency dividers may be included in a PLL. Most dividers use a digital circuit, although analog techniques dating from 1939 are available for very high frequency devision. With the availability of complete synthesizers on a single IC, fewer stand-alone divider circuits are on the market. Most dividers have a division ratio equal to a binary number or switchable from a binary to a binary +1 (e.g., divide by 64 or 65). The upper limit on the input frequency is about 3 GHz, although only a few ICs will go that high. Divide by four circuits have been demonstrated with inputs above 14 GHz, but this is a very specialized device not required by most PLLs. For a linear analysis when the loop bandwidth is much less than the reference frequency, dividers are modeled as a gain element with a value = $1/N$.

12.6 LINEAR ANALYSIS OF THE PLL [5]*

From the perspective of the time domain, the control voltage for the VCO is

$$V_{\text{tune}}(t) = V_{\text{tune}-0} + \int_0^t V_e(t) f(t-\mu)\, d\mu \tag{12.11}$$

where $f(t)$ is the impulse response of the filter and V_e is the error voltage coming from the phase detector. The Laplace transform of $f(t)$ is

$$F(s) = \int_0^\infty f(t) e^{-st} dt \qquad t > 0 \tag{12.12}$$

*This material is based on A. J. Viterbi, *Principles of Coherent Communication*, 1966, by permission of The McGraw-Hill Companies.

and the inverse transform can be obtained in principle by the integral:

$$f(t)=\frac{1}{2\pi i}\lim_{\beta\to\infty}\int_{\gamma-i\beta}^{\gamma+i\beta}F(s)e^{st}dt \qquad \Re\{s\}>\gamma \tag{12.13}$$

Substituting Eq. (12.11) into the derivative of Eq. (12.8) gives the VCO frequency:

$$\frac{d\phi_2(t)}{dt}=\omega_0+K_{\text{vco}}V_{\text{tune-0}}+\frac{K_{\text{vco}}K_{\text{m}}V_{\text{a}}V_{\text{b}}}{2}\int_0^t f(t-\mu)\cos\Delta\phi(\mu)\,d\mu \tag{12.14}$$

where $\Delta\phi(t)=\phi_1(t)-\phi_2(t)$. Consequently, a general equation describing the phase error is

$$\frac{d\Delta\phi}{dt}\triangleq\frac{d\phi_1}{dt}-\omega_0-K_{\text{vco}}V_{\text{tune-0}}-\frac{K_{\text{vco}}K_{\text{m}}V_{\text{a}}V_{\text{b}}}{2}\int_0^t f(t-\mu)\cos\Delta\phi(\mu)\,d\mu \tag{12.15}$$

For a given input phase ϕ_1, the solution of this equation describes the exact operation of the PLL. However, to avoid carrying along ω_0, a new phase variable may be defined:

$$\psi_1(t)\triangleq\phi_1(t)-(\omega_0+K_{\text{vco}}V_{\text{tune-0}})t \tag{12.16}$$

$$\psi_2(t)\triangleq\phi_2(t)-(\omega_0+K_{\text{vco}}V_{\text{tune-0}})t \tag{12.17}$$

The equation for the phase error is now given without ω_0:

$$\frac{d\Delta\phi}{dt}=\frac{d\psi_1}{dt}-\frac{K_{\text{vco}}K_{\text{m}}V_{\text{a}}V_{\text{b}}}{2}\int_0^t f(t-\mu)\cos\Delta\phi(\mu)\,d\mu \tag{12.18}$$

This suggests an alternate representation for the phase-lock loop as shown in Fig. 12.6. In this representation the multiplier is replaced by a subtracter and a cosinusoidal nonlinearity while the VCO is replaced by an integrator.

When the phase error $\Delta\phi$ deviates from 90° by a small amount, $\cos(\Delta\phi-90°)\approx\Delta\phi$. Then Eq. (12.18) becomes

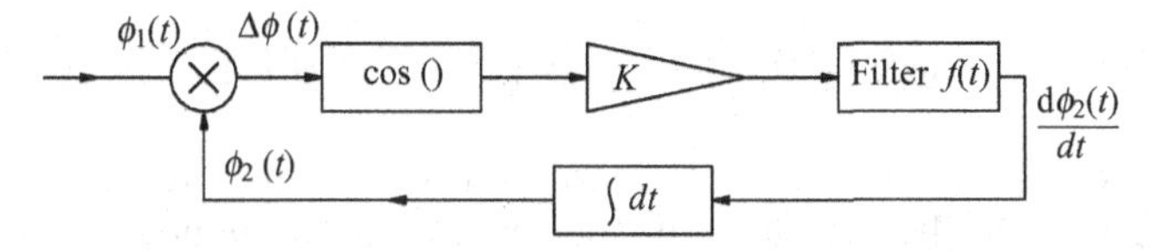

FIGURE 12.6 Time-domain nonlinear phase-lock loop.

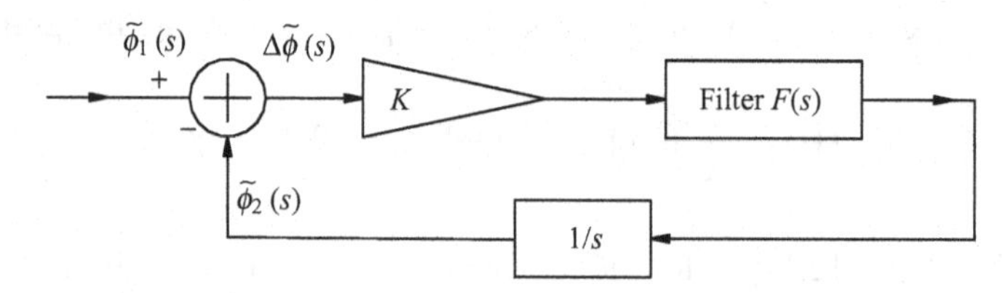

FIGURE 12.7 Frequency-domain linear phase-lock loop.

$$\frac{d\psi_1}{dt} = \frac{d\Delta\phi}{dt} + K\int_0^t f(t-\mu)\Delta\phi(\mu)\,d\mu \tag{12.19}$$

where

$$K = \frac{K_{\text{vco}}K_{\text{m}}V_{\text{a}}V_{\text{b}}}{2} \tag{12.20}$$

If the Laplace transform of $\psi_1(t)$ is represented by $\tilde{\psi}(s)$ and the Laplace transform of $\Delta\phi(t)$ is represented by $\Delta\tilde{\phi}(s)$, then the Laplace transform of Eq. (12.19) is

$$s\Delta\tilde{\phi}(s) + KF(s)\Delta\tilde{\phi}(s) = s\tilde{\psi}_1(s) \tag{12.21}$$

This linear frequency-domain equation for the PLL can be represented as shown in Fig. 12.7. The solution for the phase error gives

$$\Delta\tilde{\phi}(s) = \frac{\tilde{\psi}_1(s)}{1 + KF(s)/s} \tag{12.22}$$

so that the phase shift at the output of the PLL is

$$\begin{gathered} \tilde{\psi}_2(s) = \tilde{\psi}_1(s) - \Delta\tilde{\phi}(s) \\ \frac{\tilde{\psi}_2(s)}{\tilde{\psi}_1(s)} \triangleq H'(s) = \frac{G(s)}{1 + G(s)} \end{gathered} \tag{12.23}$$

where $G(s) = KF(s)/s$. The phase error can in turn be written in terms of this phase transfer function:

$$\Delta\tilde{\phi}(s) = \tilde{\psi}_1(s) - \tilde{\psi}_2(s) = [1 - H'(s)]\tilde{\psi}_1(s) \tag{12.24}$$

This is the difference of the two phase terms entering the phase detector. A frequency multiplier circuit will incorporate a frequency divider in the feedback loop, so that $K \to KN$. However, the output will differ from $\tilde{\psi}_2(s)$

by the frequency multiplier factor, N, and the transfer function from input to output is

$$H(s) = H'(s)N = \frac{[G(s)/N]N}{1+G(s)/N} \tag{12.25}$$

For example, let $N = 1$, the phase of the incoming signal be given by $\phi_1(t) = \omega t + \phi_0$, and the PLL has no filter so that $F(s) = 1$. Readjusting the phase reference as was done in Eqs. (12.16) and (12.17) gives

$$\psi_1(t) = \phi_1(t) - \omega_0 t \tag{12.26}$$

$$= (\omega - \omega_0)t + \phi_0 \tag{12.27}$$

In the frequency domain this becomes

$$\tilde{\psi}_1(s) = \frac{\omega - \omega_0}{s^2} + \frac{\phi_0}{s} \tag{12.28}$$

The phase error is found from Eq. (12.22):

$$\Delta\tilde{\phi}(s) = \frac{s}{s+K}\left(\frac{\omega - \omega_0}{s^2} + \frac{\phi_0}{s}\right) \tag{12.29}$$

The inverse transform in this case is straightforward and gives the phase error in the time domain:

$$\Delta\phi(t) = \frac{\omega - \omega_0}{K}(1 - e^{-Kt}) + \phi_0 e^{-Kt} \tag{12.30}$$

The steady-state phase error is found by allowing $t \to \infty$:

$$\Delta\phi(t = \infty) = \frac{\omega - \omega_0}{K} \tag{12.31}$$

Clearly, the phase will change when the incoming frequency changes, so that phase lock is not achieved.

The insertion of a low-pass filter into the PLL will produce lock. An active filter such as that shown in Fig. 12.8 is recognized as basically a noninverting amplifier. The inverting amplifier would have right half-plane poles in the PLL and is therefore unstable unless the VCO or phase detector circuit can accommodate the minus sign. For the noninverting case, the voltage transfer function can be found by writing node equations at the input nodes of the operational amplifier:

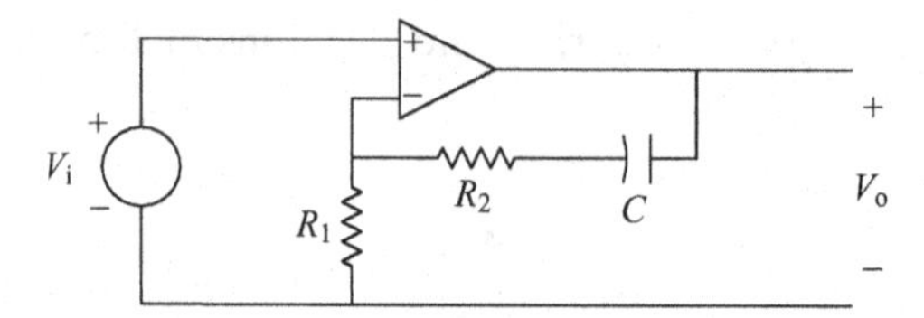

FIGURE 12.8 Possible active low-pass filter for PLL.

$$F(s) = \frac{V_o}{V_1} = 1 + \frac{R_2 + 1/sC}{R_1} = \left(1 + \frac{R_2}{R_1}\right) + \frac{1}{sCR_1} \tag{12.32}$$

The phase transfer factor is found from Eq. (12.23):

$$H'(s) = \frac{\tilde{\psi}_2(s)}{\tilde{\psi}_1(s)} = \frac{K[(1 + R_2/R_1)R_1Cs + 1]}{s^2CR_1 + CR_1K(1 + R_2/R_1)s + K} \tag{12.33}$$

Thus, using Eq. (12.24), the phase error is easily obtained:

$$\Delta\tilde{\phi}(s) = [1 - H'(s)]\psi_1(s) \tag{12.34}$$

$$\Delta\tilde{\phi}(s) = \frac{(\omega - \omega_0)CR_1 + \phi_0 CR_1 s}{s^2CR_1 + CR_1K(1 + R_2/R_1)s + K} \tag{12.35}$$

Rather than find the inverse transform this time, the final value theorem may be used to find the steady-state phase error:

$$\lim_{t \to \infty} \Delta\phi(t) = \lim_{s \to 0} s\Delta\tilde{\phi}(s) = 0 \tag{12.36}$$

In this case, the phase error is independent of frequency and in the steady state is zero.

12.7 LOCKING A PHASE-LOCK LOOP

The previous sections examined the individual elements of a PLL. A simple loop with no frequency divider will serve as an example of how these parts work together. Assume that initially the loop is not locked, and the reference frequency is 100 MHz. A tuning voltage of 5 V is required to make the VCO frequency operate at 100 MHz. The phase detector can produce a cosine wave beat note of 1 V peak to peak.

An inverting op-amp type 1 loop filter will be used with a gain of 100 at low frequency and a gain of 0.1 at high frequency (Fig. 12.9). With the loop unlocked, the VCO frequency could be anywhere within its operating limits.

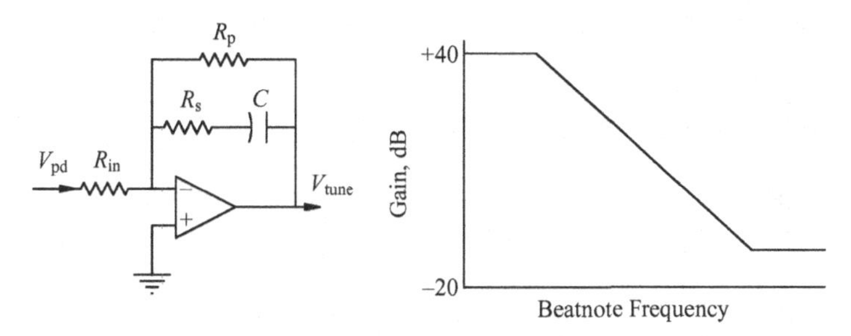

FIGURE 12.9 (*a*) Type 1 loop filter and (*b*) its frequency response.

Assume that it is operating at 101 MHz, so that there is a 1-MHz beat note at the phase detector output when the reference frequency is first applied. This beat note frequency is high enough to only be amplified with a gain of 0.1 by the loop filter. The VCO tuning voltage will be modulated by the phase detector output of 0.1-V peak to peak, but this voltage will not cause any significant change in the VCO frequency.

Since the VCO frequency is too far away from the reference frequency, there is not enough gain in the loop to bring the loop into lock. However, if the VCO frequency is 100.1 MHz when the reference frequency is applied, the beat note frequency is 100 kHz. That is well within the high-gain frequency range of the loop filter for this design. The amplified beat note voltage modulates the VCO frequency. As the VCO frequency swings closer to the reference frequency, the beat note frequency gets even lower, and it enters an even higher gain region of the loop filter. This action accelerates the VCO frequency change until it crosses the reference frequency. At this point the beat note frequency is zero. The PLL has been designed as a stable closed-loop system, and the VCO is at the same frequency as the reference. The transient phase detector output voltage and the VCO tuning voltage are shown in Figs. 12.10 and 12.11, respectively. The input voltage to the VCO is 5 V when the PLL is at frequency lock. Since the loop filter has a dc inverting gain of 100, the voltage at the phase detector output is

$$V_e = \frac{5}{-100} = -50\ \text{mV} \tag{12.37}$$

The maximum voltage from the phase detector was specified to be 1 V peak to peak or $V_e = 0.5$ V from the zero level to the peak. This would occur when $\Delta\phi = 0°$ as specified by Eq. (12.4). This equation then gives the value for $K_m V_a V_b = 1$. When $V_e = -50$ mv as given in Eq. (12.37) the value for the phase difference from Eq. (12.4) is $\Delta\phi = \arccos(2V_e/K_m V_a V_b) = \arccos(-0.1) = 95.7°$. The loop filter will keep the VCO at 100 MHz and maintain a 95.7° phase difference between the two phase detector inputs.

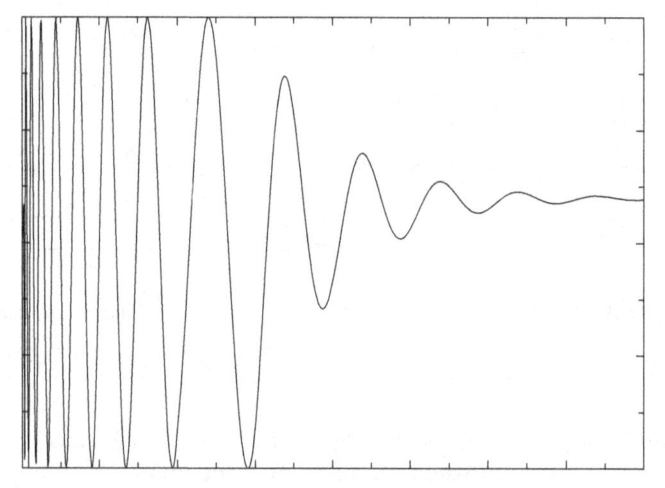

FIGURE 12.10 Phase detector voltage as PLL pulls into lock.

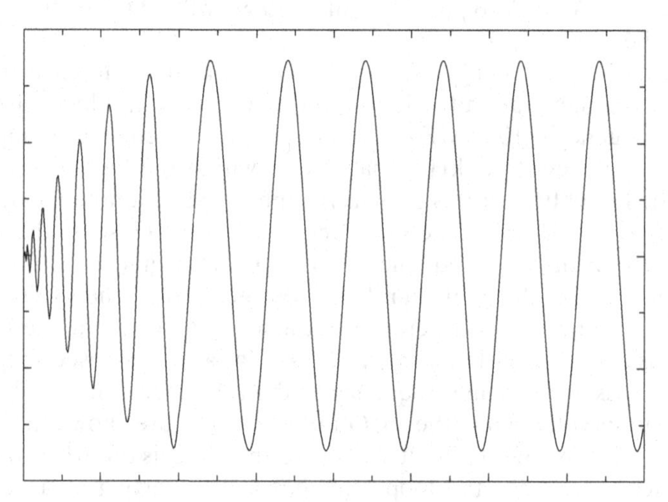

FIGURE 12.11 VCO tuning voltage as PLL pulls into lock.

An oscillator accumulates 360° of phase rotation in each cycle. If the frequency increases, it will accumulate more phase rotation in a given period of time. If the VCO tries to drift higher in frequency, it will quickly accumulate more phase rotation. The phase detector output voltage will go up, and the loop filter will amplify this change, which will lower the VCO control voltage. The VCO output frequency will drop and return to 100 MHz. The situation is similar for the VCO trying to move lower in frequency. This is the effect of the negative feedback within the loop. The battle for control goes on continuously. Small changes in the VCO due to temperature, noise, or even gravity cause small frequency changes. The PLL will not tolerate errors due to frequency or phase changes. When an error voltage develops at the phase detector output, the loop filter will amplify it and the VCO frequency and phase will return to the correct value. The corrective action of the loop will make whatever adjustments are required to hold the phase and frequency constant.

The initial beat note frequency must be well within the loop filter bandwidth to achieve lock without frequency aiding. In any loop, the first event is to bring the VCO frequency in line with the reference frequency. Once the circuit is in lock, a steady-state phase relationship that satisfies the loop feedback and dc requirements is found that will hold its lock frequency.

12.8 LOOP TYPES

The PLL is a closed-loop system controlled by negative feedback. The closed-loop gain $H(s)$ for the circuit in Fig. 12.7 is described by

$$H(s) = \frac{\tilde{\phi}_o(s)}{\tilde{\phi}_1(s)} = \frac{G(s)}{1+G(s)/N} \tag{12.38}$$

where $G(s)$ is called the open loop or forward gain, and $G(s)/N$ is called the loop gain. The forward gain is the product of the phase detector gain, the loop filter gain, and the VCO gain, that is, $G(s) = KF(s)/s$. The frequency divide ratio is N.

At every point in the loop, the signal has a voltage amplitude and a phase. The function, $H'(s)$, from Eq. (12.23) represents a ratio of two phase terms. The $G(s)$ function can be seen as converting a phase to a voltage in the phase detector and converting voltage to a phase in the VCO. The frequency divider simply multiplies the output of $G(s)$ (a phase or frequency) by $1/N$. The functions $G(s)$ and N are both dimensionless quantities.

The number of pure integrators (or number of poles at the frequency origin) in the loop gain $G(s)/N$ determines the *type* of the system. A VCO is a pure phase integrator that will contribute one pole to the type determination. Therefore, a PLL will be at least type 1. A loop filter with a finite dc gain will not increase the type number. A loop filter with an integrator such as that in Fig. 12.5*a*, will increase the type to 2.

The *order* of the PLL is the degree of the denominator polynomial of Eq. (12.38). The loop filter operational amplifier has at least two significant break points: typically one at a frequency between 1 Hz and 100 kHz and a second above 10 MHz. In the example given in Section 12.7 that used a type 1 loop, the only pure integrator is the VCO, so there is only one pole at dc. The loop filter has a dc inverting gain of 100. If the VCO gain is 1 MHz/V and the reference frequency is changed to 103 MHz, the VCO tuning voltage will now be 8 V. This is found from $f_{ref} = f_{out} = K_{vco}V_{tune} + f_0$ where $V_{tune} = 5$ gives $f_0 = 95$ MHz.

Thus, when $f_{ref} = 103$ MHz, V_{tune} must be 8 V. With a gain of −100, the phase detector voltage must be $V_e = 8/(-100) = -80$ mV.

This represents an angular difference of $\Delta\phi = \arccos(2V_e/K_m V_a V_b) = 99.2°$ in contrast to 95.7° found earlier when the reference frequency was 100 MHz. If the reference frequency continues to change, the VCO frequency will change to match it, which in turn will change the phase detector output voltage. As the reference frequency changes in a type 1 loop, the phase difference changes. This is an important characteristic that is sometimes desirable and other times unacceptable.

If the dc gain of the loop filter is increased to 1000, the phase detector output voltage for a 100-MHz lock is only −5 mV. For phase lock at 103 MHz the phase detector output voltage is −8 mV. These values represent phase differences, $\Delta\phi$, of 90.57° and 90.92°, respectively. If the dc gain is further increased, the change of $\Delta\phi$ with frequency will further decrease. Finally, the dc feedback resistor, R_p, will approach an open circuit and the loop filter dc gain will increase to infinity. The loop filter in Fig. 12.5*b* is transformed to that shown in Fig. 12.5*a*.

This loop filter is now a pure integrator, thereby producing a type 2 loop. The total number of integrators for the PLL with this loop filter is two: one for the VCO and one for the loop filter. Among the features of this loop is the constant phase shift between the VCO and reference frequency that is maintained with a change in frequency.

Type 1 and type 2 loops constitute the majority of applications. Type 3 and higher loops are required to solve frequency change problems in unusual situations. For example, a ground-launched missile must track an orbiting satellite during its own launch and orbital insertion. During the launch phase, the rocket is consuming fuel and thus reducing its mass. With a constant force, its acceleration will increase at an increasing rate. As the satellite comes overhead, its transmit frequency is shifted due to the relative motion with the rocket. This shift is changing at an increasing rate. Then the booster separates from the rocket and the force goes to zero during coast. To track the satellite frequency with no phase error requires a PLL type of at least 4. Most high-type loops are used to solve complicated motional problems.

12.9 NEGATIVE FEEDBACK IN A PLL

A frequency change that generates a change in the phase of a stable negative feedback loop generates a correction for the phase error. In the previous example, the type 1 loop filter was described as having a dc inverting gain of 100. The VCO requires 5 V to produce a 100-MHz output. An open-loop connection of the PLL components will demonstrate what is called the "sense" of the loop. For open-loop testing, the VCO is connected to a manually adjustable power supply. With the power supply set at 5 V, there will be a low-frequency beat note observed at the phase detector output. If the voltage is changed to either 4 or 6V, the beat note will be 1 MHz. The frequency of the VCO cannot be determined from the beat note, since the beat note shows only the frequency difference between the two signals and not which signal is the higher or lower frequency. A complete description of the difference frequency between the VCO and reference requires both a direction and a magnitude. With the loop out of lock, this type of phase detector can only determine the magnitude, $|\Delta\phi|$. The VCO frequency must be forced close enough to the reference frequency for the beat note to be inside the loop bandwidth for a PLL with this type of phase detector to pull into phase lock.

In the previous example with the inverting filter, the VCO frequency increased as the tuning voltage decreased as seen from Eq. (12.7) where $K_{vco} < 0$. Many VCOs have the opposite characteristic; that is, the frequency increases with increasing tuning voltage. Stability is assured by the appropriate choice of an inverting or noninverting filter. If the loop locked up at 90° difference between the two inputs with the positive slope VCO, it will lock up at 270° with the negative slope VCO. The phase detector output in either case will be correct to adjust the VCO to match the reference input frequency and phase.

Most synthesizer ICs and PLLs using frequency dividers or logic ICs have a different type of phase detector. Using flip-flops to count the input edges, these phase detectors produce an error voltage that has not only a magnitude but also a sense of the direction between the two inputs. The output is a series of voltage or current pulses. The loop filter averages these pulses to form the control voltage for the VCO. A pulse duty cycle above 50% indicates that the VCO frequency is higher than the reference frequency, and a duty cycle of less than 50% indicates that the VCO frequency is lower. If the VCO is running higher than the reference frequency, the control voltage will force it toward the correct value. If the VCO is running too low, the error voltage will drive the frequency higher. This type of phase detector can drive a PLL into lock even when the VCO and reference frequencies are a great distance apart, far outside the loop bandwidth. The typical IC synthesizer will have a pin available to reverse the sense of the error voltage to accommodate VCOs of either positive or negative tuning slope.

12.10 PLL DESIGN EQUATIONS

Each block of the normal phase-lock loop model (Fig. 12.12) is described by a gain value that may be a constant or a function of frequency. The components are carefully selected to ensure a locking mechanism and stable loop operation. The frequency response of the closed loop is typically displayed as a Bode plot with a minimum frequency of 1 Hz and a maximum frequency between 10 kHz and 10 MHz. The Bode plot describes the filtering bandwidth and in turn the transient response of the PLL to the input voltage.

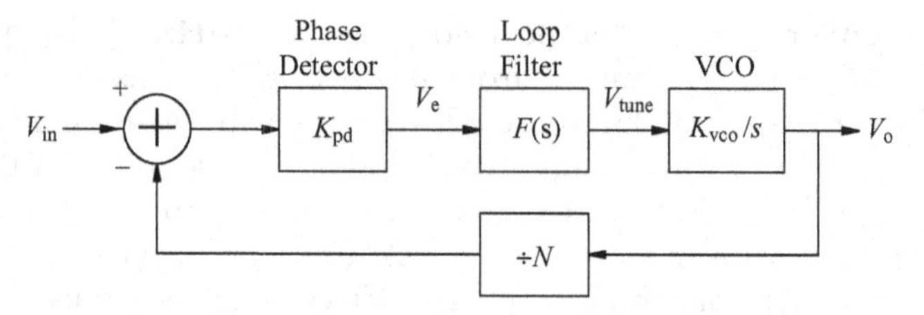

FIGURE 12.12 Frequency-domain closed-loop model for PLL.

12.10.1 Inverting Loop Filter

An inverting filter, such as that shown in Fig. 12.5*b*, produces a 180° phase shift between its input and output. The VCO would have to match the filter so as to produce an increase in frequency with a decrease in voltage. Thus, with either an inverting filter or a noninverting filter, the phase transfer function given by Eq. (12.25) remains:

$$H(s) = \frac{G(s)}{1 + G(s)/N} \tag{12.39}$$

The example that follows makes use of a second-order type 1 PLL. The transfer function for the filter in Fig. 12.5*b* is

$$\begin{aligned} F(s) &= -\frac{R_p \| (R_s + 1/Cs)}{R_{in}} \\ &= -\frac{(R_p R_s / R_{in}) sC + (R_p / R_{in})}{(R_p + R_s) sC + 1} \end{aligned} \tag{12.40}$$

The open-loop gain is

$$G(s) = \frac{F(s) K_{pd} K_{vco}}{s} \triangleq \frac{KF(s)}{s} \tag{12.41}$$

When Eqs. (12.40) and (12.41) are substituted into the expression for the the closed-loop PLL [Eq. (12.39)], the result is clearly of second order in the denominator. This can be written in terms of notation commonly used in control theory:

$$H(s) = \frac{-KR_p R_s}{R_{in}(R_p + R_s)} \left[\frac{s + 1/(CR_s)}{s^2 + 2\zeta\omega_n s + \omega_n^2} \right] \tag{12.42}$$

The variable, ζ, is the damping ratio and ω_n is called the natural frequency. The roots of the denominator are

$$s_{1,2} = -\zeta\omega_n \pm \omega_n\sqrt{\zeta^2 - 1} \tag{12.43}$$

For the second-order type 1 loop under consideration,

$$2\zeta\omega_n = \frac{NR_{in} - KR_pR_sC}{R_{in}NC(R_p + R_s)} \tag{12.44}$$

$$\omega_n = \sqrt{\frac{-KR_p}{NR_{in}C(R_p) + R_s}} \tag{12.45}$$

so that

$$\zeta = \frac{1/C - KR_pR_s/(NR_{in})}{2\omega_n(R_p + R_s)} \tag{12.46}$$

The design specification for a PLL is typically given in terms of a damping ratio and a natural frequency. The design task is to determine circuit values that will meet the specification. The filter response at dc is

$$F_{dc} = -\frac{R_p}{R_{in}} \tag{12.47}$$

where the minus sign is to be absorbed by K. Thus, Eq. (12.45) can be rearranged to give

$$R_p + R_s = \frac{KF_{dc}}{NC\omega_n^2} \tag{12.48}$$

and this substituted into Eq. (12.46) to give

$$R_p + R_s = \frac{KF_{dc}}{NC\omega_n^2} = \frac{1}{2C\omega_n\zeta} + \frac{KF_{dc}R_s}{N2\omega_n\zeta} \tag{12.49}$$

Using Eq. (12.48) to replace R_s above gives,

$$\frac{KF_{dc}}{NC\omega_n^2} = \frac{1}{2C\omega_n\zeta} + \frac{KF_{dc}}{N2\omega_n\zeta}\left(\frac{F_{dc}K}{NC\omega_n^2} - R_p\right) \tag{12.50}$$

If, in addition to the damping ratio and the natural frequency, values for C and the dc gain are chosen, then the required resistance values can be found as summarized below. Solution of Eq. (12.50) gives the value for R_p, then R_s, and finally R_{in}:

$$R_p = \frac{K}{NC}\left[\frac{F_{dc}}{\omega_n^2} + \frac{N^2}{K^2F_{dc}} - \frac{2\zeta N}{K\omega_n}\right] \tag{12.51}$$

$$R_s = \frac{KF_{dc}}{NC\omega_n^2} - R_p \quad (12.52)$$

$$= \frac{K}{NC}\left[\frac{2\zeta N}{K\omega_n} - \frac{N^2}{K^2 F_{dc}}\right] \quad (12.53)$$

$$R_{in} = -\frac{R_p}{F_{dc}} \quad (12.54)$$

To guarantee $R_p > 0$, the first two terms in the brackets of Eq. (12.51) must be more negative than the third term. Since $K < 0$ for a PLL with an inverting filter, R_p would be positive. Thus

$$\frac{F_{dc}}{\omega_n^2} + \frac{N^2}{K^2 F_{dc}} < \frac{2\zeta N}{K\omega_n} \quad (12.55)$$

To guarantee $R_s > 0$, the first term in Eq. (12.53) must be more negative than the second terms:

$$\frac{2\zeta N}{K\omega_n} < \frac{N^2}{K^2 F_{dc}} \quad (12.56)$$

These two inequalities lead to the following conclusion:

$$\frac{F_{dc}}{\omega_n^2} + \frac{N^2}{k^2 F_{dc}} < \frac{2\zeta N}{K\omega_n} < \frac{N^2}{K^2 F_{dc}} \quad (12.57)$$

Comparing the first and last inequalities of Eq. (12.57) gives $F_{dc} < 0$, which was assumed at the outset. However, Eqs. (12.55) and (12.56) give the explicit requirements for R_p and R_s.

Figure 12.13 illustrates the expected PLL bandwidth versus frequency for several values of damping ratio in a typical second-order circuit. These results are calculated for a natural frequency of 1 Hz. The results can be easily scaled for loops requiring higher natural frequencies. When $\zeta < 1$, the PLL is underdamped and peaking occurs. The response of such a loop to a disturbance will be a damped oscillation that finally converges to the final answer. When $\zeta > 1$, the system is overdamped. The –3-dB frequency is found by setting the transfer function to $1/\sqrt{2}$ and solving for ω_3/ω_n. It is assumed in doing this that only the denominator terms are frequency dependent. Thus,

$$|H(j\omega)|^2 = \frac{1}{\left|-\omega_3^2 + j\omega_3 2\zeta\omega_n + \omega_n^2\right|^2} = \left|\frac{1}{\sqrt{2}\omega_n^2}\right|^2 \quad (12.58)$$

Therefore,

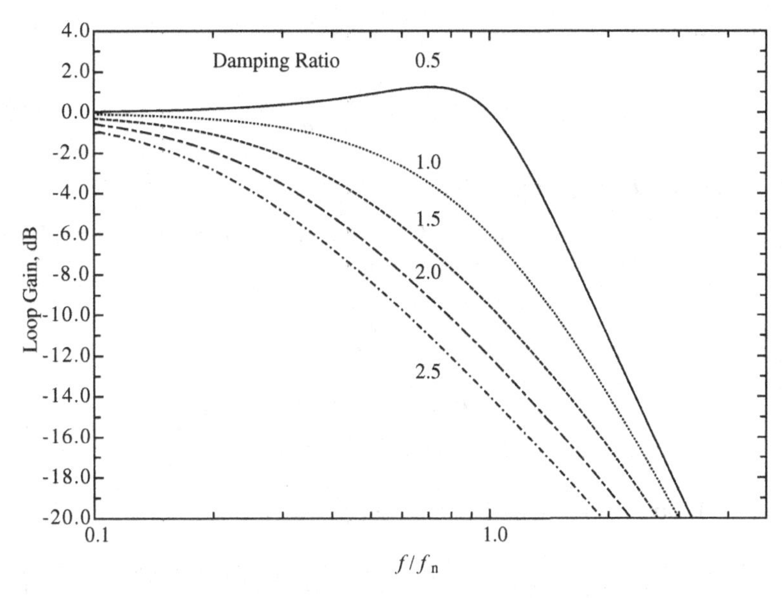

FIGURE 12.13 PLL response with natural frequency of 1 Hz and various damping ratios.

$$\omega_3^2 = \omega_n^2\left[\sqrt{(2\zeta^2-1)^2+1}-(2\zeta^2-1)\right] \tag{12.59}$$

The –3-dB gain frequency, f_3, for a damping ratio of 1.0 is 0.6436 times the natural frequency. If a –3-dB frequency of 50 kHz were required with a damping ratio of 1.0, then a natural frequency of 77.889 kHz would be chosen.

A type 2 PLL can be formed by allowing $R_p \to \infty$ to give the filter shown in Fig. 12.5*a*. Equation (12.44) becomes

$$2\zeta\omega_n = -\frac{K}{NR_{in}C} \tag{12.60}$$

and from Eq. (12.45)

$$\omega_n^2 = -\frac{K}{NR_{in}C} \tag{12.61}$$

which gives

$$R_{in} = -\frac{K}{NC\omega_n^2} \tag{12.62}$$

Consequently,

$$R_s = -\frac{2\zeta N R_{in}}{K} = \frac{2\zeta}{C\omega_n} \tag{12.63}$$

12.10.2 Noninverting Loop Filter

Design equations can be developed for a noninverting loop filter like that shown in Fig. 12.5*d*. The filter transfer function is

$$\begin{aligned} F(s) &= 1 + \frac{R_p \| [R_s + 1/(sC)]}{R_1} \\ &= \frac{1 + R_p/R_1 + sC(R_p R_s/R_1 + R_p + R_s)}{1 + sC(R_p + R_s)} \end{aligned} \tag{12.64}$$

The closed-loop gain is found by substituting Eq. (12.64) into Eq. (12.38):

$$H(s) = K\left(1 + \frac{R_s R_p}{R_1(R_p + R_s)}\right)\left[\frac{s + 1/[C(R_p \| R_1 + R_s)]}{s^2 + 2\zeta\omega_n + \omega_n^2}\right] \tag{12.65}$$

From this the loop natural frequency and damping ratio can be identified:

$$\omega_n = \sqrt{\frac{(R_1 + R_p)K}{N(R_p + R_s)R_1 C}} \tag{12.66}$$

$$\zeta = \frac{1 + C(R_s R_p/R_1 + R_p + R_s)K/N}{2\omega_n C(R_p + R_s)} \tag{12.67}$$

The typical synthesis procedure is to design a PLL for a given natural frequency and damping ratio using a specified capacitance, C. Solving Eq. (12.66) for $R_p + R_s$ and substituting this into Eq. (12.67) gives an equation in terms of one unknown, R_p. First, from Eq. (12.66),

$$R_s = \frac{K(R_p + R_1)}{N R_1 C \omega_n^2} - R_p \tag{12.68}$$

then substitution gives

$$\frac{K(R_p + R_1)}{N R_1 C\omega_n^2} = \frac{1}{2\zeta\omega_n C R_1}\left[R_1 + \frac{C R_p K}{N}\left(\frac{K(R_1 + R_p)}{N R_1 C\omega_n^2} - R_p\right) + \left(\frac{CKR_1}{N}\right)\frac{K(R_1 + R_p)}{N R_1 C\omega_n^2}\right] \tag{12.69}$$

This has one unknown, R_p, which can be solved by the quadratic formula as follows:

$$0 = R_p^2 a + R_p b + c \tag{12.70}$$

where

$$a = \frac{K}{N}\left(\frac{K}{N} - R_1 C \omega_n^2\right) \tag{12.71}$$

$$b = \frac{2KR_1}{N}\left(\frac{K}{N} - \zeta\omega_n\right) \tag{12.72}$$

$$c = R_1^2\left(\omega_n^2 + \frac{K^2}{N^2} - \frac{2\zeta K\omega_n}{N}\right) \tag{12.73}$$

so that

$$R_p = \frac{-b + \sqrt{b^2 - 4ac}}{2a} \tag{12.74}$$

The value of R_1 is associated with the dc voltage gain of the noninverting loop filter:

$$R_1 = \frac{R_p}{F_{dc} - 1} \tag{12.75}$$

The value for R_s is obtained from Eq. (12.68). Physically realizable solutions would require the discriminant of Eq. (12.74) to be positive. This can be found in terms of the PLL parameters:

$$b^2 - 4ac = \frac{4R_1^2\omega_n^2 K}{N}\left[\frac{K}{N}(\zeta^2 - 1) + R_1 C\left(\omega_n^2 + \frac{K^2}{N^2}\right) - \frac{R_1 K C\omega_n}{N}\right] > 0 \tag{12.76}$$

The type 2 PLL parameters with the noninverting loop filter can be found by letting $R_p \to \infty$. Thus, the design equations for a given natural frequency, damping ratio, and capacitance for a type 2 PLL are

$$\omega_n^2 = \frac{K}{NR_1C} \tag{12.77}$$

or

$$R_1 = \frac{K}{NC\omega_n^2} \tag{12.78}$$

and the damping ratio is

$$2\omega_n\zeta = 0 + \frac{R_s K}{N R_1} + \frac{K}{N} \tag{12.79}$$

or

$$R_s = R_1\left(\frac{2\omega_n\zeta N}{K} - 1\right) \tag{12.80}$$

The value for K is given by Eq. (12.41). A second-order loop can be built either as a type 1 or a type 2 with either an inverting or noninverting loop filter. The actual loop order may be several orders higher than 2 when all the extraneous poles are considered. A good design procedure initially ignores these poles and assumes ideal VCOs, phase detectors, operational amplifiers, and the like and determines a set of loop filter values based on the second-order model. Subsequently, nonideal parts can then be added and computer simulation used to refine the analysis.

12.11 PHASE DETECTOR TYPES

Previous sections have introduced both the mixer and flip-flop-based phase detectors. These two widely used configurations have many specialized variations. A sampling phase detector is a third type that is frequently used in RF and microwave applications. The design requirements for a specific application will usually point to the correct choice.

12.11.1 Mixer Phase Detectors

Mixers with a dc-coupled output make an excellent phase detector. At high frequencies, a mixer may be either active (with transistors) or passive (with diodes). The diode versions usually provide the best dc stability. It is the best choice for low-noise designs where the PLL reference input is a low-level signal. The beat note output is typically 100 mV to 1 V peak to peak, depending on the mixer type and application. The mixer type phase detector is the best choice when the input signal is pulsed or noncontinuous. The ability of this type of detector to resolve an angular difference is limited to ±90°. It has no ability to determine which input is the higher frequency, so it is not capable of frequency discrimination.

12.11.2 Sampling Phase Detectors

Sampling phase detectors (SPD) can be used in a phase-lock loop to produce an output frequency that is an integer multiple of the reference frequency. This mixer relies on a device to generate a comb of frequencies at multiples of the reference. The VCO then uses the correct spectral line to produce an error signal. An SPD is used in a phase-lock loop where the output frequency is an integer multiple of the input frequency. If $f_{out} = Nf_{in} + \Delta f$, the SPD output is a cosine wave of frequency Δf. If f_{out} is exactly N times f_{in}, the SPD output is a dc level proportional to the phase difference between its two inputs. The input frequency, f_{in} is typically between 50 and 200 MHz at a power level of +20 dBm or higher, and f_{out} can be at any harmonic of f_{in} up to about 18 GHz or up to about the 150th harmonic of f_{in}.

A sampling phase detector is one of the best choices for a very high frequency PLL where excellent phase noise is a requirement. However, an external circuit is usually required to bring the VCO into lock range. Additionally, provisions must be made to ensure that the VCO will be locked to the correct multiple of f_{in}.

12.11.3 Flip-Flop Phase Detector with Frequency Acquisition Aiding

All of the phase detectors previously discussed have a major drawback. They produce an output equal to the difference between the two input frequencies, either a cosine wave or a triangle wave. However, this output does not have information about whether the VCO is too high in frequency or too low. A PLL using these phase detectors must also include a sweep or search circuit to initially bring the VCO frequency close enough to lock. This can involve a substantial amount of circuitry. The phase detector circuit shown in Fig. 12.14*a* uses positive edge triggered *D*-type flip-flops to overcome this problem. The *D* inputs are connected to a logic 1. Figure 12.14*b* illustrates the timing sequence. The signal, f_1, positive edge arrives first, causing Q_1 to clock high. Later the positive edge of f_2 causes Q_2 to clock high. Two 1's at the NAND gate's input cause its output to go low and clear both Q_1 and Q_2. The output at Q_2 is a pulse whose duty cycle represents the time delay between f_1 and f_2. The pulse at Q_2 is very short since it is the sum of the propagation times through the flip-flops and gates. Of course, if f_2 arrives before f_1, then the output pictures are reversed.

The outputs, Q_1 and Q_2, turn on the current sources. These current sources either source or sink current to the capacitor, which ramps up or down the phase detector output voltage. The action of this circuit is identical to the op-amp integrator. So the phase detector has added another integrator to the PLL. A PLL using this circuit will be at least a type 2 loop.

The advantage of this circuit is the self-searching capability. If f_1 is higher than f_2, the output voltage will go to the positive voltage limit. If f_1 is lower than f_2, the output voltage will go to the negative voltage limit. If f_1 equals f_2, the output voltage will be proportional to the phase difference. Thus, this circuit can sense which input frequency is higher. The output voltage can then be used to drive the VCO in the correct direction to bring the loop into lock. Once the two frequencies are the same, this circuit becomes a phase detector and drives the VCO for no phase error.

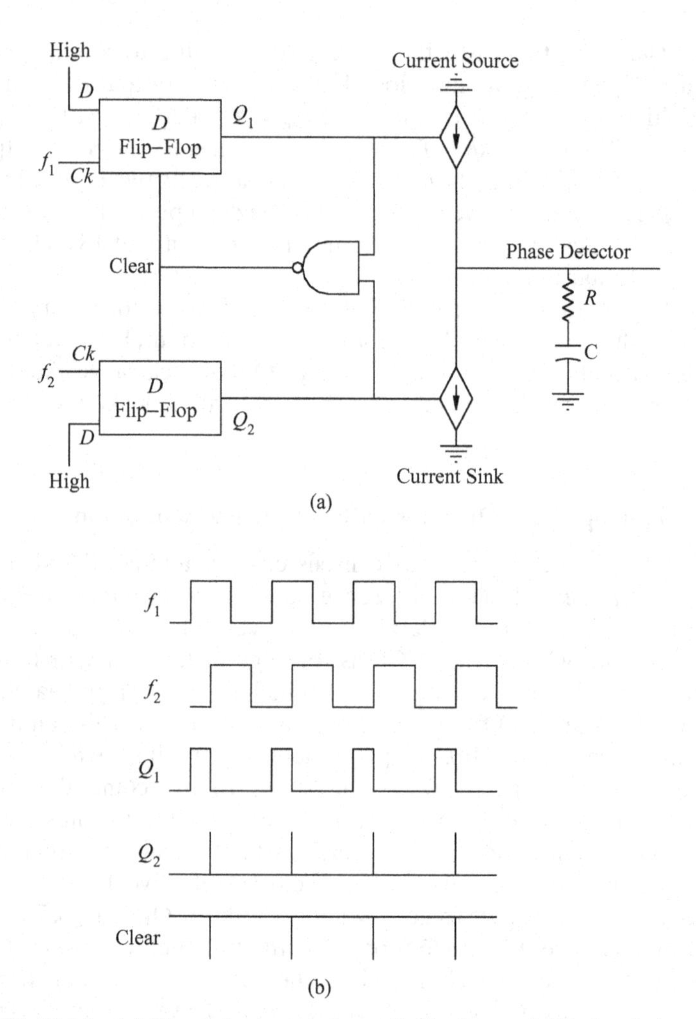

FIGURE 12.14 (*a*) Phase detector using a *D* flip-flop and (*b*) the timing chart.

This phase detector circuit is used in many present-day frequency synthesizer ICs, where its built-in search capability makes it ideal for a variety of applications. However, there are at least two drawbacks to this circuit that limits its usage. The largest problem is the short pulse on one of the flip-flop outputs. With high-speed logic, this pulse is only a few nanoseconds long. If f_1 and f_2 are high in frequency, this pulse width may be a significant part of their period. The pulse dead time due to propagation delays results in a nonlinear phase detector transfer curve. In older versions, there were flat spots with zero gain and regions where the gain reversed its slope. The pulse also contributes heavily to the output noise, easily adding 20dB of noise to the PLL output

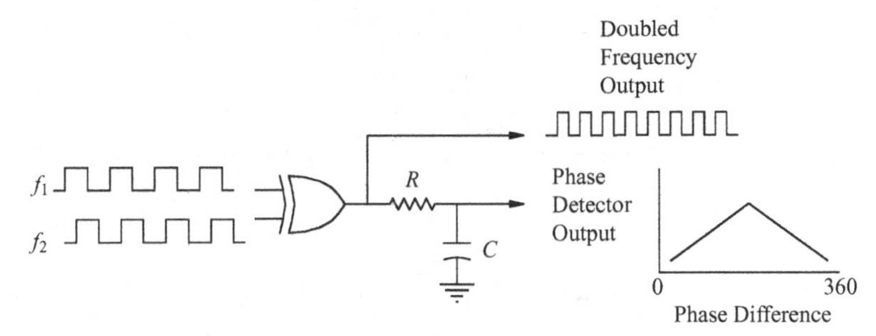

FIGURE 12.15 Exclusive OR phase detector.

even in the most modern devices. The second problem is that the searching capability can become confused if there is any interruption in either f_1 or f_2. Therefore, this circuit should be used in applications where very low phase noise is not required and the inputs are continuous. For this phase detector the gain is K_{pd} volts/rad.

12.11.4 Exclusive OR Phase Detector

An exclusive OR gate works as a frequency doubler and phase detector. Figure 12.15 illustrates a typical example for the phase detector. For correct operation, both inputs, f_1 and f_2, must be at the same frequency and both must have 50% duty cycles. The XOR output will be a logic level waveform at twice the input frequency. The duty cycle of the output depends on the phase difference between the two inputs. Phase shifts of 90° or 270° produce a 50% duty cycle output. The *RC* low-pass filter produces a dc value proportional to the duty cycle. For a 90° or a 270° phase difference, the filter output is one half the difference between the logic high- and logic low-output voltages.

The XOR gate is the functional equivalent of the balanced mixer. This circuit is useful for PLL applications requiring a high-frequency VCO to be divided down and locked to a low-frequency logic-level frequency reference. This phase detector is suitable for low phase noise applications, but it frequently requires an external search circuit to initially achieve lock.

12.11.5 Charge Pump PLL

The charge pump phase-lock loop described by Gardner [6] incorporates a digital phase/frequency detector (PFD) plus a charge pump that basically converts the digital output voltage to an analog current. This in turn is fed to the loop filter. The block diagram is shown in Fig. 12.16 where the PFD has a three-state digital logic output of up, down, and neutral (UP, DW, and N).* This is fed into a charge pump that charges or discharges the loop filter capacitance

*See, for example, the Max9382 PFD from Maxim.

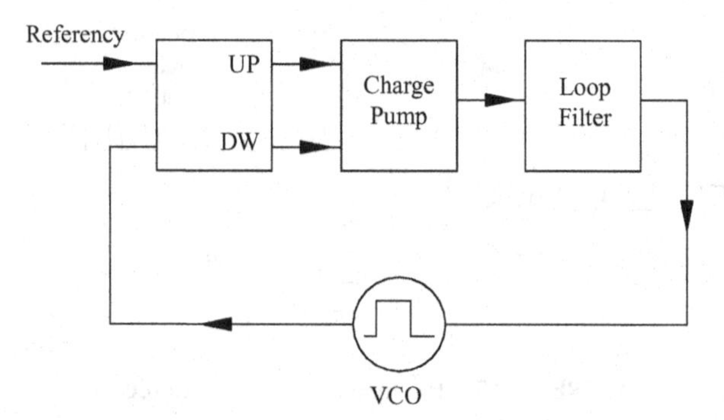

FIGURE 12.16 Charge pump phase-lock loop.

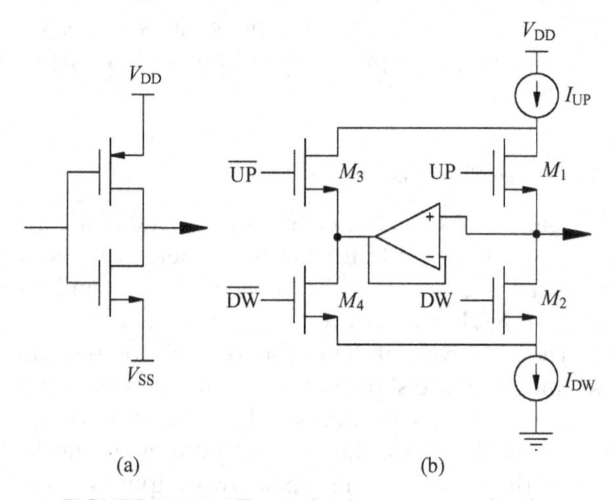

FIGURE 12.17 Example charge pump circuits.

depending on whether the output logic is UP or DW. If the phase detector is N, the loop is locked and no current flows. The loop filter can be as simple as a shunt capacitance to ground or something like that shown in Fig. 12.5*c*.

There are a wide variety of charge pump circuits, two of which are shown in Fig. 12.17. The first of these uses positive and negative voltage logic states that charges or discharges the loop filter, depending on the state of the PFD. It is possible for both transistors to be off so that the voltage on the loop filter is floating. The operational amplifier in Fig. 12.17*b* helps to control this voltage. This circuit, as drawn, uses positive and 0 voltage logic states. When UP is high and DW is 0, transistors M_1 and M_4 are on and M_2 and M_3 are off. Current I_{UP} charges the loop filter and I_{DW} draws current through M_4 and the low-impedance output of the operational amplifier. The opposite transistors turn on when DW is high and UP is 0, and the loop filter is discharged. The operational amplifier must be capable of handling the current levels of the current sources. Several advanced designs have been proposed that address the floating voltage, the additional complexity of the operational amplifier, and nonideal components [7–9].

12.12 DESIGN EXAMPLES

Example 1 A phase-lock loop design shown in Fig. 12.18 requires an output frequency of 1600 MHz where the reference oscillator is 100 MHz. The design approach chosen is to use an inverting type 2 loop filter with a frequency divider and a mixer phase detector. The VCO chosen shows a typical tuning slope of 1 MHz/V. Measurement of the phase detector output shows a 100-mV peak to peak cosine wave. The filter is to have a 3-dB bandwidth of 100 kHz with a damping ratio of 1.

a. Using a 100-pF capacitor, find the remaining loop filter values.
b. Using a 10-kΩ R_{in}, find the remaining loop filter values.

Solution 1 From the graph in Fig. 12.13 the 3 dB frequency for a damping ratio $\zeta = 1$ is $f/f_n = 0.6435$ so that $f_n = 100\,\text{kHz}/0.6436 = 155.38\,\text{kHz}$ and $\omega_n = 976.26 \cdot 10^3\,\text{rad/s}$. The type 2 circuit for this example is shown in Fig. 12.16. The output frequency is 16 times the input frequency so $N = 16$. The value for K_{vco} is specified to be at 1 MHz/V. The phase detector output is a cosine wave. If the loop locks at 90° or 270°, the phase detector output voltage is zero. For a positive R_{in}, the slope is the first derivative evaluated at 270°, so that from Eq. (12.6)

$$K_{pd} = -\frac{K_m V_a V_b}{2} \sin \Delta\phi = -50\ \text{mV/rad}$$

The 50 mV/rad comes from the 100-mV peak to peak specification. The value for K/N is

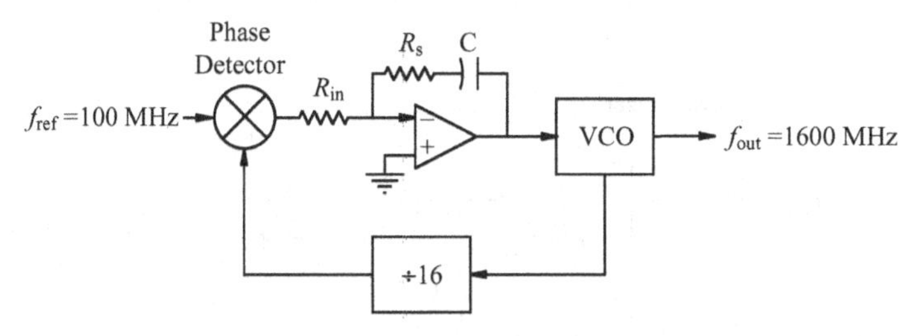

FIGURE 12.18 PLL for Example 1.

$$\frac{K}{N} = -\frac{50[\text{MV/rad}]\times 10^{6}[\text{Hz/v}]2\pi[\text{rad/Hz}]\times 10^{-3}[\text{v/mV}]}{16}$$
$$= -19.635\times 10^{3} \tag{12.81}$$

For part (a). where C is set at 100 pF, Eqs. (12.62) and (12.63) are to be used:

$$R_{\text{in}} = +\frac{19.635\times 10^{3}}{100\times 10^{-12}(976.26\times 10^{3})^{2}}$$
$$= 206\ \Omega \tag{12.82}$$

and

$$R_{\text{s}} = \frac{2\times 1}{(100\times 10^{-12})(9.7626\times 10^{5})}$$
$$= 20.48\ \text{k}\Omega \tag{12.83}$$

Example 2 A synthesizer design shown in Fig. 12.19 requires an output frequency from 900 to 920 MHz. The output frequency can be changed in 1 kHz steps by changing the divide ratio. Design a PLL using a synthesizer IC and an external VCO. The synthesizer IC data sheet lists the current mode phase detector output as 5 mA/rad. The VCO data sheet lists the tuning rate at 10 MHz/V.

Solution 2 The output frequency must be an integer multiple of the reference frequency so the reference frequency is 1 kHz. The circuit diagram is shown in Fig. 12.19. The divide ratio must change from 900 MHz/1 kHz or 9×10^5 to 920 MHz/1 kHz or 9.2×10^5. The midpoint value, 9.1×10^5 can be used for the design. A damping ratio of 1 is chosen for a rapid settling time when the divide ratio changes. The loop filter must attenuate the pulses from the phase detector output running at 1 kHz. Figure 12.13 shows that 14 dB of attenuation can be expected at 10 times the natural frequency. With a slope of −20 dB/decade, 34 dB attenuation can be expected at 100 times the natural frequency. Choosing $f_n = 10$ Hz or $\omega_n = 62.83$ rad/s will work with

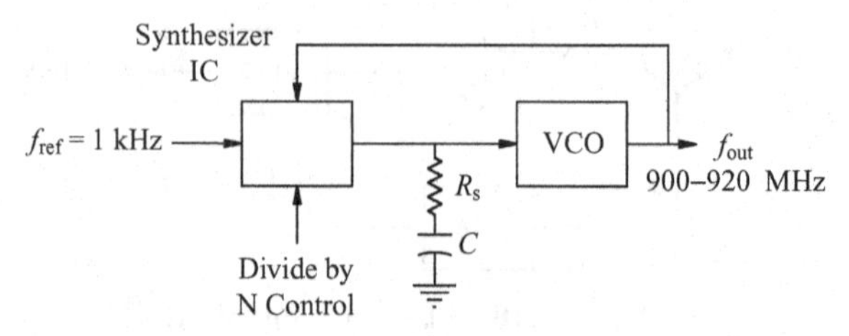

FIGURE 12.19 Synthesizer design for Example 2.

a 1-kHz reference frequency. Here the value of K is in dimensions of ampere/volt:

$$\frac{K}{N}=\frac{K_{\text{vco}}K_{\text{pd}}}{N}$$
$$=10\frac{\text{MHz}}{\text{volt}}\times 5\frac{\text{ma}}{\text{rad}}\times 2\pi\frac{\text{rad}}{\text{Hz}}\times 10^{-3}\frac{\text{A}}{\text{ma}}\times 10^{6}\frac{\text{Hz}}{\text{MHz}}$$
$$=3.1416\times 10^{5}\frac{\text{A}}{\text{V}}$$

The VCO is assumed to have a high input impedance so that the ratio of the voltage, V_e, seen at the input of the VCO to the input current i is

$$\frac{V_e}{i}=F(s)=R_2+\frac{1}{sC}$$

The phase transfer function is found from $G(s)$:

$$H(s)=\frac{G(s)}{1+G(s)}$$

where

$$G(s)=\frac{KF(s)}{s}=K\frac{R_2Cs+1}{s^2}$$

Thus,

$$H(s)=\frac{K(R_2Cs+1)}{s^2+sKCR_2+K} \tag{12.84}$$

From Eq. (12.84) $K=\omega_2$ and $2\zeta\omega_n=KCR_2$ so that

$$R_2=\frac{2\zeta}{\omega_n C}$$

which gives for $C=100\,\mu\text{F}$ and $\zeta=1$, $R_2=318\,\Omega$.

Example 3 A frequency synthesizer contains a phase-lock loop circuit that uses a type 1 noninverting filter. The data sheets for the phase detector show that the output waveform has a slope of 100 mV/rad. The VCO nominal output frequency is 3 GHz with a tuning rate of 100 MHz/V. The reference is a 100-MHz crystal oscillator. If $R_{in}=620\,\Omega$, $R_s=150\,\Omega$, $R_p=56\,k\Omega$, and $C=1\,\text{nF}$, what is the expected 3-dB bandwidth and damping ratio for this PLL?

Solution 3 With a 3-GHz output and a 100-MHz reference, the frequency divide ratio N must be 30:

$$
\begin{aligned}
K &= K_{\text{vco}} K_{\text{pd}} \\
&= 100 \frac{\text{MHz}}{\text{V}} \times 10^{8} \frac{\text{Hz}}{\text{V}} \times 10^{-3} \frac{\text{V}}{\text{V}} \times 2\pi \\
&= 6.2822 \times 10^{7}
\end{aligned}
$$

and

$$\frac{K}{N} = 2.0944 \times 10^{6}$$

For the noninverting loop, Eq. (12.66) gives the natural frequency:

$$\omega_{\text{n}} = 1.847 \times 10^{6} \text{ rad/s}$$

and from Eq. (12.67):

$$\zeta = 0.7091$$

Finally, the 3-dB frequency for the loop is found from Eq. (12.59):

$$\omega_3 = 1.8404 \times 10^{6} \text{ rad/s}$$

Thus, the 3-dB frequency is $f_3 = 292.9\,\text{kHz}$. This can be confirmed from the curve in Fig. 12.13. A linear interpolation for the 3-dB normalized frequency when $\zeta = 0.709$ is found from Fig. 12.13 to be approximately 1. The natural frequency for this PLL is approximately $\omega_{\text{n}}/2\pi = 293.7\,\text{kHz}$.

12.13 CONCLUSIONS

Chapter 1 began with a description of a generic superheterodyne transmitter–receiver pair. Each of the components in that block diagram were analyzed using a variety of circuits often requiring special RF design techniques. Passive circuit elements had to be chosen with care because of their stray inductance, parasitic capacitance, skin effect losses, and the like. A large amount of attention was given to filters and impedance transformers. Filters control the bandwidth and are necessary to avoid transmitting unwanted signals and receiving extraneous interference or noise. Impedance matching played a crucial role in determining the gain, stability, and noise figure of amplifiers as well as design of oscillators. The discussion of transmitter power amplifiers described the pros and cons of different types of amplifiers and provided analytical guidelines for their design. Low-phase noise oscillators supply the transmitter carrier frequency by exciting the LO port of a mixer. The modulator and demodulator scheme takes on a wide variety of forms in communication circuits, both analog and digital. The basic forms given in Chapter 11 provide the basic theory for mixers. Not found explicitly on the transceiver block diagram is the phase-lock loop described in this final chapter. Nevertheless, they are pervasive in RF designs and, for example, often are used to lock the frequency of the transmitter and receiver together. The journey does not conclude here, for there are a multitude of applications and designs that were left out, and others yet to be discovered. What was provided in this book is the background for further study and invention.

PROBLEMS

12.1. A phase-lock loop can be described in the frequency domain in terms of the input and output phase angles shown in Fig. 12.7. The input phase is $\tilde{\phi}_1(s) = a + b/s$. The filter transfer function is

$$F(s) = \frac{1 + sCR_2}{1 + sC(R_1 + R_2)}$$

a. What is the steady-state phase error?

b. What is the steady-state phase error if the capacitance $C = \infty$?

12.2. The circuit in Fig. 12.7 has an input signal with a phase that varies as $\tilde{\phi}_1(s) = (\omega - \omega_0)/s^3 + \phi_0/s^2$, where ω_0 and ϕ_0 are offset constant values. The filter has a transfer function given by $F(s) = 2 + 10/s^2$. Determine if this loop provides phase lock. If not, determine if it provides frequency lock.

REFERENCES

1. U. L. Rhode, *Microwave and Wireless Synthesizers, Theory and Design*, New York: Wiley, 1997.
2. J. A. Crawford, *Frequency Synthesizer Design Handbook*, Norwood, MA: Artech House, 1994.
3. U. L. Rhode and J. C. Whitaker, *Communications Receivers*, 3rd ed., New York: McGraw-Hill, 1996.
4. B. Razavi, *Monolithic Phase-Locked Loops and Clock Recovery Circuits*, New York: IEEE Press, 2001.
5. A. J. Viterbi, *Principles of Coherent Communication*, New York: McGraw-Hill, 1966.
6. F. M. Gardner, "Charge-Pump Phase-Lock Loops," *IEEE Trans. Commun.*, **28**, pp. 1849–1858, Nov. 1980.

7. W. Rhee, "Design of High-Performance CMOS Charge Pumps in Phase-Locked Loops," *Proc. of the 1999 IEEE International Symposium on Circuits and Systems*, Orlando, Florida, pp. 545–548, May 1999.
8. R. C. Chang and L-C. Kuo, "A New Low-Voltage Charge Pump Circuit for PLL," *Proc. Of the 2000 IEEE International Symposium on Circuits and Systems*, Geneva, Switzerland, pp. V-701–V-704, May 2000.
9. H. Yu, Y. Inoue, and Y. Han, "A New High-Speed Low-Voltage Charge Pump for PLL Applications," *2005 6th International Conference on ASIC*, Shanghai, China, pp. 387–390, Oct. 2005.

APPENDIX A

Example of a Solenoid Design

The design of a solenoid inductor was described in Chapter 2. An example for a 100-nH solenoid design where the Q is evaluated at 200 MHz is shown below using the program SOLENOID where the **bold** type indicates user inputs. After initial values for the solenoid are given, the form length and number of integral turns are modified, but the form diameter remains unchanged. The interwire capacitance, C, the self-resonant frequency, F_{res}, and the Q are calculated.

```
Frequency in Hz for Q calculation = 200.E6
Geometry -> Inductance, type <G>
Inductance -> n, the number of turns, type <L>

L
Desired inductance, L (H), initial form Length,
Form Diameter - inches, Initial value for no. turns
100.E-9, 0.5, 0.1, 20
Number of turns = 13.00000
AWG = 21.0 Wire Diameter = .284625E-01
Final Length = .576708E+00 inches
Pitch = .44362E-01 Wire Diameter = .28462E-01 in.
Turn Diameter = .12846E+00 in. Form Diameter = .10000E+00
in.
L = .10002E+00 µH
C = .203350E+00 pF Fres = .111600E+04 MHz
Q = .649760E+02 at F = .200000E+03 MHz
```

Radio Frequency Circuit Design, Second Edition, by W. Alan Davis

APPENDIX B

Analytical Spiral Inductor Model

Modeling equations for spiral inductors given in [1, 2] are collected here for convenience. A straight line of length l, width w, thickness t, resistivity ρ, and in a material of permeability μ_0 has the following inductance:

$$L = \frac{\mu_0 l}{2\pi}\left[\ln\left(\frac{2l}{\rho}\right) - 1\right] \tag{B.1}$$

where

$$\ln\left(\frac{\rho}{2c}\right) = -\frac{25}{12} - \frac{1}{6}\left[\left(\frac{w}{t}\right)^2 \ln\sqrt{1+\left(\frac{t}{w}\right)^2} + \left(\frac{t}{w}\right)^2 \ln\sqrt{1+\left(\frac{w}{t}\right)^2}\right] + \frac{2}{3}\left[\frac{w}{t}\arctan\left(\frac{t}{w}\right) + \frac{t}{w}\arctan\left(\frac{w}{t}\right)\right] \tag{B.2}$$

$$2c = \sqrt{w^2 + t^2} \tag{B.3}$$

If inhomogeneous current density across the conductor cross section is to be considered, an additional expression is found in [1].

The single-loop inductor is illustrated in Fig. B.1 where the total angular rotation, ϕ_0, is somewhat less than 360°. The inductance is found by numerical integration of the following equation:

$$L = \frac{\mu_0}{2\pi w^2}\int_0^{\phi_0/\sqrt{2}} \left(\phi_0\sqrt{2} - 2\phi\right)\cos\left(\sqrt{2}\phi\right) F(\phi)\, d\phi \tag{B.4}$$

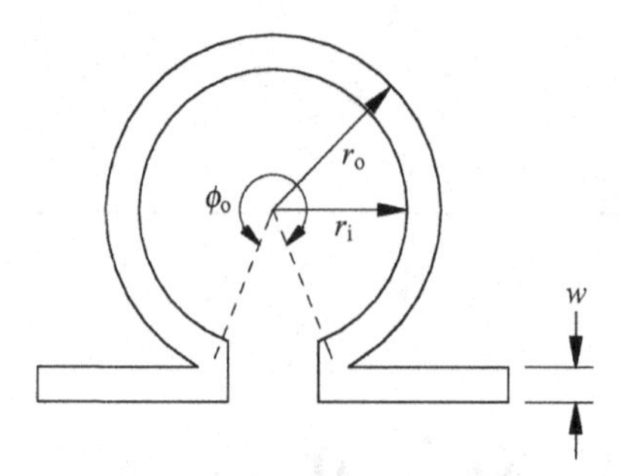

FIGURE B.1 Single-loop inductor.

where

$$F(\phi) = G(r_o, r_o) - G(r_i, r_o) - G(r_o, r_i) + G(r_i, r_i) \tag{B.5}$$

The values for r_1 and r_2 below are replaced by r_i and r_o as specified by Eq. (B.5):

$$G(r_1, r_2) = \tfrac{1}{3}R^3 + \tfrac{2}{3}r_1 r_2 R\cos\left(\sqrt{2}\phi\right) + \frac{2}{3}r_1^3\cos\left(\sqrt{2}\phi\right)\arcsin\left[\frac{r_2 - r_1\cos\left(\sqrt{2}\phi\right)}{r_1\left|\sin\left(\sqrt{2}\phi\right)\right|}\right] \tag{B.6}$$

$$R = \sqrt{r_1^2 + r_2^2 - 2r_1 r_2\cos\left(\sqrt{2}\phi\right)} \tag{B.7}$$

The inductance of a circular spiral with n turns (with air bridge) consists of n static inductances, L_i, $i = 1, \ldots ,n$ as found by Eq. (B.4) plus mutual inductance terms between the ith and jth line segments. This mutual inductance is

$$M_{ij} = \mu\sqrt{ab}\left[\left(\frac{2}{k_{ij}} - k_{ij}\right)K(k_{ij}) - \frac{2}{k_{ij}}E(k_{ij})\right] \tag{B.8}$$

where

$$k_{ij} = \frac{4ab}{(a+b)^2} \tag{B.9}$$

$$a = r_i + (i - 0.5)(w + s) \tag{B.10}$$

$$b = r_i + (j - 0.5)(w + s) \tag{B.11}$$

In this expression for the mutual inductance, r_i is the inner radius of the innermost turn of the circular spiral, w is the conductor width, and s is the spacing between turns. The outermost radius of the outermost turn is determined by these parameters together with the number of turns, n. The $K(k_{ij})$ and $E(k_{ij})$

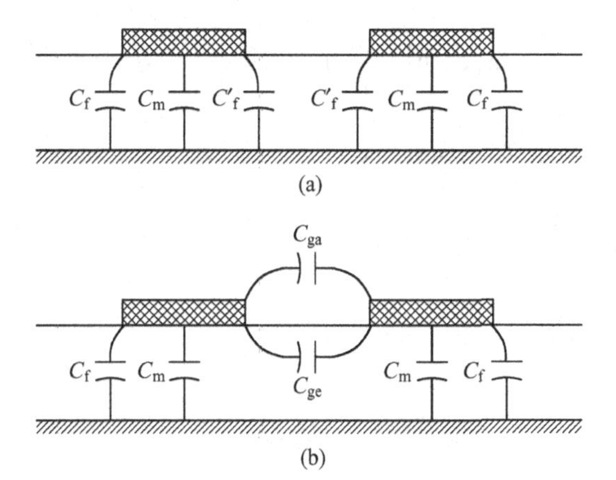

FIGURE B.2 Capacitances associated with coupled microstrip: (*a*) even mode and (*b*) odd mode.

are the complete elliptic integrals of the first and second kind, respectively. If there is a ground plane underneath the spiral conductor a distance h away, there is an additional mirrored mutual inductance, M_{ij}^{m}, given by Eq. (B.8) where

$$k_{ij} = \frac{4ab}{4h^2 + (a+b)^2} \tag{B.12}$$

and where a and b are given by Eqs. (B.10) and (B.11). The inductance of the multiple-turn circular spiral is then

$$L = \sum_{i=1}^{n} L_i + 2\sum_{i=1}^{n-1} \sum_{j=i+1}^{n} M_{ij} + \sum_{i=1}^{n} \sum_{j=1}^{n} M_{ij}^{m} \tag{B.13}$$

The associated capacitances are shown in Fig. B.2, which were given by [3–5]

$$C_{ga} = \frac{\varepsilon_0}{2} \frac{K(k')}{K(k)} \tag{B.14}$$

The arguments of the elliptic integrals are

$$k = \frac{s}{h}\left(\frac{s}{h} + \frac{2w}{h}\right) \tag{B.15}$$

$$k' = \sqrt{1-k^2} \tag{B.16}$$

The dielectric coupling capacitance is

$$C_{\mathrm{ge}} = \frac{\varepsilon_0 \varepsilon_{\mathrm{r}}}{\pi} \ln\left[\coth\left(\frac{\pi}{4}\frac{s}{h}\right)\right] + 0.65 C_{\mathrm{f}} \left[\frac{0.02}{s/h}\sqrt{\varepsilon_{\mathrm{r}}} + 1 - \frac{1}{\varepsilon_{\mathrm{r}}^2}\right] \tag{B.17}$$

The main capacitance to ground is

$$C_{\mathrm{m}} = \frac{\varepsilon_0 \varepsilon_{\mathrm{r}} w}{h} \tag{B.18}$$

The fringing capacitances are

$$C_{\mathrm{f}} = \frac{1}{2}\left[\frac{\sqrt{\varepsilon_{\mathrm{eff}}}}{c Z_0} - \frac{\varepsilon_0 \varepsilon_{\mathrm{r}} w}{h}\right] \tag{B.19}$$

$$C_{\mathrm{f}}' = \frac{C_{\mathrm{f}}}{1 + A(h/s)\tanh(8s/h)} \tag{B.20}$$

$$A = \exp\left[-0.1 \exp\left(2.33 - 2.53\frac{w}{h}\right)\right] \tag{B.21}$$

The capacitance, C_{f}, is the fringing capacitance of a single microstrip line of width w/h, characteristic impedance, Z_0, and effective dielectric constant $\varepsilon_{\mathrm{eff}}$, in which the velocity of light in a vacuum is c. The microstrip parameters can be calculated based on Section 4.7.4. Hence, the total even-mode capacitance is

$$C_{\mathrm{e}} = C_{\mathrm{m}} + C_{\mathrm{f}} + C_{\mathrm{f}}' \tag{B.22}$$

and the odd-mode capacitance is

$$C_{\mathrm{o}} = C_{\mathrm{m}} + C_{\mathrm{f}} + C_{\mathrm{ga}} + C_{\mathrm{ge}} \tag{B.23}$$

REFERENCES

1. E. Pettenkpaul, H. Kapusta, A. Weisgerber, H. Mampe, J. Luginsland, and I. Wolff, "CAD Models of Lumped Elements on GaAs up to 18 GHz," *IEEE Trans. Microwave Theory Tech.*, **36**, pp. 294–304, Feb. 1988.
2. C. Hentschel, "Die Analyse von Schaltungen mit Dünnfilmschichtspulen," *Arch Elek. Übertragung.*, **26**, pp. 319–328, 1972.
3. K. C. Gupta, R. Garg, and I. J. Bahl, *Microstrip Lines and Slotlines*, Norwood MA: Artech, Chapter 8, 1979.
4. R. Garg and I. J. Bahl, "Characteristics of Coupled Microstrips," *IEEE Trans. Microwave Theory Tech.*, **MTT-27**, pp. 700–705, July 1979.
5. R. Garg and I. J. Bahl, "Correction to 'Characteristics of Coupled Microstriplines," *IEEE Trans. Microwave Theory Tech.*, **MTT-28**, pp. 272, March 1980.

APPENDIX C

Double-Tuned Matching Circuit Example

Assume that an impedance transformation is required between a 50-Ω source and a 15-Ω load. The matching is to be done using the double-tuned matching circuit described in Chapter 3 using the program DBLTUNE. The center frequency is at 4 MHz, the bandwidth is 100 kHz, and the pass-band ripple is 0.5 dB. The capacitances and transformer parameters are to be determined. In the following computer output, the **bold** characters are the responses the program expects from the user. Furthermore, in this example, the verbose mode is chosen by choosing to display the intermediate results. An analysis of this circuit using SPICE is shown in Fig. C.1.

```
Display intermediate results? < Y/N > Y
Center Freq, Bandwidth (Hz) = ? 4.E6, 100.E3
Fm1 = .396480E+07 Fm2 = .403551E+07
GTMIN = .99992E+00
Passband ripple in dB = ? 0.5
Resistance Ratio r = .19841E+01
Q2_m1 = .97432E+00 Q2_m2 = .10097E+01
Generator and Load resistances values = 50., 15.
L2' = .56259E+02 μH C2' = .28140E+02pF
RL' = .79332E+05 Bm1 = .19480E-01 Bm2 = -.20193E-01
Given terminal resistances: RG = .500E+02 RL = .150E+02
Input Circuit: C1 = .446554E+05pF L11 = .354637E-01μH
Output Circuit: C2 = .148828E+06pF L22 = .106441E-01μH
Transformer coupling coefficient k = .250991E-01
```

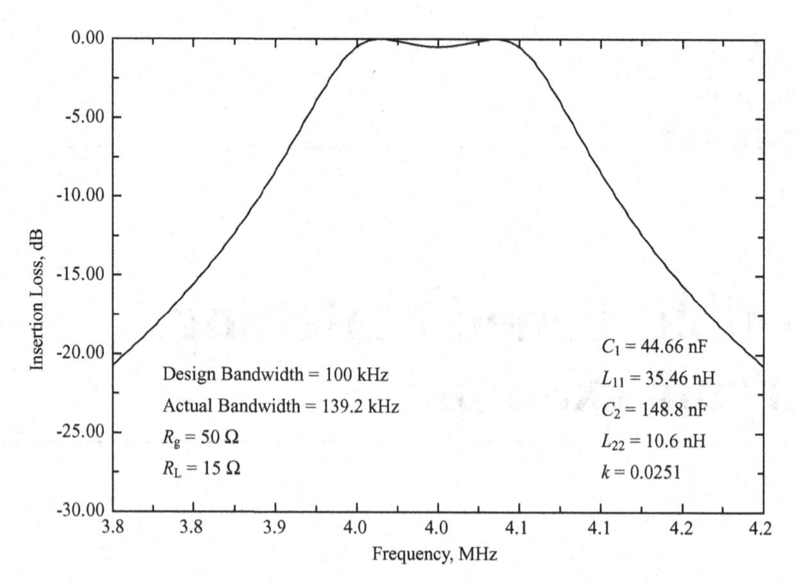

FIGURE C.1 Double-tuned matching circuit example.

APPENDIX D

Two-Port Parameter Conversion

Conversion between the z, y, h, and g two-port voltage–current parameters is simply rearrangement of two linear equations relating voltages and currents at the two ports. Converting between these and the S parameters requires relating the voltage waves to voltages and currents. This latter relationship always includes the characteristic impedance, Z_0, by which the S parameters are referenced. Typically, this value is $50\,\Omega$. Table D.1 shows this conversion. The program PARCONV is basically a code of many of the conversions in Table D.1.

The definitions of the various two-port parameters are described below. In each case, it is assumed that the current is flowing into the port terminal:

$$\begin{bmatrix} v_1 \\ v_2 \end{bmatrix} = \begin{bmatrix} z_{11} & z_{12} \\ z_{21} & z_{22} \end{bmatrix} \begin{bmatrix} i_1 \\ i_2 \end{bmatrix} \tag{D.1}$$

$$\begin{bmatrix} i_1 \\ i_2 \end{bmatrix} = \begin{bmatrix} y_{11} & y_{12} \\ y_{21} & y_{22} \end{bmatrix} \begin{bmatrix} v_1 \\ v_2 \end{bmatrix} \tag{D.2}$$

$$\begin{bmatrix} v_1 \\ i_2 \end{bmatrix} = \begin{bmatrix} h_{11} & h_{12} \\ h_{21} & h_{22} \end{bmatrix} \begin{bmatrix} i_1 \\ v_2 \end{bmatrix} \tag{D.3}$$

$$\begin{bmatrix} i_1 \\ v_2 \end{bmatrix} = \begin{bmatrix} g_{11} & g_{12} \\ g_{21} & g_{22} \end{bmatrix} \begin{bmatrix} v_1 \\ i_2 \end{bmatrix} \tag{D.4}$$

$$\begin{bmatrix} v_1 \\ i_1 \end{bmatrix} = \begin{bmatrix} A & B \\ C & D \end{bmatrix} \begin{bmatrix} v_2 \\ -i_2 \end{bmatrix} \tag{D.5}$$

Radio Frequency Circuit Design, Second Edition, by W. Alan Davis

TABLE D.1 *S*-Parameter Conversion Chart

	S	z	y	$ABCD$
S_{11}	S_{11}	$\frac{(z_{11}-Z_0)(z_{22}+Z_0)-z_{12}z_{21}}{(z_{11}+Z_0)(z_{22}+Z_0)-z_{12}z_{21}}$	$\frac{(Y_0-y_{11})(Y_0+y_{22})+y_{12}y_{21}}{(Y_0+y_{11})(Y_0+y_{22})-y_{12}y_{21}}$	$\frac{A+B/Z_0-CZ_0-D}{A+B/Z_0+CZ_0+D}$
S_{12}	S_{12}	$\frac{2z_{12}Z_0}{(z_{11}+Z_0)(z_{22}+Z_0)-z_{12}z_{21}}$	$\frac{-2y_{12}Y_0}{(Y_0+y_{11})(Y_0+y_{22})-y_{12}y_{21}}$	$\frac{2(AD-BC)}{A+B/Z_0+CZ_0+D}$
S_{21}	S_{21}	$\frac{2z_{12}Z_0}{(z_{11}+Z_0)(z_{22}+Z_0)-z_{12}z_{21}}$	$\frac{-2y_{12}Y_0}{(Y_0+y_{11})(Y_0+y_{22})-y_{12}y_{21}}$	$\frac{2}{A+B/Z_0+CZ_0+D}$
S_{22}	S_{22}	$\frac{(z_{11}+Z_0)(z_{22}-Z_0)-z_{12}z_{21}}{(z_{11}+Z_0)(z_{22}+Z_0)-z_{12}z_{21}}$	$\frac{(Y_0+y_{11})(Y_0-y_{22})+y_{12}y_{21}}{(Y_0+y_{11})(Y_0+y_{22})-y_{12}y_{21}}$	$\frac{-A+B/Z_0-CZ_0+D}{A+B/Z_0+CZ_0+D}$
z_{11}	$Z_0\frac{(1+S_{11})(1-S_{22})+S_{12}S_{21}}{(1-S_{11})(1-S_{22})-S_{12}S_{21}}$	z_{11}	$\frac{y_{22}}{y_{11}y_{22}-y_{12}y_{21}}$	$\frac{A}{C}$
z_{12}	$Z_0\frac{2S_{12}}{(1-S_{11})(1-S_{22})-S_{12}S_{21}}$	z_{12}	$\frac{-y_{12}}{y_{11}y_{22}-y_{12}y_{21}}$	$\frac{AD-BC}{C}$
z_{21}	$Z_0\frac{2S_{21}}{(1-S_{11})(1-S_{22})-S_{12}S_{21}}$	z_{21}	$\frac{-y_{21}}{y_{11}y_{22}-y_{12}y_{21}}$	$\frac{1}{C}$
z_{22}	$Z_0\frac{(1-S_{11})(1+S_{22})+S_{12}S_{21}}{(1-S_{11})(1-S_{22})-S_{12}S_{21}}$	z_{22}	$\frac{y_{11}}{y_{11}y_{22}-y_{12}y_{21}}$	$\frac{D}{C}$

y_{11}	$Y_0\frac{(1-S_{11})(1+S_{22})+S_{12}S_{21}}{(1+S_{11})(1+S_{22})-S_{12}S_{21}}$	$\frac{z_{22}}{z_{11}z_{22}-z_{12}z_{21}}$	y_{11}	$\frac{D}{B}$
y_{12}	$Y_0\frac{-2S_{12}}{(1+S_{11})(1+S_{22})-S_{12}S_{21}}$	$\frac{-z_{12}}{z_{11}z_{22}-z_{12}z_{21}}$	y_{12}	$\frac{BC-AD}{B}$
y_{21}	$Y_0\frac{-2S_{21}}{(1+S_{11})(1+S_{22})-S_{12}S_{21}}$	$\frac{-z_{21}}{z_{11}z_{22}-z_{12}z_{21}}$	y_{21}	$\frac{-1}{B}$
y_{22}	$Y_0\frac{(1+S_{11})(1-S_{22})+S_{12}S_{21}}{(1+S_{11})(1+S_{22})-S_{12}S_{21}}$	$\frac{z_{11}}{z_{11}z_{22}-z_{12}z_{21}}$	y_{22}	$\frac{A}{B}$
A	$\frac{(1+S_{11})(1-S_{22})+S_{12}S_{21}}{2S_{21}}$	$\frac{z_{11}}{z_{21}}$	$\frac{-y_{22}}{y_{21}}$	A
B	$Z_0\frac{(1+S_{11})(1+S_{22})-S_{12}S_{21}}{2S_{21}}$	$\frac{z_{11}z_{22}-z_{12}z_{21}}{z_{21}}$	$\frac{-1}{y_{21}}$	B
C	$\frac{1}{Z_0}\frac{(1-S_{11})(1-S_{22})-S_{12}S_{21}}{2S_{21}}$	$\frac{1}{z_{21}}$	$-\frac{y_{11}y_{22}-y_{12}y_{21}}{y_{21}}$	C
D	$\frac{(1-S_{11})(1+S_{22})+S_{12}S_{21}}{2S_{21}}$	$\frac{z_{22}}{z_{21}}$	$\frac{-y_{11}}{y_{21}}$	D

$$\begin{bmatrix} b_1 \\ b_2 \end{bmatrix} = \begin{bmatrix} S_{11} & S_{12} \\ S_{21} & S_{22} \end{bmatrix} \begin{bmatrix} a_1 \\ a_2 \end{bmatrix} \tag{D.6}$$

For conversion to and from S parameters for circuits with more than two ports, the following formulas may be used [1]. Each variable is understood to be a matrix representing the S, z, or y parameters. The conversion formulas are

$$S = F(Z - G^*)(Z + G)^{-1}F^{-1} \tag{D.7}$$

$$Z = F^{-1}(I - S)^{-1}(SG + G^*)F \tag{D.8}$$

$$S = F(I - G^*Y)(I + GY)^{-1}F^{-1} \tag{D.9}$$

$$Y = F^{-1}G^{-1}(I + S)^{-1}(I - S)F \tag{D.10}$$

where

$$F = \begin{bmatrix} \frac{1}{2\sqrt{Z_{01}}} & 0 & \cdots & 0 \\ 0 & \frac{1}{2\sqrt{Z_{02}}} & \cdots & 0 \\ \vdots & \vdots & \ddots & \vdots \\ 0 & 0 & \cdots & \frac{1}{2\sqrt{Z_{0n}}} \end{bmatrix} \tag{D.11}$$

and

$$G = \begin{bmatrix} Z_{01} & 0 & \cdots & 0 \\ 0 & Z_{02} & \cdots & 0 \\ \vdots & \vdots & \ddots & \vdots \\ 0 & 0 & \cdots & Z_{0n} \end{bmatrix} \tag{D.12}$$

The I in Eqs. (D.8) through (D.10) is the square identity matrix, and the Z_{0i}, $i = 1 \dots n$, are the characteristic impedances associated with each of the ports. An example of the usage of PARCONV is shown below. In using the program, make sure to include the decimals with the input data. Boldface values represent user inputs to the program. To exit the program use Ctrl. C.

```
TYPE SOURCE AND LOAD REFERENCE IMPEDANCE Z01,Z02 =
50., 50.Y --> S = YS OR S --> Y = SY
Z --> S = ZS OR S --> Z = SZ
ABCD --> S = AS OR S --> ABCD = SA
H --> S = HS OR S --> H = SH
H --> Z = HZ OR Z --> H =ZH
SY
INPUT S11, MAG. AND PHASE (deg)
.9, -80.
```

TABLE D.2 S-Parameter to Hybrid Parameter Conversion Chart

	S	h
S_{11}	S_{11}	$\frac{(h_{11}-Z_0)(h_{22}Z_0+1)-h_{12}h_{21}Z_0}{(h_{11}+Z_0)(h_{22}Z_0+1)-h_{12}h_{21}Z_0}$
S_{12}	S_{12}	$\frac{2h_{12}Z_0}{(h_{11}+Z_0)(h_{22}Z_0+1)-h_{12}h_{21}Z_0}$
S_{21}	S_{21}	$\frac{-2h_{12}Z_0}{(h_{11}+Z_0)(h_{22}Z_0+1)-h_{12}h_{21}Z_0}$
S_{22}	S_{22}	$\frac{(h_{11}+Z_0)(1-h_{22}Z_0)+h_{12}h_{21}Z_0}{(h_{11}+Z_0)(h_{22}Z_0+1)-h_{12}h_{21}Z_0}$
h_{11}	$Z_0\frac{(1+S_{11})(1+S_{22})-S_{12}S_{21}}{(1-S_{11})(1+S_{S22})+S_{12}S_{21}}$	h_{11}
h_{12}	$\frac{2S_{12}}{(1-S_{11})(1+S_{S22})+S_{12}S_{21}}$	h_{12}
h_{21}	$\frac{-2S_{21}}{(1-S_{11})(1+S_{S22})+S_{12}S_{21}}$	h_{21}
h_{22}	$\frac{1}{Z_0}\frac{(1-S_{11})(1-S_{22})-S_{12}S_{21}}{(1-S_{11})(1+S_{S22})+S_{12}S_{21}}$	h_{22}

```
INPUT S21, MAG. AND PHASE (deg)
1.9, 112.
INPUT S12, MAG. AND PHASE (deg)
0.043, 48.
INPUT S22, MAG. AND PHASE (deg)
0.7, -70.
Y(1,1) = .162912E-02 J .156482E-01
Y(1,2) = .304363E-03 J -.759390E-03
Y(2,1) = .360540E-01 J -.262179E-02
Y(2,2) = .483468E-02 J .123116E-01
Y --> S = YS OR S --> Y = SY
Z --> S = ZS OR S --> Z = SZ
ABCD --> S = AS OR S --> ABCD = SA
H --> S = HS OR S --> H = SH
H --> Z = HZ OR Z --> H =ZH
```

Table D.2 provides a direct conversion between two-port S parameters and two-port h parameters. This can be convenient with transistor models that are given in terms of h parameters.

REFERENCE

1. K. Kurokawa, "Power Waves and the Scattering Matrix," *IEEE Trans. Microwave Theory Tech.*, **MTT-11**, pp. 194–202, March 1965.

APPENDIX E

Termination of a Transistor Port with a Load

In the three-port circuit in Fig. E.1 one of the three ports is terminated with an impedance that has a reflection coefficient relative to the reference impedance Z_{ref}:

$$r_i = \frac{Z_i - Z_{\text{ref}}}{Z_i + Z_{\text{ref}}} \tag{E.1}$$

In this expression the subscript i represents s, g, or d, depending on whether the device connection is common source, gate, or drain terminated with Z_s, Z_g, or Z_d. For example, $r_s = a_2/b_2$ in Fig. E.1 or $b_2 = a_2/r_s$. This is substituted in the appropriate place in the following equations. The incident and scattered waves from the three-port circuit are

$$b_1 = S_{11}a_1 + S_{12}a_2 + S_{13}a_3 \tag{E.2}$$

$$b_2 = S_{21}a_1 + S_{22}a_2 + S_{23}a_3 \tag{E.3}$$

$$b_3 = S_{31}a_1 + S_{32}a_2 + S_{33}a_3 \tag{E.4}$$

When one of the ports is terminated with r_i, then the circuit really is a two port. The scattering parameters for the common source, gate, and drain connection follow:

Radio Frequency Circuit Design, Second Edition, by W. Alan Davis

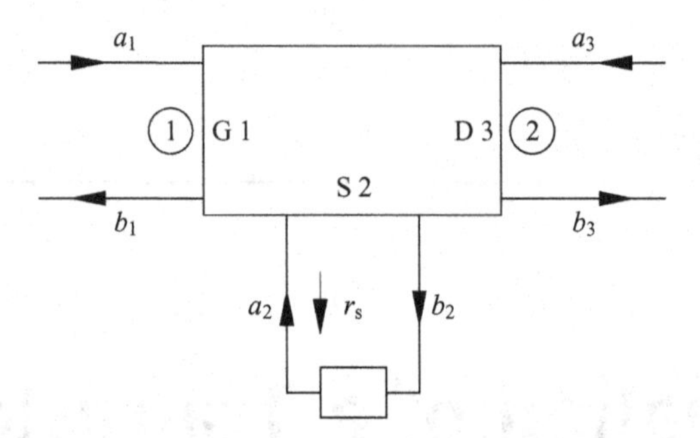

FIGURE E.1 Three-port with source terminated with r_s.

Common Source

$$S_{11s} = S_{11} + \frac{S_{12}S_{21}}{1/r_s - S_{22}} \tag{E.5}$$

$$S_{12s} = S_{13} + \frac{S_{12}S_{23}}{1/r_s - S_{22}} \tag{E.6}$$

$$S_{21s} = S_{31} + \frac{S_{32}S_{21}}{1/r_s - S_{22}} \tag{E.7}$$

$$S_{22s} = S_{33} + \frac{S_{23}S_{32}}{1/r_s - S_{22}} \tag{E.8}$$

Common Gate

$$S_{11g} = S_{22} + \frac{S_{12}S_{21}}{1/r_g - S_{11}} \tag{E.9}$$

$$S_{12g} = S_{23} + \frac{S_{21}S_{13}}{1/r_g - S_{11}} \tag{E.10}$$

$$S_{21g} = S_{32} + \frac{S_{31}S_{12}}{1/r_g - S_{11}} \tag{E.11}$$

$$S_{22g} = S_{33} + \frac{S_{31}S_{13}}{1/r_g - S_{11}} \tag{E.12}$$

Common Drain

$$S_{11d} = S_{11} + \frac{S_{13}S_{31}}{1/r_d + S_{33}} \tag{E.13}$$

$$S_{12d} = S_{12} + \frac{S_{13}S_{32}}{1/r_d - S_{33}} \tag{E.14}$$

$$S_{21d} = S_{21} + \frac{S_{23}S_{31}}{1/r_d - S_{33}} \tag{E.15}$$

$$S_{22d} = S_{22} + \frac{S_{23}S_{32}}{1/r_d - S_{33}} \tag{E.16}$$

A numerical example illustrates the process. A transistor with a set of common source S parameters at 2 GHz is given below:

$$S_{11} = 0.136 \angle 86$$

$$S_{21} = 3.025 \angle 6$$

$$S_{12} = 0.085 \angle -164$$

$$S_{22} = 0.304 \angle -136$$

These are then converted to two-port y parameters. These will be called y_{11}, y_{31}, y_{13}, and y_{33}. The indefinite admittance matrix is formed by adding a third row and column such that the sum of each row and the sum of each column is zero. The resulting 3×3 set of y parameters are obtained:

$$y_{11} = 9.681 \times 10^{-3} - 7.695 \times 10^{-3}$$

$$y_{12} = -2.77 \times 10^{-3} + 6.776 \times 10^{-3}$$

$$y_{13} = -3.086 \times 10^{-3} + .9194 \times 10^{-3}$$

$$y_{21} = 104.2 \times 10^{-3} + 20.85 \times 10^{-3}$$

$$y_{22} = -82.89 \times 10^{-3} + 14.39 \times 10^{-3}$$

$$y_{23} = -21.28 \times 10^{-3} + 6.452 \times 10^{-3}$$

$$y_{31} = -113.8 \times 10^{-3} + 13.15 \times 10^{-3}$$

$$y_{32} = -95.65 \times 10^{-3} + 7.618 \times 10^{-3}$$

$$y_{33} = -18.19 \times 10^{-3} + 5.533 \times 10^{-3}$$

These are then converted to three-port S parameters using Eq. (10.32) as originally given in [1]:

$$S_{11} = 1.6718 \angle -168.12°$$

$$S_{12} = 1.6573 \angle 3.639°$$

$$S_{13} = 1.0103 \angle 13.684°$$

$$S_{21} = 3.1794 \angle -157.77°$$

$$S_{22} = 2.0959 \angle 14.185^{\circ}$$

$$S_{23} = 0.7156 \angle 74.511^{\circ}$$

$$S_{31} = 1.6455 \angle 70.181^{\circ}$$

$$S_{32} = 2.7564 \angle -167.02^{\circ}$$

$$S_{33} = 2.1085 \angle -153.87^{\circ}$$

At this point it is desired to transform these parameters to common gate parameters in which the gate is connected to ground through a short circuit. The resulting common gate two-port S parameters are found from Eqs. (E.9) through (E.12):

$$S_{11g} = 5.317 \angle 170.925^{\circ}$$

$$S_{21g} = 10.772 \angle -14.852^{\circ}$$

$$S_{12g} = 2.496 \angle 177.466^{\circ}$$

$$S_{22g} = 6.250 \angle -7.553^{\circ}$$

With the transistor now characterized in the orientation that it is to be used in the oscillator, a choice is made for the impedance at the generator side. If this impedance is chosen to be a 5-nH inductor at 2 GHz, the output reflection coefficient is

$$\Gamma_{o} = 1.7775 \angle -30.35^{\circ}$$

This shows that oscillation is possible under these loading conditions. The above expressions for the revised S parameters can be found in [2] using slightly different notation.

REFERENCES

1. K. Kurokawa, "Power Waves and the Scattering Matrix," *IEEE Trans. Microwave Theory Tech.*, **13**, pp. 194–202, March 1965.
2. R. M. Dougherty, "Feedback Analysis and Design Techniques," *Microwave J.*, pp. 133–150, April 1985.

APPENDIX F

Transistor and Amplifier Formulas

The following formulas are meant as a reminder of the fundamentals given in most standard electronics textbooks. Notation for the formulas have the traditional meanings. Depletion capacitances are all given with a negative sign in the denominator as in $C = C_0/(1 - V/\phi)^{\gamma}$. Consequently, when the junction is reverse biased, the minus sign turns into a positive sign. Figure F.1 presents the basic FET features and symbols. The transistor body terminal typically is connected to the source or to ground. When the source is not grounded, the body effect modifies the transistor properties. The equations using the Early voltage, V_A, should conform with the sign convention used by SPICE ($V_A > 0$).

BIPOLAR TRANSISTOR PARAMETERS (BJT)

Description	Formula
Collector current	$I_C = I_S \exp\left(\dfrac{qV_{BE}}{kT}\right)$
Transconductance	$g_m = \dfrac{qI_C}{kT}$
Input resistance	$r_\pi = \dfrac{\beta_0}{g_m}$

(continued)

Radio Frequency Circuit Design, Second Edition, by W. Alan Davis

(*continued*)

Description	Formula
Output resistance	$r_o = \dfrac{V_A}{I_C}$
Base charging capacitance	$C_D = \tau F_{gm}$
Emitter–base junction	$C_{je} = A_E\left(\dfrac{q\varepsilon N_B}{V_j}\right)^{1/3}$
Input capacitance	$C_\pi = C_b + C_{je}$
Collector base	$C_\mu = \dfrac{C_{\mu o}}{(1 - V_{BC}/\psi_{oc})^{1/3}}$
Collector substrate	$C_{cs} = \dfrac{C_{CSO}}{(1 - V_{sc}/\psi_{os})^{1/2}}$
Transition frequency	$f_T = \dfrac{1}{2\pi}\dfrac{g_m}{(C_\pi + C_\mu)}$
Thermal voltage	$V_T = \dfrac{kT}{q} = 0.0259\text{ V}$

JUNCTION FIELD-EFFECT TRANSISTOR PARAMETERS (JFET)

Description	Formula
Saturated drain current	$I_D = I_{DSS}\left(1 - \dfrac{V_{GS}}{V_P}\right)^2\left(1 + \dfrac{V_{GS}}{V_A}\right) \quad V_A > 0$
	$V_{DS} \geq V_{GS} - V_P$
Ohmic region drain current	$I_D = G_o\left[V_{DS} + \dfrac{3}{2}\dfrac{(\psi_0 + V_{GS} - V_{DS})^{3/2} - (\psi_0 + V_{GS})^{3/2}}{(\psi_0 + V_P)^{1/2}}\right]$
	$V_{DS} < V_{GS} - V_P$
	$G_o = \dfrac{2aW}{L}\sigma_c$
	$I_D \approx K[2(V_{GS} - V_P)V_{DS} - V_{DS}^2]$
Transconductance	$g_m = \dfrac{2I_{DSS}}{V_P}\left(1 - \dfrac{V_{GS}}{V_P}\right)$

(*continued*)

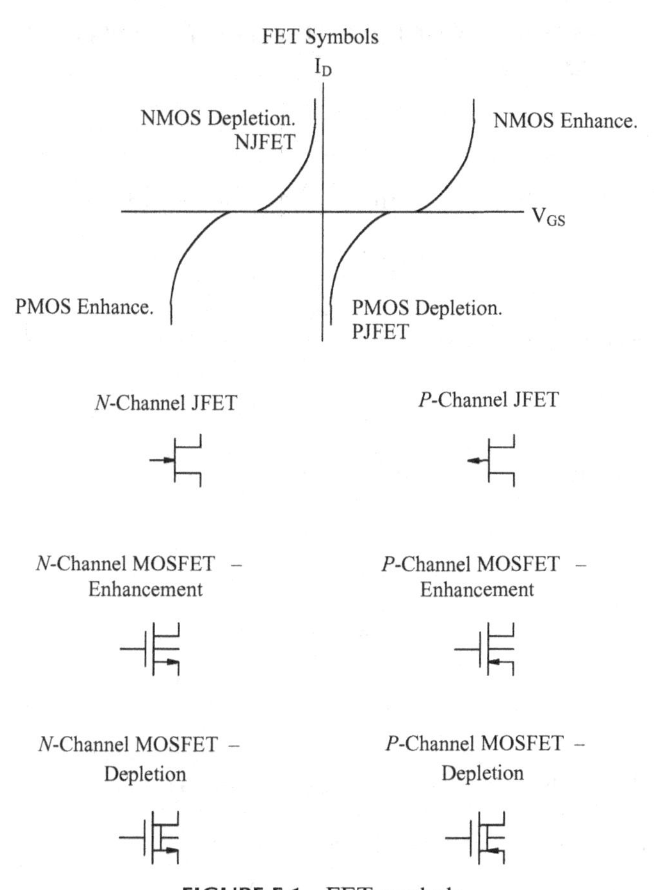

FIGURE F.1 FET symbols.

(*continued*)

Description	Formula
Output resistance	$r_o = \dfrac{V_A}{I_D}$
Gate–source capacitance	$C_{gs} = \dfrac{C_{gs0}}{(1 - V_{GS}/\psi_0)^{1/3}}$
Gate–drain capacitance	$C_{gd} = \dfrac{C_{gd0}}{(1 - V_{GD}/\psi_0)^{1/3}}$
Gate–substrate capacitance	$C_{gss} = \dfrac{C_{gss0}}{(1 - V_{GSS}/\psi_0)^{1/2}}$
n-Channel JFET	$V_P < 0$
p-Channel JFET	$V_P > 0$

METAL–OXIDE SEMICONDUCTOR FIELD-EFFECT TRANSISTOR (MOSFET) PARAMETERS

Description	Formula
Saturation region drain current	$I_D = \dfrac{\mu C_{ox} W}{2L}(V_{GS} - V_t)^2\left(1 + \dfrac{V_{DS}}{V_A}\right) \quad V_A > 0$
	$V_{DS} \geq V_{GS} - V_t$
Ohmic region drain current	$I_D = \dfrac{\mu C_{ox} W}{2L}[2(V_{GS} - V_t)V_{DS} - V_{DS}^2] \times \left(1 + \dfrac{V_{DS}}{V_A}\right)$
	$V_{DS} < V_{GS} - V_t$
Oxide capacitance	$C_{ox} = \dfrac{\varepsilon_{ox}}{t_{ox}}$
Transconductance	$g_m = \mu C_{ox} \dfrac{W}{L}(V_{GS} - V_t)$
Output resistance	$r_o = \dfrac{\lvert V_A \rvert}{I_{D0}}$
Input capacitance	$C_{in} = C_{GS} + C_{GD} = C_{ox} LW$
Transition frequency	$f_c = \dfrac{g_m}{2\pi C_{in}} = \dfrac{\mu_s(V_{GS} - V_t)}{2\pi L^2}$
Surface mobility holes	$\mu_s = 200\,\text{cm}^2/\text{V–s}$
Surface mobility electrons	$\mu_s = 450\,\text{cm}^2/\text{V–s}$

n-Channel JFET	*p*-Channel JFET
$I_{DSS} > 0$	$I_{DSS} < 0$
$V_P < 0$	$V_P > 0$
$g_{ms} = \dfrac{-2I_{DSS}}{V_P}$	$g_{ms} = \dfrac{-2I_{DSS}}{V_P}$
$K \triangleq \dfrac{I_{DSS}}{V_P^2} > 0$	$K \triangleq \dfrac{I_{DSS}}{V_P^2} < 0$
$V_P < V_{GS}$ for $\lvert I_{DS} \rvert > 0$	$V_{GS} < V_P$ for $\lvert I_{DS} \rvert > 0$

(continued)

(*continued*)

n-Channel JFET	*p*-Channel JFET
NMOS Enhancement	PMOS Enhancement
$V_t > 0$	$V_t < 0$
$V_{GS} > V_t$	$V_{GS} < V_t$
$K \triangleq \frac{\mu_n C_{ox} W}{2L} > 0$	$K \triangleq \frac{\mu_p C_{ox} W}{2L} < 0$
NMOS Depletion	PMOS Depletion
$V_t < 0$	$V_t > 0$
$V_{GS} > V_t < 0$ for $\lvert I_{DS}\rvert > 0$	$V_{GS} < V_t$ for $\lvert I_{DS}\rvert$
$K \triangleq \frac{\mu_n C_{ox} W}{2L} > 0$	$K \triangleq \frac{\mu_p C_{ox} W}{2L} < 0$

SMALL-SIGNAL SINGLE-TRANSISTOR AMPLIFIER CONFIGURATIONS

MOSFET	BJT
Common Source	Common Emitter
$R_{in} = R_B = R_1 \| R_2$	$R_{in} = (r_\pi + r_b) \| R_B \approx r_\pi$
$R_{out} = R_D \| r_o$	$R_{out} = R_c \| r_o$
$A_V = -g_m(r_0 \| R_D \| R_L) \propto \frac{1}{\sqrt{I_D}}$	$A_V = -g_m (R_c \| r_o \| R_L)$
Source Degeneration	Emitter Degeneration
$R_{in} = R_B = R_1 \| R_2$	$R_{in} = R_B \| [r_\pi + R_E (\beta + 1)]$ $\approx r_\pi (1 + g_m R_E)$
$R_{out} = r_o[1 + (g_m + g_{mb})R_S] + R_S$	$R_{out} = R_E \| r_\pi + r_o [1 + g_m (r_\pi \| R_E)]$ $\approx r_o (1 + g_m R_E)$
$G_m = \frac{g_m}{1 + (g_m + g_{mb})R_S + R_S/r_o}$	$G_m = \frac{g_m}{1 + g_m R_E(1 + 1/\beta)}$

(*continued*)

(*continued*)

MOSFET	BJT
Common Gate	Common Base
$R_{in} = \dfrac{r_0 + R_D \| R_L}{1 + (g_m + g_{mb}) r_0} \approx \dfrac{1}{g_m + g_{mb}} + \dfrac{R_D \| R_L}{(g_m + g_{mb}) r_0}$	$R_{in} = \dfrac{r_0 + R_C \| R_L}{1 + g_m [R_C \| R_L + (\beta_0 + 1) r_0] / \beta_0} \approx \dfrac{\alpha_0}{g_m} + \dfrac{\alpha_0 R_C \| R_L}{g_m r_0}$
$R_{out} = R_D \| [r_o + R_{gen} + R_{gen} r_o (g_m + g_{mb})]$	$R_{out} = R_C \left\| \left[\dfrac{r_0 + R_{gen}(1 + r_0 g_m / \alpha_0)}{1 + R_{gen}/r_\pi} \right] \right.$
$G_m = g_m + g_{mb}$	$G_m = g_m \left(\dfrac{1}{1 + r_b / r_\pi} \right) \approx g_m$
	$A_V = g_m (R_C \| R_L)$
	$A_I = \dfrac{g_m R_C}{R_C + R_L} \dfrac{R_{gen}}{1 + g_m R_{gen}} \approx g_m r_e = \alpha_0$
Common Drain (Source Follower)	Common Collector (Emitter Follower)
$R_{in} = R_1 \| R_2$	$R_{in} = R_B \| [r_\pi + r_b + (\beta + 1)(r_o \| R_E)]$
$R_{out} = \dfrac{r_0}{1 + r_0 (g_m + g_{mb})} \approx \dfrac{1}{g_m + g_{mb}}$	$R_{out} = \dfrac{r_\pi + R_{gen} + r_b}{1 + \beta} \approx \dfrac{1}{g_m} + \dfrac{R_{gen} + r_b}{1 + \beta}$
$A_V = \dfrac{g_m r_0}{1 + r_0 (g_m + g_{mb}) + r_0 / R_L} \approx \dfrac{1}{1 + g_{mb}/g_m} \approx 1$	$A_V = \dfrac{1}{1 + [(R_{gen} + r_b + r_\pi)]/(R_E \| r_0)(\beta + 1)} \approx 1$

APPENDIX G

Transformed Frequency-Domain Measurements Using SPICE

G.1 INTRODUCTION

Time-domain measurements taken on an automatic network analyzer can be easily replicated theoretically for dispersionless transmission lines using the SPICE time-domain program simulation. A technique is presented here that gives the type, value, and position of a measured discontinuity directly from time-domain measurements. Furthermore, the effect of multiple discontinuities can be replicated by SPICE.

Microwave impedance measurements have evolved from the slotted line to the computer-controlled automatic network analyzer. One of the major innovations of the network analyzer made possible by its broadband frequency capability was the simulation of time-domain measurements. Though fundamentally a frequency-domain machine, the network analyzer has been very useful in doing time-domain analysis of broadband circuits. This was demonstrated by the pioneering work of Hines and Stinehelfer [1]. Fundamentally, time-domain equipment such as the time-domain reflectometer have lagged because of the requirement for fast switching devices. However, that has not been true of the software side, as the widespread use of SPICE can readily attest. Here, it will be demonstrated how time-domain measurements on the network analyzer can be simulated using the transient analysis in SPICE.

There have been several studies that have shown how S parameters can be plotted using SPICE [2–4]. SPICE can be used to model physical directional

Radio Frequency Circuit Design, Second Edition, by W. Alan Davis

couplers by means of interconnection of three transmission lines [5]. By this means, a SPICE model was developed for the characteristics of a network analyzer in the frequency domain. However, the above-mentioned applications are not using the time-domain capability of SPICE.

First of all, the SPICE *S* parameter simulation can be expanded to include unequal input and output impedance levels. This is useful in analyzing impedance steps and transformers with two different resistance levels. Second, rather than using an ac steady-state voltage source, a time-domain pulse can be used. The automatic network analyzer is made to simulate a time-domain reflectometer by mathematically producing an "impulse" that can be replicated in SPICE approximately by using the PULSE function or more accurately by the piecewise linear (PWL) transient function. Rather than use an ideal impulse, the SPICE analysis uses the "impulse"-like waveform produced by the network analyzer. This SPICE replica of the actual network analyzer impulse can be used to show how various circuit elements react to time-domain measurements. Simple formulas are given that can be used to obtain the value of the inductance, capacitance, or impedance step directly from time-domain impulse data.

G.2 FREQUENCY-DOMAIN *S* PARAMETERS

In [4] the *S* parameters were obtained for a given circuit from SPICE for a two-port circuit in which the input and output resistances were both $50\,\Omega$. However, the conversion circuit used to obtain the *S* parameters can be modified so that the two ports of the circuit are at different impedance levels (Fig. G.1). This is done by the addition of an ideal transformer whose turns ratio is

$$n = \sqrt{\frac{R_{02}}{R_{01}}} \tag{G.1}$$

where R_{01} and R_{02} are the reference characteristic impedance levels of the input and output sides, respectively. The input impedance, Z_1, looking into the transformer from the port 1 side is

$$Z_1 = \frac{Z_L}{n^2} \tag{G.2}$$

which is equal to R_{01} when $Z_L = R_{02}$. The voltage drop caused by the independent current source of value $1/R_{01}$ is

$$V_1 = \frac{1}{R_{01}}(R_{01} \| Z_1) = \frac{Z_1}{R_{01} + Z_1} \tag{G.3}$$

The voltage at node 11 in Fig. G.1 is numerically the same as S_{11} since

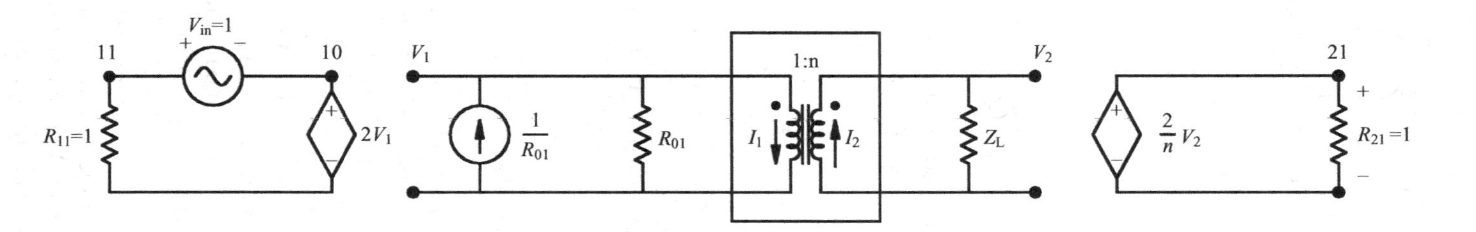

FIGURE G.1 SPICE circuit for finding S parameters with different input and output impedance levels.

$$V(11) = 2V_1 - V_{\text{in}} = \frac{Z_1 - R_{01}}{Z_1 + R_{01}} \tag{G.4}$$

For the output side, the secondary current in the transformer is $I_2 = I_1 / n$. The portion of current going into the transformer primary from the current source side is

$$I_1 = \frac{1}{R_{01}}(R_{01} \| Z_1)\frac{1}{Z_1} = \frac{1}{Z_1 + R_{01}} \tag{G.5}$$

The voltage at node 2 is

$$V_2 = I_2 Z_{\text{L}} = \frac{I_1}{n} Z_{\text{L}} = \frac{Z_{\text{L}}}{(Z_1 + R_{01})n} \tag{G.6}$$

The voltage across R_{21} with the marked polarity is $V(21) = 2V_2/n$, which upon replacing V_2 with the above value gives a value for $V(21)$ that is numerically equal to S_{21}:

$$V(21) = \frac{2Z_{\text{L}}}{(Z_1 + R_{01})n^2} = \frac{2Z_{\text{L}}}{Z_{\text{L}} + R_{02}} = S_{21} \tag{G.7}$$

The right-hand side is obtained using Eqs. (G.1) and (G.2). For a matched load when $Z_{\text{L}} = R_{02}$, $S_{21} = 1$ as expected. Finding the S_{22} and S_{12} for a circuit is achieved by direct analogy. A suggested SPICE listing for finding the S parameters with unequal source and load impedance levels is shown in Section G.6.

G.3 TIME-DOMAIN REFLECTOMETRY ANALYSIS

Time-domain reflectometer (TDR) measurements from an automatic network analyzer can be directly compared with a theoretical circuit model in the time domain in SPICE. All that is required is to make the two modifications described in Section G.7. Of course, a near ideal impulse can be implemented in SPICE, but the point of view taken here is to calculate time-domain data that would actually be measured using the time-domain feature on a network analyzer. Two actual time-domain impulses were measured on a network analyzer under the conditions in which the maximum frequency was 18 and 26 GHz, respectively. A third impulse with a maximum frequency of 50 GHz was calculated from the precursor of Agilent's Advanced Design System (ADS) called Microwave Development System (MDS). This was justified on the basis that (1) the program and the network analyzer use the same algorithm for the chirp-Z transform, and (2) the measured and calculated 18-GHz impulses were nearly identical. Circuit responses to impulses with different

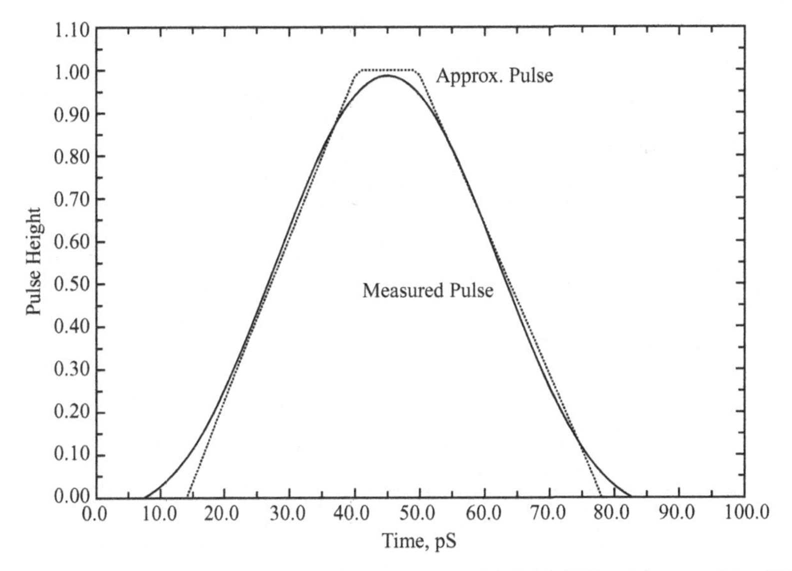

FIGURE G.2 Measured and approximate trapezoidal 26-GHz pulse used in SPICE analysis.

maximum frequency content will of course differ, so the time -domain response must be always coupled to the frequencies used in producing the impulse.

Two options for representing the network analyzer impulse are illustrated in Fig. G.2. The approximate impulse is modeled in SPICE as a PULSE, while the more accurate approach is achieved by using the PWL SPICE function. In the present case, the PWL function uses 77 different (x,y) coordinate pairs to represent the 26-GHz impulse. The result is indistinguishable from the measured impulse within the resolution of the graph in Fig. G.2. A similar fit was made for the 18- and 50-GHz impulses. The piecewise linear fit of the network analyzer impulses for use in SPICE are found in Section G.7. The approximate trapezoidal pulse for a 50-Ω source impedance has the advantage of providing results similar to the PWL approximation with a lot less data entry. However, to closely represent the actual time-domain response of the network analyzer, the PWL approach should be used.

G.4 TIME-DOMAIN IDENTIFICATION OF CIRCUIT ELEMENTS

Time-domain analysis of discontinuities in broadband circuits enables determining the type of circuit element, the position of the circuit element, and the size of the circuit element causing the discontinuity. For example, if a shunt capacitance is causing the discontinuity, the time-domain impulse reflection first goes below the baseline, then rises above the baseline, and then settles

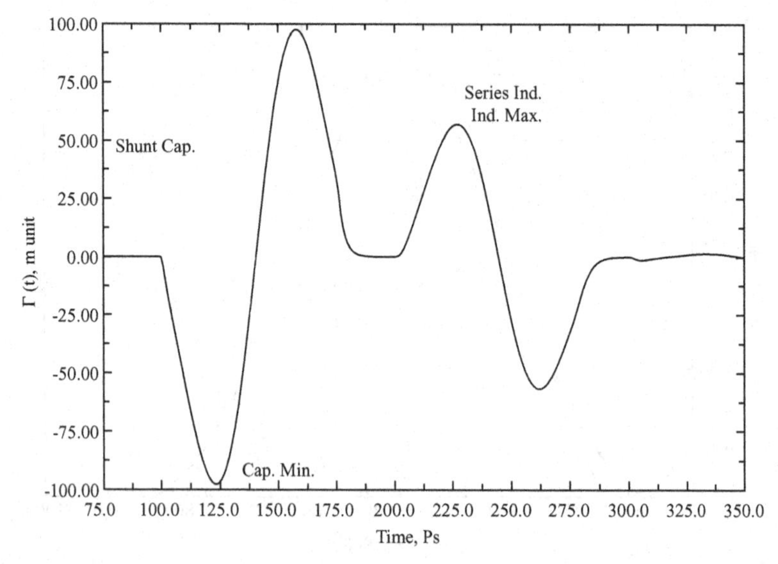

FIGURE G.3 Time-domain response from shunt capacitor separated from series inductor by 100-pS 50-Ω transmission line.

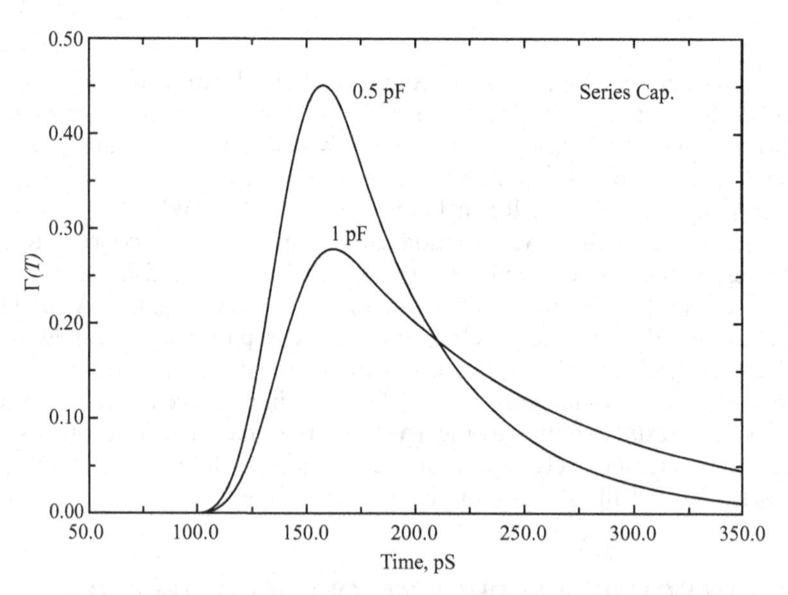

FIGURE G.4 Typical time-domain response for series capacitor.

back to the baseline, looking roughly like a single-period sine wave. Figure G.3 shows a typical response to a shunt capacitance and series inductance that are a half wavelength apart at 10 GHz. The typical response for a series capacitance is shown in Fig. G.4. A shunt inductance response would look like the

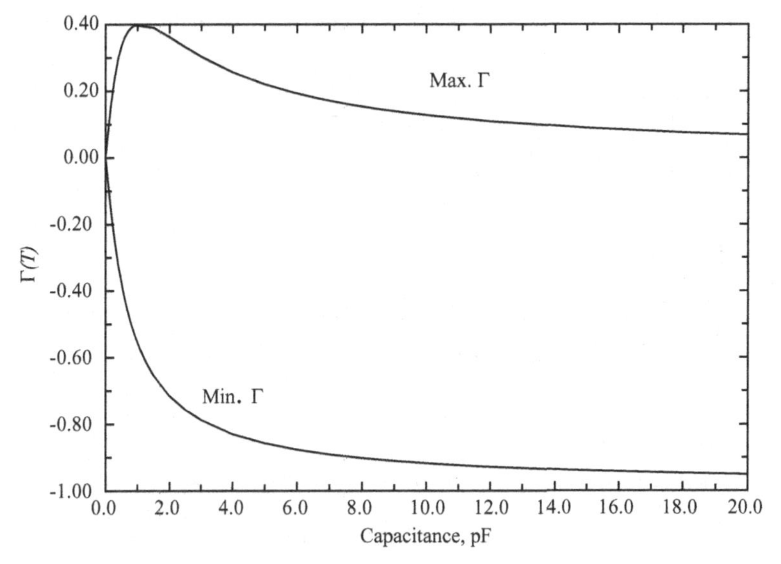

FIGURE G.5 Maximum and minimum values of time-domain response for shunt capacitance using piecewise linear approximation to measured impulse.

negative of the series capacitance. The shunt capacitance, series inductance, series capacitance, shunt inductance, and step in characteristic impedance all have their peculiar time-domain signature.

In general, the larger the discontinuity, the larger is the time-domain reflection, $|\Gamma(t)|$. For the larger shunt capacitances, the magnitude of the dip below the baseline is not equal to the magnitude of the rise above the baseline. So, there is some ambiguity in choosing what part of the curve to use to predict the value of the shunt capacitance. Various capacitance values were tested using the PWL representation for the network analyzer time-domain impulse. The results in Fig. G.5 show that if the algebraic maximum of $\Gamma(t)$ (the second extremum) is chosen, there are two possible values of capacitance for $\Gamma(t)$. However, the maximum value of the negative excursion of $\Gamma(t)$ (the first extremum) is monotonic and hence gives a unique value for the capacitance. The series inductance response is a mirror image about the $\Gamma(t) = 0$ line. Thus, the first maximum of $\Gamma(t)$ is monotonically related to the value of the series inductance. It is this first maximum of $|\Gamma(t)|$ that can be used to find the value of these two types of discontinuities.

The time-domain not only allows determination of the discontinuity type and size but also the discontinuity position in time. Here again, there is some ambiguity in what part of the impulse response curve should be used to find the position. For shunt capacitances, a position midway between the lower and upper part of the curve could be used, that is, where $\Gamma(t) = 0$. From a comparison of this choice with the theoretical distance in Fig. G.6 it is seen that this is

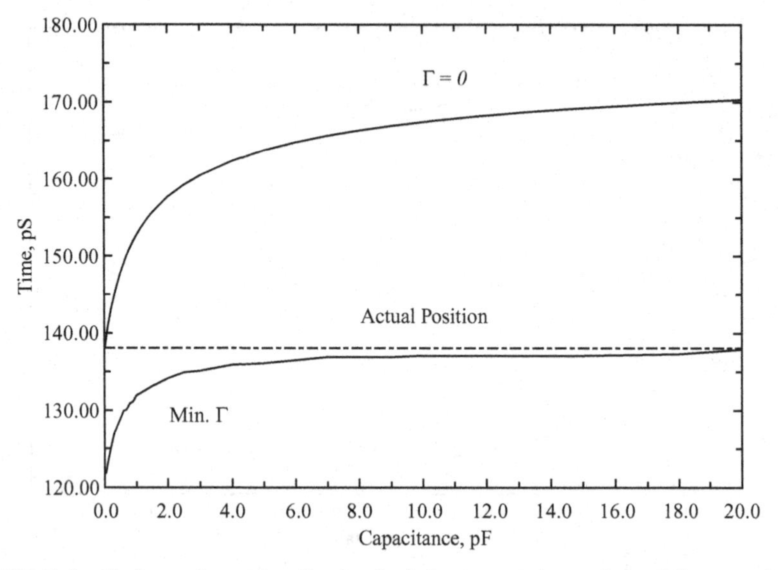

FIGURE G.6 Estimated position (in time) of shunt capacitor using minimum value of $\Gamma(t)$ and position where $\Gamma(t) = 0$.

TABLE G.1 Expressions for Reactance Element Values in Terms of Γ

Max. Freq.	Shunt C, pF $C = -\dfrac{\Gamma A}{1+\Gamma B}$	Series L, nH $L = -\dfrac{\Gamma A}{1+\Gamma B}$	Series C, pF $C = \dfrac{1-\Gamma A}{\Gamma B}$	Shunt L, nH $L = \dfrac{1-\Gamma A}{\Gamma B}$
18-GHz	$A = 1.212$	$A = -2.965$	$A = 0.9379$	$A = -0.9239$
pulse	$B = 0.9887$	$B = -0.9908$	$B = 1.691$	$B = -0.6892$
26-GHz	$A = 0.8216$	$A = -2.061$	$A = 0.9068$	$A = -0.9275$
pulse	$B = 1.004$	$B = -1.010$	$B = 2.624$	$B = -1.036$
50-GHz	$A = 0.4275$	$A = -1.091$	$A = 0.9279$	$A = -0.8770$
pulse	$B = 0.9875$	$B = -1.010$	$B = 4.801$	$B = -1.9885$

accurate if the discontinuity is very small. However, for larger discontinuities, a better choice would be the location of the first extremum of $|\Gamma(t)|$.

The minimum $\Gamma(t)$ of the shunt capacitor, the maximum $\Gamma(t)$ of the series inductor, the maximum $\Gamma(t)$ of the series capacitance, the minimum $\Gamma(t)$ of the shunt inductance, and the peak $\Gamma(t)$ (either negative or positive) of the impedance step can be each described by a simple formula that could be readily stored in a handheld calculator. These formulas are listed in Table G.1. The parameters for these expressions are listed for the 18, 26, and 50 GHz impulses.

Shunt capacitance:

$$C = -\frac{\Gamma A}{1+\Gamma B} \tag{G.8}$$

Series inductance:

$$L = -\frac{\Gamma A}{1+\Gamma B} \tag{G.9}$$

Series capacitance:

$$C = \frac{1-\Gamma A}{B\Gamma} \tag{G.10}$$

Shunt inductance:

$$L = \frac{1-\Gamma A}{B\Gamma} \tag{G.11}$$

Step in characteristic impedance:

$$Z_0' = Z_0\frac{1+\Gamma}{1-\Gamma} \tag{G.12}$$

Measurements of a given series-mounted chip capacitor on a LCR meter at say 1 MHz would not necessarily give accurate correlation with a time-domain model of a simple series capacitance as modeled by Eq. (G.10) or direct measurements on a network analyzer. However, a better fit with the time-domain data can be obtained by using a model of a high-frequency capacitor as described in Chapter 2.

G.5 MULTIPLE DISCONTINUITIES

The preceding formulas are correct when there is only one significant discontinuity in the circuit that is being measured. When there are multiple discontinuities, the SPICE analysis will also display the results that would be expected in a real time-domain reflectometer measurement. The gating error in measuring a discontinuity in the presence of other discontinuities was analyzed in [6]. That analysis used a rectangular gating function with a depth of 40 dB rather than the chirp-Z transform used in the network analyzer software. Four sources of error are identified by [6]. The first is out-of-gate attenuation associated with incomplete suppression of reflections outside the gating function. The

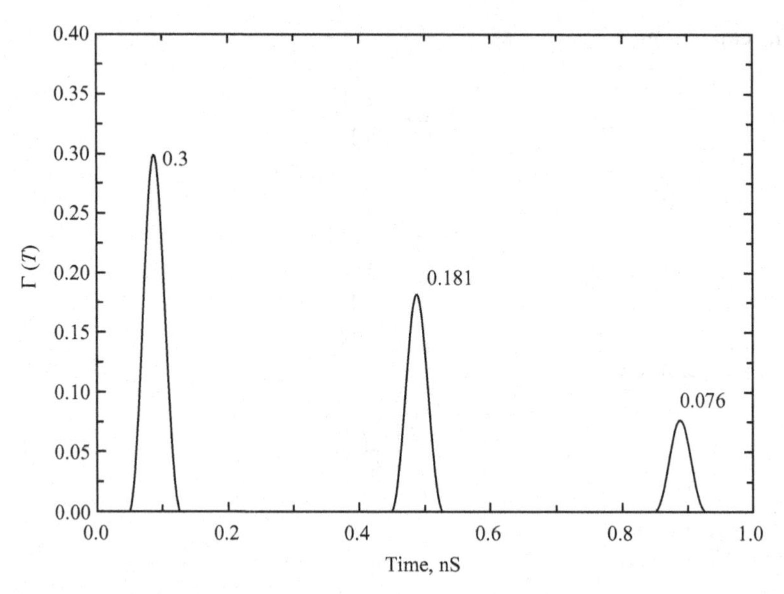

FIGURE G.7 Predicted response from three discontinuities whose actual reflection coefficients are 0.3, 0.2, and 0.1.

second is a truncation error where the gate is made too narrow to pick up all the response due to the discontinuity in question. The third is a masking error where the transmission coefficients of previous discontinuities reduces the signal getting to the discontinuity under investigation. The fourth is a multireflection aliasing error that occurs when the circuit has commensurate line lengths, and the residual reflection of one discontinuity adds to or subtracts from the reflection of the discontinuity under investigation.

When two discontinuities are sufficiently separated in time, each can be analyzed separately. However, the accuracy of predicting the later discontinuities from the formulas in Table G.1 are less accurate than when there is only one discontinuity. Three impedance steps were set up such that if they were alone, they would have had a $\Gamma(t)$ of 0.3, 0.2, and 0.1, respectively (Fig. G.7), as was done by [6]. However, when the three steps were put in one circuit suitably separated from one another, the SPICE analysis showed that $\Gamma(t) = 0.3$ (0% error), $\Gamma(t) = 0.181$ (9.6% error), and $\Gamma(t) = 0.087$ to 0.076 (12.9 to 24% error depending on whether line lengths are commensurate or not). The estimated errors in [6] were 0, 12, and 20%, respectively, which correlate well with the SPICE results given the approximations that are employed.

In addition, the technique described here can be used in a wide variety of circuits to model two elements that are too close together to be resolved separately in time. Adjustment of a theoretical time-domain model to make its response match that of the measured time-domain measurements gives a method to extract an equivalent circuit in the time domain.

G.6 SAMPLE SPICE LIST

The SPICE listing below can be used to find the $S_{11}[=V(11)]$ and $S_{21}[=V(21)]$ parameters when the source and load resistances are $50\,\Omega$, and the circuit is terminated with a 50-Ω load. The PSPICE specific functions have been commented out, but they may be useful if PSPICE is used. For example, the .PARAM statement can be used to set the input and output impedance levels. The circuit to be analyzed is entered in the .SUBCKT section. The number of frequencies and the frequency range must also be added to the .AC statement.

```
Analysis of a circuit for S11 and S21
*
* R01 and R02 are input and output resistance levels.
* RL is the load resistance.The load may be supplemented
* with additional elements.
* Statements below with ** may be used for PSPICE
**.PARAM R01=50, R02=50. RLOAD=50.IIN={-1/R01}
**.FUNC N(R01,R02) {SQRT(R02/R01)}
** R0110{R01}
R0110''value''
VIN1011AC1
** GI110VALUE={-V(10,11)/R01}
GI1101011''-1/R01''
E1110  0102
R1111  01
Xcircuit12netname
** RL20{RLOAD}
RL20''value''
** E21210VALUE={V(2)*2/N(R01,R02)}
*n = SQRT(R02/R01)
E2121020''2/n''
R212101
*
.SUBCKT netname''first node''''last node''
* Input side
*.
*.
*.
* Output side
.ENDSnetname
* Code for S11 and S21
*.AC DEC ''num''''f1''''f2''
** PROBE V(11)V(21)
.END
```

G.7 IMPULSE RESPONSE SPICE NET LIST MODIFICATION

Time-domain analysis with SPICE requires replacing the `.AC` statement with a `.TRAN` statement similar to the following:

```
.TRAN .01ps 250ps 0 .2ps
```

In addition the `VIN` statement should be replaced with one containing either the `PULSE` or `PWL` transient function. The pulse statement has the form `PULSE` (initial volt, pulse volt, delay time, rise time, fall time, pulse width, period). The measured impulse from the time-domain measurements from the network analyzer is approximated by the following `PULSE` statements. The 18-GHz pulse, with base width of 94.019 pS, is approximated as follows:

```
VIN 10 11 PULSE(0 1 0 39.482p 39.491p 15.05p)
```

The 26-GHz pulse, with base width of 64.150 pS, is approximated as follows:

```
VIN 10 11 PULSE(0 1 0 26.395p 28.477p 9.278p)
```

Alternately, the more accurate PWL fit could be used. The 18-GHz impulse is approximated using 81 points:

```
VIN 10 11 PWL(0ps 0, 1.5ps 0.0037, 3.0ps 0.0089, 4.5ps 0.0156,
+ 6.0ps 0.0237, 7.5ps 0.0336, 9.0ps 0.0452, 10.5ps 0.0589,
+ 12.0ps 0.0745, 13.5ps 0.0922, 15.0ps 0.1121, 16.5ps 0.1342,
+ 18.0ps 0.1584, 19.5ps 0.1848, 21.0ps 0.2134, 22.5ps 0.2441,
+ 24.0ps 0.2767, 25.5ps 0.3111, 27.0ps 0.3472, 28.5ps 0.3848,
+ 30.0ps 0.4236, 31.5ps 0.4634, 33.0ps 0.5038, 34.5ps 0.5448,
+ 36.0ps 0.5858, 37.5ps 0.6267, 39.0ps 0.6671, 40.5ps 0.7065,
+ 42.0ps 0.7448, 43.5ps 0.7815, 45.0ps 0.8162, 46.5ps 0.8489,
+ 48.0ps 0.8789, 49.5ps 0.9062, 51.0ps 0.9304, 52.5ps 0.9513,
+ 54.0ps 0.9687, 55.5ps 0.9824, 57.0ps 0.9922, 58.5ps 0.9983,
+ 60.0ps 1.0002, 61.5ps 0.9982, 63.0ps 0.9922, 64.5ps 0.9823,
+ 66.0ps 0.9686, 67.5ps 0.9512, 69.0ps 0.9303, 70.5ps 0.9061,
+ 72.0ps 0.8788, 73.5ps 0.8487, 75.0ps 0.8161, 76.5ps 0.7813,
+ 78.0ps 0.7446, 79.5ps 0.7064, 81.0ps 0.6669, 82.5ps 0.6266,
+ 84.0ps 0.5857, 85.5ps 0.5446, 87.0ps 0.5037, 88.5ps 0.4633,
+ 90.0ps 0.4235, 91.5ps 0.3847, 93.0ps 0.3471, 94.5ps 0.3111,
+ 96.0ps 0.2767, 97.5ps 0.2441, 99.0ps 0.2135, 100.5ps 0.1849,
+ 102.0ps 0.1585,103.5ps 0.1342,105.0ps 0.1122,106.5ps 0.0923,
+ 108.0ps 0.0746,109.5ps 0.0591,111.0ps 0.0455,112.5ps 0.0338,
+ 114.0ps 0.0240,115.5ps 0.0158,117.0ps 0.0092,118.5ps 0.0040,
+ 120.0ps 0)
```

The PWL fit for the 26-GHz impulse is approximated using 77 points:

```
VIN 10 11 PWL(0ps 0,1ps .005, 2ps .015, 3ps .0267, 4ps .0402,
+ 5ps .0556, 6ps .0731, 7ps .0925, 8ps .1140, 9ps .1375,
+ 10ps .1632, 11ps .1909, 12ps .2204, 13ps .2519, 14ps .2850,
+ 15ps .3198, 16ps .3560, 17ps .3933, 18ps .4318, 19ps .4709,
+ 20ps .5106, 21ps .5505, 22ps .5904, 23ps .6299, 24ps .6688,
+ 25ps .7067, 26ps .7433, 27ps .7784, 28ps .8116, 29ps .8427,
+ 30ps .8713, 31ps .8972, 32ps .9202, 33ps .9401, 34ps .9566,
+ 35ps .9697, 36ps .9792, 37ps .9850, 38ps .9871, 39ps .9855,
+ 40ps .9801, 41ps .9710, 42ps .9584, 43ps .9423, 44ps .9227,
+ 45ps .9001, 46ps .8745, 47ps .8462, 48ps .8155, 49ps .7825,
+ 50ps .7477, 51ps .7112, 52ps .6734, 53ps .6346, 54ps .5952,
+ 55ps .5553, 56ps .5154, 57ps .4756, 58ps .4363, 59ps .3979,
+ 60ps .3604, 61ps .3240, 62ps .2891, 63ps .2557, 64ps .2240,
+ 65ps .1942, 66ps .1663, 67ps .1404, 68ps .1166, 69ps .0948,
+ 70ps .0751, 71ps .0575, 72ps .0418, 73ps .0279, 74ps .0160,
+ 75ps .0058, 76ps 0)
```

The PWL fit for the 50-GHz impulse is approximated using 46 points:

```
VIN 10 11 PWL(0ps -4.530E-03, 1ps -611.4E-06, 2ps 6.941E-03,
+ 3ps 0.019, 4ps 0.037, 5ps 0.061, 6ps 0.091, 7ps 0.130,
+ 8ps 0.175, 9ps 0.229, 10s 0.289, 11ps 0.355, 12ps 0.425,
+ 13ps 0.500, 14ps 0.576, 15ps 0.651, 16ps 0.724, 17ps 0.792,
+ 18ps 0.854, 19ps 0.906, 20ps 0.948, 21ps 0.979, 22ps 0.996,
+ 23ps 1.000, 24ps 0.990, 25ps 0.967, 26ps 0.931, 27ps 0.884,
+ 28ps 0.828, 29ps 0.763, 30ps 0.693, 31ps 0.618, 32ps 0.542,
+ 33ps 0.467, 34ps 0.394, 35ps 0.325, 36ps 0.262, 37ps 0.204,
+ 38ps 0.155, 39ps 0.112, 40ps 0.077, 41ps 0.049, 42ps 0.028,
+ 43ps 0.013, 44ps 3.141E-03, 45ps -2.711E-03)
```

Acknowledgment

The author wishes to acknowledge the help of Terry Jamison in the analysis of the impulse response data.

REFERENCES

1. M. E. Hines and H. E. Stinehelfer, Sr., "Time-Domain Oscillographic Microwave Network Analysis Using Frequency-Domain Data," *IEEE Trans. Microwave Theory Tech.*, **MTT-22**, pp. 276–282, March 1974.
2. T. B. Mills, "S-Parameters in Spice," *RF Design*, pp. 45–48, June 1989.
3. K. B. Kumar and T. Wong, "Methods to Obtain Z, Y, H, G and S Parameters from the SPICE Program," *IEEE Circuits Devices Mag.*, pp. 30–31, Sept. 1988.

4. R. Goyal, "S-Parameter Output from the SPICE Program," *IEEE Circuits Devices Mag.*, pp. 28–30, March 1988.
5. C. E. Smith, "Frequency Domain Analysis of RF and Microwave Circuits Using SPICE," *IEEE Trans. Microwave Theory Tech.*, **MTT-42**, pp. 1904–1909, Oct. 1994.
6. K. Lu and T. J. Brazil, "A Systematic Error Analysis of HP8510 Time-Domain Gating Techniques with Experimental Verification," *1993 IEEE MTT-S Digest*, pp. 1259–1262, May 1993.

APPENDIX H

Single-Tone Intermodulation Distortion Suppression for Double-Balanced Mixers

An expression is given in [1] for the suppression for single-tone intermodulation distortion in a double-balanced mixer. This is repeated below as well as coded in the program IMSUP. The intermodulation suppression in dBc (dB below the carrier) is S_{nm} for a set of frequencies $nf_p \pm mf_1$:

$$S_{nm} = (|m| - 1)\Delta P + 20\log(|A_{nm}|) \tag{H.1}$$

The difference in dB between the RF signal and LO power is ΔP.

$$\begin{aligned} A'_{nm} = {} & \frac{\Gamma\left(\frac{|n|+|m|-1}{2}\right)}{\Gamma\left(\frac{|n|-|m|+3}{2}\right)} \frac{1}{2}\left[\sin\frac{|n|\pi}{2}\sin\frac{|m|\pi}{2}B_{oo} + \cos\frac{|n|\pi}{2}\cos\frac{|m|\pi}{2}B_{ee}\right] \\ & + \frac{\Gamma\left(\frac{|n|+|m|}{2}\right)}{\Gamma\left(\frac{|n|-|m|+2}{2}\right)} V_f\left[\sin\frac{|n|\pi}{2}\sin\frac{|m|\pi}{2}B_{oe} + \cos\frac{|n|\pi}{2}\sin\frac{|m|\pi}{2}B_{eo}\right] \end{aligned} \tag{H.2}$$

$$-A_{nm} = \frac{1}{B_{IF}|m|!} A'_{nm} \tag{H.3}$$

Radio Frequency Circuit Design, Second Edition, by W. Alan Davis

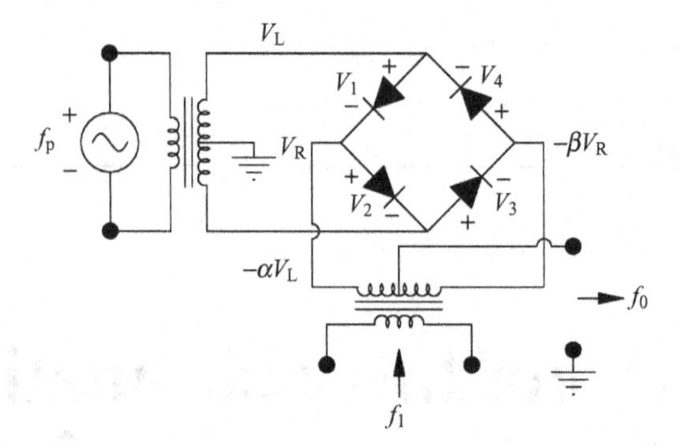

FIGURE H.1 Double-balanced mixer with transformer and diode imbalanced.

$$
\begin{aligned}
B_{oo} &= 1+\delta_4+\alpha(\delta_3+\delta_2)-|m|[\delta_4-\delta_2-\alpha(\delta_3+\delta_2)-\beta(\delta_3+\delta_4)]\\
B_{ee} &= -1+\delta_4-\alpha(\delta_3-\delta_2)-|m|[\delta_4-\delta_2-\alpha(\delta_3-\delta_2)+\beta(\delta_3-\delta_4)]\\
B_{oe} &= |m|[-\delta_4-\delta_2+\alpha(\delta_3+\delta_2)+\beta(\delta_4-\delta_3)]\\
B_{eo} &= |m|[\delta_4+\delta_2+\alpha(\delta_3-\delta_2)-\beta(\delta_4+\delta_3)]\\
B_{IF} &= B_{oo}(m=1)
\end{aligned}
$$

The value for V_f in Eq. (H.2) is the ratio of the forward saturation diode voltage, V_{sat} and the peak LO voltage, V_L. Thus,

$$V_f = \frac{V_{sat}}{V_L} \tag{H.4}$$

The α and β represent the isolation in the LO and RF transformers resulting from their imbalance. The imbalance is illustrated in Fig. H.1, which is the same as used in Fig. 11.5 except the diode numbering convention has been made here to conform to that used in [1]:

$$
\begin{aligned}
\text{LO isolation} &= 20\log(1-\alpha)\\
\text{RF isolation} &= 20\log(1-\beta)
\end{aligned}
$$

The values for δ are a measure of the inequality of the forward voltages across the diodes:

$$\delta_2 = \frac{V_2}{V_1}$$

$$\delta_3 = \frac{V_3}{V_1}$$

$$\delta_4 = \frac{V_4}{V_1}$$

Under ideal conditions

$$\alpha = \beta = \delta_2 = \delta_3 = \delta_4 = 1$$

Typical values for isolation by the transformers are 10 to 15 dB while the values for δ range from 0.85 to 1.15.

The equations have been implemented in a program called IMPSUP, which determines the single-tone intermodulation suppression for a given set of frequency harmonics of the RF signal and LO oscillator, the relative RF signal and LO power levels, the peak value of the LO voltage, imbalances resulting in finite isolation in the transformers, and imbalances in the diode forward voltage drops. A sample run of IMSUP shows the intermodulation suppression for a variety of frequency harmonics.

```
LO and RF Signal Transformer Isolation (typ. 10 to 15 dB)
10.,10.
Ring diode voltage ratios: V2/V1, V3/V1, V4/V1 = ?
Typically .85 to 1.15 (ideally =1)
0.85, 0.90, 1.15
Difference in LO and RF power in dB (typ. -20.)
-20.
Peak LO voltage = ?
3.
Forward diode saturation voltage (typ. 0.1)
IM product n x FL +- m x Frf: n,m = ?
1, 1
For intermodulation product n x m = 1 1
IM Suppression = 0.000000E+00 dBc
New n,m values only? <Y/N>
y
IM product n x FL +- m x Frf: n,m = ?
2, 1
For intermodulation product n x m = 2 1
IM Suppression = -0.437641E+02 dBc
New n,m values only? <Y/N>
y
IM product n x FL +- m x Frf: n,m = ?
3, 1
For intermodulation product n x m = 3 1
IM Suppression = -0.954243E+01 dBc
New n,m values only? <Y/N>
y
IM product n x FL +- m x Frf: n,m = ?
3, 2
```

```
For intermodulation product n x m = 3 2
IM Suppression = -0.602423E+02 dBc
New n,m values only? <Y/N>
n
Completely new mixer specs? <Y/N>
n
fin
```

REFERENCE

1. B. C. Henderson, "Reliably Predict Mixer IM Suppression," *Microwaves RF*, **22**, pp. 63–66, 68–70, 132, Nov. 1983. Also reprinted in *RF and Microwave Designer's Handbook*, Watkins-Johnson 97–98 Catalog.

INDEX

Radio Frequency Circuit Design, Second Edition, by W. Alan Davis

Printed in the United States
By Bookmasters